Select Ideas in Partial Differential Equations

PETER J. COSTA

Synthesis Lectures on Mathematics and Statistics

Editor
***Steven G. Krantz**, Washington University, St. Louis*

Select Ideas in Partial Differential Equations
Peter J. Costa
June 2021

Crowd Dynamics by Kinetic Theory Modeling: Complexity, Modeling, Simulations, and Safety
Bouchra Aylaj, Nicola Bellomo, Livio Gibelli, and Damián Knopoff
October 2020

Probability and Statistics for STEM: A Course in One Semester
E.N. Barron and J.G. Del Greco
July 2020

An Introduction to Proofs with Set Theory
Daniel Ashlock and Colin Lee
June 2020

Discrete Distributions in Engineering and the Applied Sciences
Rajan Chattamvelli and Ramalingam Shanmugam
June 2020

Affine Arithmetic Based Solution of Uncertain Static and Dynamic Problems
Snehashish Chakraverty and Saudamini Rout
March 2020

Time-Fractional Order Biological Systems with Uncertain Parameters
Snehashish Chakraverty, Rajarama Mohan Jena, and Subrat Kumar Jena
March 2020

Fast Start Advanced Calculus No Access
Daniel Ashlock
September 2019

Fast Start Integral Calculus No Access
Daniel Ashlock
September 2019

Fast Start Differential Calculus No Access
Daniel Ashlock
September 2019

Introduction to Statistics Using R No Access
Mustapha Akinkunmi
July 2019

Inverse Obstacle Scattering with Non-Over-Determined Scattering Data No Access
Alexander G. Ramm
June 2019

Analytical Techniques for Solving Nonlinear Partial Differential Equations s
Daniel J. Arrigo
June 2019

Aspects of Differential Geometry IV
Esteban Calviño-Louzao, Eduardo García-Río, Peter Gilkey, JeongHyeong Park, and Ramón Vázquez-Lorenzo
April 2019

Symmetry Problems. The Navier--Stokes Problem
Alexander G. Ramm
March 2019

An Introduction to Partial Differential Equations
Daniel J. Arrigo
December 2017

Numerical Integration of Space Fractional Partial Differential Equations: Vol 2 - Applications from Classical Integer PDEs
Younes Salehi and William E. Schiesser
December 2017

Numerical Integration of Space Fractional Partial Differential Equations: Vol 1 - Introduction to Algorithms and Computer Coding in R
Younes Salehi and William E. Schiesser
November 2017

Aspects of Differential Geometry III
Esteban Calviño-Louzao, Eduardo García-Río, Peter Gilkey, JeongHyeong Park, and Ramón Vázquez-Lorenzo
May 2017

The Fundamentals of Analysis for Talented Freshmen
Peter M. Luthy, Guido L. Weiss, and Steven S. Xiao
August 2016

Aspects of Differential Geometry II
Peter Gilkey, JeongHyeong Park, and Ramón Vázquez-Lorenzo
May 2015

Aspects of Differential Geometry I
Peter Gilkey, JeongHyeong Park, and Ramón Vázquez-Lorenzo
February 2015

An Easy Path to Convex Analysis and Applications
Boris S. Mordukhovich and Nguyen Mau Nam
December 2013

Applications of Affine and Weyl Geometry
Eduardo García-Río, Peter Gilkey, Stana Nikčević, and Ramón Vázquez-Lorenzo
May 2013

Essentials of Applied Mathematics for Engineers and Scientists, Second Edition
Robert G. Watts
February 2012

Chaotic Maps: Dynamics, Fractals, and Rapid Fluctuations No Access
Goong Chen and Yu Huang
August 2011

Matrices in Engineering Problems
Marvin J. Tobias
June 2011

The Integral: A Crux for Analysis
Steven G. Krantz
January 2011

Statistics is Easy! Second Edition
Dennis Shasha and Manda Wilson
August 2010

Lectures on Financial Mathematics: Discrete Asset Pricing
Greg Anderson and Alec N. Kercheval
August 2010

Jordan Canonical Form: Theory and Practice
Steven H. Weintraub
2009

The Geometry of Walker Manifolds
Miguel Brozos-Vázquez, Eduardo García-Río, Peter Gilkey, Stana Nikčević, and Ramón Vázquez-Lorenzo
2009

An Introduction to Multivariable Mathematics
Leon Simon
2008

Jordan Canonical Form: Application to Differential Equations
Steven H. Weintraub
2008

Statistics is Easy!
Dennis Shasha and Manda Wilson
2008

A Gyrovector Space Approach to Hyperbolic Geometry
Abraham Albert Ungar
2008

Select Ideas in Partial Differential
Equations Peter J. Costa

ISBN: 978-3-031-01306-5 print
ISBN: 978-3-031-02434-4 ebook
ISBN: 978-3-031-00280-9 hardcover

DOI 10.1007/978-3-031-02434-4

A Publication in the Springer series
SYNTHESIS LECTURES ON MATHEMATICS AND STATISTICS, #40
Series Editor: *Steven G. Krantz, Washington University, St. Louis*

Series ISSN: 1946-7680 Print 1946-7699 Electronic

Select Ideas in Partial Differential Equations

Peter J. Costa

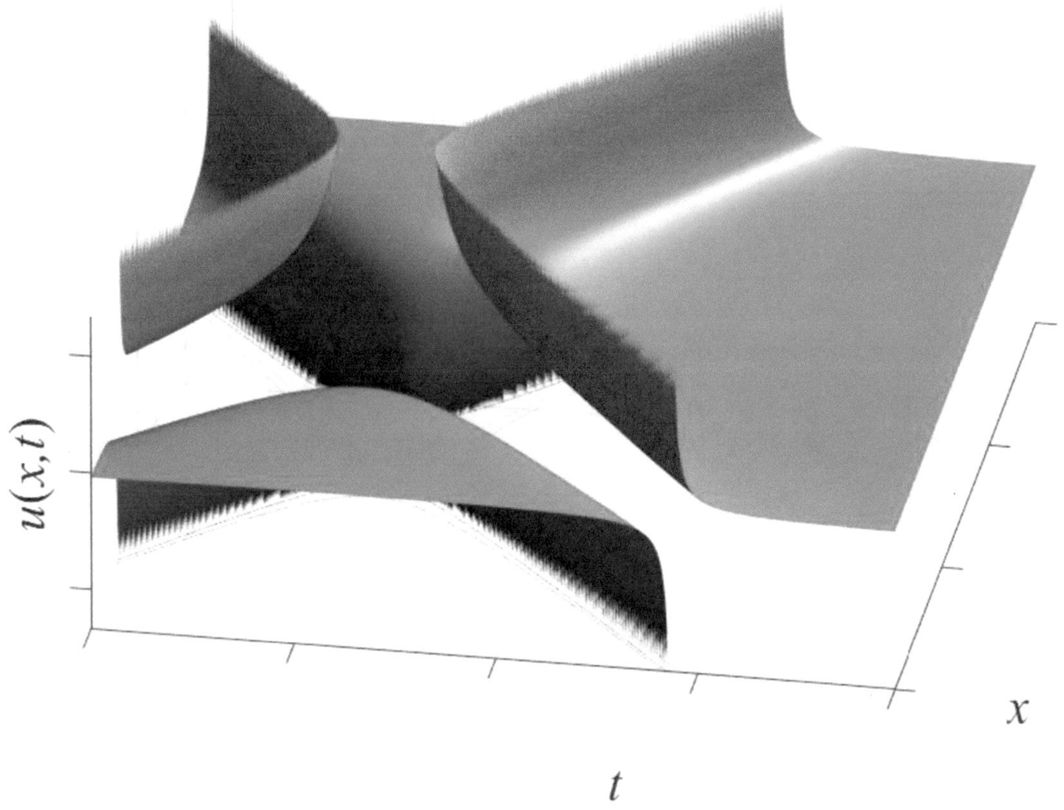

SYNTHESIS LECTURES ON MATHEMATICS AND STATISTICS #40

Dedication

For Anne
Always and forever

Contents

Preface

As an undergraduate math major, I had the good fortune to be assigned a gregarious and sharp-witted professor as my advisor. His mathematical acumen was even keener than his insightful and razor-edged sense of humor. At the close of my freshman year, I had completed the calculus sequence (differential, integral, and multivariate calculus) and an introductory course in linear algebra. When asked what he recommended for my next term, my advisor responded *ordinary differential equations*. Thinking that I was up to the challenge and now being semi-conversant in partial derivatives, I asked about the difference between *ordinary* and *partial differential equations*. My advisor looked at me and responded.

Ordinary differential equations model the rate of change of physical systems. Partial differential equations are too hard for human beings.

The mischief in his remark was softened by the glint in his eye. *Ordinary differential equations* it would then so be. After winding my way through the standard undergraduate program in mathematics (abstract algebra, *ODE*s, real analysis, advanced calculus, advanced linear algebra, topology, complex variables, Euclidean and non–Euclidean geometry, and other related classes), I was set to graduate. To my surprise and pleasure, my advisor was teaching a class in "engineering mathematics" during my final semester. I thought it would be a fitting and uplifting way to end my undergraduate studies by taking this class with my beloved professor. When I asked about the content of the course, he mentioned Fourier series, the Laplace transform, surface and volume integrals, and *partial differential equations*.

I learned the basic notions of Laplace's equation and separation of variables. In addition, I began to see that with a bit of discipline and more than a bit of hard work, I could figure out how to solve the rudimentary *PDE*s. This gave me sufficient confidence to apply to graduate school and eventually earn a doctorate in applied mathematics. My specialty? Partial differential equations.

I do not pretend to know everything about this vast and rich subject area. Indeed, I have spent my career *primarily* as an industrial mathematician doing data analysis, statistical modeling, and algorithm development. But partial differential equations followed me from time-to-time, giving me considerable challenges and a genuine measure of satisfaction when I could bring mathematical knowledge to glean some insight from these wily creations.

This text is a mathematical diary of my encounters and adventures with partial differential equations. Some elements will be familiar to the general mathematical audience; some will be different; some will be novel; and some will reflect work heretofore unpublished. It is my hope that the reader will find the contents interesting, clearly presented, and intellectually satisfying. If I have been true to the lessons my advisor taught me, then the material contained herein will show the reader that partial differential equations are not so hard after all.

Acknowledgments

In October of 1981, the University of Massachusetts' Department of Mathematics and Statistics hosted a special conference to celebrate the 175th birthday of James Clerk Maxwell. *The Maxwell Symposium* was envisioned and developed by Professor Melvyn S. Berger, a prominent faculty member within the Department. Professor Berger taught graduate-level courses in applied mathematics which focused on partial differential equations. I was one of his students in a first-year sequence in applied mathematics. He asked me to organize a group of graduate students to act as ambassadors, assistants, and all-purpose "gophers" to attend to the distinguished scientists and guests invited to participate in the symposium.

As the "troubleshooter coordinator" for the conference, it fell upon me to insure all of the attendees and speakers had access to overhead projectors, microphones, seating, writing materials, hotel rooms, and anything else they required during the conference. Among others, I met C. N. Yang (1957 Noble Laureate in Physics), Leon Cooper (1972 Noble Laureate in Physics), and Roger Penrose (long-time collaborator of Stephen Hawking and 2020 Noble Laureate in Physics). Heady times for a very young man.

Following the successful conclusion of the conference, Professor Berger asked whether I wished to pursue a doctoral dissertation under his guidance. After some hesitation, I agreed and specialized in partial differential equations (*PDE*s). Maxwell's equations held a particularly exalted status in my worldview. Three years later, I finished my degree. The debt I owe Professor Berger is more than I can ever hope to repay. It saddens me to write that he passed away in 2005 and could not see this work come to fruition.

This book is, in part, a consequence of his mentorship and that remarkable conference. It begins with *Maxwell's equations* and ends with a revised and updated version of my doctoral dissertation. In-between these two end points are a mix of the essential elements of *PDE*s along with some novel work in nonlinear optics, numerical analysis, and functional analysis applied to *PDE*s. If it gives the reader encouragement to explore and insight into this deep and elegant branch of applied mathematics, then I will be well satisfied as to the value of this work.

I do not work alone.

No one was more instrumental in the writing of this text than my lovely, beloved, and aptly named wife Anne. Derived from *Hannah/Anna* the name *Anne* means *grace*. Her skillful proofreading and copyediting graces these pages. My friend and "unindicted co-conspirator" Dr. William J. Satzer read through the entire manuscript and made numerous insightful suggestions. Professor Rüdiger Seydel (Köln Unisersität, retired) helped me develop this material by extending invitations to present my latest results to his many students. For his friendship, hospitality, and generosity, I am profoundly grateful. My former colleagues Professors Douglas Dokken, Yongzhi Yang, and Uwe Scholz read through earlier versions of Chapters 1–5 and made numerous helpful recommendations. Many thanks to my former students, especially Michelle Mitchell, Michael Wackerfuss, and Dana Willeke, for their patience and enthusiasm in learning the ideas presented in this text. Dr. Vladimir Krapchev first introduced me to the problem of Raman scattering and gave keen discernment into the physics of the nonlinear interaction. Dr. Alexander Eydeland helped me to develop the proof that the numerical scheme for the Raman scattering problem is convergent and stable. Professor Stephen Krone (University of Idaho) provided clarity. Special thanks go to my Hologic colleagues in the Scientific Product Support Group for their assistance, kindness, and good cheer, especially Paul MacLean, Steven Hecht, Michael Hopkins, Steven Guertin, Wendy Stanick, and Allyson Mood.

For their encouragement, support, and tolerance of my myriad idiosyncrasies, I wish to thank Alexander and Alla Eydeland, Sylvan Elhay and Jula Szuster, Alfonso and Franca Farina, Vladimir and Tania Krapchev, Stephen and Claudia Krone, William and Carol Link, John and Lanette McGovern, William and Amy Moore, William Satzer and Joyce Mitchell, Rüdiger and Friederike Seydel, John and Ellen Zupa, and the "lads and lasses." To my family (all 36 of you and counting) despite being familiar with my "unique perspective" you still abide my presence and occasionally even appear to enjoy it. Heaven only knows why. Thank you.

Finally, I wish to thank Susanne Filler and Melanie Carlson of Morgan & Claypool for their patient support of this book. Their efforts helped make this work successful. To all who worked to convert the manuscript into a text, thank you.

All mistakes are my responsibility and I solicit your corrections and patience where they occur.

Peter J. Costa
Hudson, MA
2021

Introduction

A Brief Overview of Partial Differential Equations

Intended Audience

The first five chapters plus §§6.1–6.3 of Chapter 6 served as a one-semester course in applied mathematics for senior-level undergraduates. It was given by the author and others at least three times with positive student response. Chapters 1–7 could be used for an *introduction to applied mathematics/partial differential equations/mathematical physics* course for first- and second-year graduate students. Chapter 8 is primarily of interest to upper-level graduate students as an aid to advanced work in partial differential equations (*PDE*s) theory.

The book, as a whole, is written for the curious mind. It is a combination of theory and practical results which try to inform the reader *how to solve PDE*s and what methods will be successful for various problems. Rather than ignoring or making a passing comment about Maxwell's equations, this seminal accomplishment of mathematical physics is the book's starting point. Indeed, one focus of this work is to develop Maxwell's equations and then explain how to find solutions. This is something that practitioners of electromagnetic wave theory (such as radar antenna engineers) find of value.

Most of all, this work records and presents a sampling of the inherent challenge, utility, and beauty of partial differential equations.

Condensed Summary by Chapter

Chapter 1: Review the definitions of the divergence, gradient, and curl in three spatial dimensions. Use Green's, Stokes', and Gauss's Divergence Theorem to help establish Faraday's Law, Coulomb's Law, and Ampere's Law. Use these results to present Maxwell's equations. This is the starting point for the book's approach to *PDE*s.

Chapter 2: Develop classical (separation of variables) and modern methods (Green's functions) solution of Laplace's equation. Introduce the Dirichlet, Neumann, and Robin boundary value variations of Laplace's equation along with Poisson's equation. Prove Green's Representation Theorem for smooth functions on general domains in $\mathbb{R}^n$. Provide examples of analytic solutions over regular bounded domains in $\mathbb{R}^2$.

Chapter 3: Introduce the idea of analytic functions (namely those functions which have convergent Taylor series) and their rôle in solving *PDE*s. Detail Fourier series, its application in representing periodic functions that need not be continuous, and its use in the solution of Laplace's equation over varying geometry. Begin the discussion of how functional analysis, through function spaces, help solve *PDE*s. Show how Green's functions solve Poisson's equation. Introduce the heat equation and the use of Bessel functions in its solution over circular geometries.

Chapter 4: Introduce the Fourier transform and show how it is used to solve the heat equation in $\mathbb{R}^n$. Use the Fourier transform to derive d'Alembert's solution for the homogeneous and inhomogeneous wave equations in $\mathbb{R}$. Construct the various properties of the Fourier transform and establish a key set of Fourier transform pairs. Solve the string equation (i.e., the one–dimensional wave equation over a finite interval), the circular vibrating drum equation, and the vibrating volume equation over finite domains.

Chapter 5: Use the Fourier transform to solve the wave equation over $\mathbb{R}^3$. Detail Huygen's and Duhamel's principles to arrive at spherical means solutions to the three–dimensional inhomogeneous wave equation. Employ the method of descent to arrive at the solution of the two–dimensional wave equation. Develop solutions of Maxwell's equations via the three–dimensional wave solutions.

Chapter 6: Introduce examples of nonlinear *PDE*s which are solved via "transformations." The first example, the nonlinear Klein–Gordon (*nKG*) equation, can be solved by the classical transformation method *separation of variables*. Burgers' equation is reduced to the heat equation via a linearizing change of variables: *The Hopf–Cole Transformation*. The well-studied Korteweg de Vries (*KdV*) equation is solved using the *inverse scattering transform* which reduces solutions of the *KdV* to solutions of the linear Airy's equation.

Chapter 7: Numerical methods. The system of nonlinear *PDE*s which model stimulated Raman scattering are established. Functional analysis is utilized to show that a particular finite difference scheme is both stable and convergent. Moreover, *a priori* conditions are placed on the temporal step size to insure stability of the finite difference method. It is demonstrated that the system is a conserved Hamiltonian. A series of computations contrasting the effects of varying initial conditions are illustrated. Discussion of the physical implications of the calculations is presented.

Chapter 8: The examples in Chapter 6 illustrate that a variety of nonlinear *PDE*s can be solved by a change of variables which transform the original nonlinear equation into a linear *PDE*. Chapter 8, conversely, presents a general theory for the existence of a change of variables $u = H(v)$

which transforms nonlinear evolution equations $\frac{\partial u}{\partial t} = Lu + N(u)$ into linear *PDE*s $\frac{\partial v}{\partial t} = Lv$ is presented. This is achieved by generalizing the *Hartman–Grobman Theorem.*

Purpose and Intent

Mathematical analysis of problems and applications in engineering, physics, and applied science can arguably be divided into three areas: statistics, estimation and optimization, and differential equations. Whenever data are collected and analyzed, statistical techniques play a major role in interpreting trends, estimating parameters, or modeling mean behaviors. In instances in which physical parameters must be estimated from data collected with underlying physical characteristics, then estimation methods (i.e., optimal filtering) are utilized. Similarly, for establishing efficient schedules, use of resources, income yields, and the like, optimization algorithms are the mathematical tools of choice. To model systems that evolve in time and/or space, however, differential equations are the boulevards upon which the applied scientist travels.

For systems which evolve only in time (e.g., equations of motion), ordinary differential equations are used to model such phenomena. Fluid flow, electromagnetic theory, optics, heat conduction, and wave diffusion are examples of physical processes that change in both space and time. These processes are modeled via partial differential equations.

In this text, the fundamental equations of heat transfer, wave motion, electromagnetic theory, shallow water waves, and Raman scattering are detailed. Some of these equations are *linear* and can be solved analytically. Some are *nonlinear* which require either very sophisticated analytic techniques or numerical methods. Direct solutions (when possible) and numerical methods of solution (when required) of these equations are presented. The aim of this work is to give the reader exposure to some key elements of the expansive topic that is partial differential equations.

The book begins with *Maxwell's equations*. This is done for two reasons. First, Maxwell's equations are central to the understanding of electromagnetic theory and Newtonian physics. Second, they are also a crowning achievement of human intellect and dedication. Chapter 1 develops these famous equations (which can regularly be seen on T-shirts adorned by MIT's industrious students) with a promise to reveal methods leading to their solution. To arrive at this, the solution of the inhomogeneous wave equation is needed. This is produced at the end of Chapter 5.

It is recommended that the reader be familiar with differential, integral, and multivariate calculus; have a working understanding of ordinary differential equations; and have some background in real analysis and advanced calculus. From the reservoirs of these subject areas, the first five chapters of the text are developed. Chapter 6 begins with a nonlinear *PDE* which can be solved via the same technique normally applied to linear *PDE*s. Once that example (the nonlinear Klein–Gordon equation) has been presented, then one of the fundamental ideas of *PDE*s and this text come into focus: transformations/changes of variables.

One of the most successful methods used to solve linear *PDE*s is the *separation of variables* transformation. Specifically, if a *PDE* models a physical phenomenon which evolves in space and time, then the solution $u(\boldsymbol{x}, t)$ of the equation is written as a product of functions of each of its independent variables. That is, for $\boldsymbol{x} = (x_1, x_2, \dots, x_n)$ then the separation of variables transform sets $u(\boldsymbol{x},t) = f_1(x_1) \cdot f_2(x_2) \cdot \cdots \cdot f_n(x_n) \cdot g(t) = g(t) \cdot \prod_{j=1}^{n} f_j(x_j)$. For many linear *PDE*s, this transformation will change the partial differential equation into a set of independent *ordinary differential equations* (*ODE*s). These *ODE*s, in turn, will admit analytic solutions; the product of which solves the corresponding *PDE*. Associated with the separation of variables methods are series solutions; the most famous of which are *Fourier series*.

While the separation of variables method can be used successfully on a wide number of *PDE*s, it is limited. For more complex underlying geometries and/or initial conditions, more sophisticated approaches are required. This is where multivariate integrals and the theorems of Green, Stokes, and Gauss come into play. As is detailed in Chapter 2, integral representation of the solutions of certain *PDE*s are required. These representations rely upon Green's Theorem of multi–dimensional integrals as well as certain mathematical objects called *Green's functions*.

When the domain of definition of the equation changes from a finite to an infinite domain (e.g., from the rectangle $\Omega = [a_1, b_1] \times [a_2, b_2]$ to the entire plane $\mathbb{R}^2$), then separation of variables and Fourier series are replaced by the *Fourier transform*. Though more difficult to implement across high spatial dimensions (i.e., $n \geq 2$), the Fourier transform can reduce linear *PDE*s to *ODE*s.

This trio of techniques, *separation of variables*, *Green's functions*, and the *Fourier transform* remind the author of two (late 19th/early 20th century) approaches to the game of chess: *classical* and *modern*. Each theory of the game is best applied to particular situations and scenarios. So too with the aforementioned *PDE* solution methods which were developed during the 19th century.

These methods are remarkably successful at solving the most important *linear PDE*s of physics and engineering. For *nonlinear PDE*s, however, these techniques almost always prove inadequate.

The idea of a change of variables or some manner of nonlinear transform as a substitute for the separation of variables and Fourier transform methods became the focus in the 20th century. Indeed, Chapter 6 gives an example of each of these techniques. In 1950, the determination of the *Hopf–Cole Transformation* [40] as a change of variables which maps a nonlinear *PDE* (Burgers' equation) into a linear *PDE* (the heat equation) was made. This is a substantial achievement and led to the "discovery" of linearizing changes of variables for other *PDE*s. A general theory of such a change of variables proved elusive. Not 20 years later (1967), Gardner, Greene, Kruskal, and Miura [27, 28] developed the *inverse scattering transform* (*IST*) which led to an analytic solution of the Korteweg de Vries (*KdV*) equation and was later extended to the nonlinear Schrödinger equation. The solution of the *KdV* equation by inverse scattering transform was so widely acclaimed that this equation rarely appeared in print without the word "celebrated" preceding it.

As a consequence, some mathematicians joke that the equation should be known hereafter as the *cKdV*.

The *IST* is generally restricted to equations with one spatial dimension though Cheney [17, 18], Newton [54], Faddeev [24], and many others extended inverse scattering to the Schrödinger equation in dimensions 2 and 3.

A general theory for a linearizing change of variables was developed by extending the *Hartman–Grobman Theorem* [19, 30, 33–35] from ordinary to partial differential equations. That work is the focus of Chapter 8.

Alas, not all *PDE*s of scientific interest admit analytic solutions. For such equations, numerical solutions are required. Chapter 7 examines the system of nonlinear *PDE*s which model *stimulated Raman scattering* and a finite difference scheme which is used to compute a solution. Most notably, Chapter 7 uses functional analysis to demonstrate that the finite difference scheme associated with this nonlinear system is stable, consistent, and convergent. This chapter can be used as a starting point to examine the convergence of finite difference schemes for other nonlinear equations.

To continue with the chess analogy, the *IST* and general theory of a linearizing change of variables could be categorized as *hypermodern* methods. The numerical methods might fit into the various efforts to program computers to play chess or the famous game *Go*. Gratefully, this book does not require advanced computer programming skills to calculate any of the solutions, analytic or numeric. Indeed, MATLAB® and its accompanying Symbolic Math Toolbox can be used to great effect in producing these solutions and computer codes. The Appendix of this text lists all of the MATLAB files used to produce the numeric and analytic calculations presented herein. MATLAB commands and M–files used to produce the figures are contained on the website associated with this text.

http://www.morganclaypoolpublishers.com/Costa/

For all of these methods, a bit more than multivariate calculus is required to arrive at a solution. The methods provided in Chapters 2–6 use real and functional analysis to show convergence of Fourier series, the construction of Fourier transform pairs, the development of the *IST*, and the *superposition principle*. More specifically, the superposition principle notes that if a *PDE* has two solutions u_1 and u_2, then the sum of those solutions $u = u_1 + u_2$ is also a solution. Hence, if $\langle u_j \rangle_{j=1}^{\infty}$ is a sequence of solutions to a *PDE*, then the method of superposition states that $u = \sum_{j=1}^{\infty} c_j \cdot u_j$ is also a solution. Here $\langle c_j \rangle_{j=1}^{\infty}$ is a sequence of constants selected so that the series for u converges and satisfies the initial/boundary conditions.

Functional analysis also plays a crucial rôle in the proof that the numerical scheme developed for the Raman scattering system in Chapter 7 converges. For the proof of the Hartman–Grobman Theorem for infinite dimensional systems (i.e., *PDE*s), functional analysis is essential.

Therefore, this book takes the reader on a select and guided tour of partial differential equations ranging from the well-known (Laplace, Poisson, heat, wave, *KdV*), to the rarely treated but nevertheless important (Maxwell's equations), to an example which requires functional and numerical analysis (Raman scattering), to the purely theoretical (Hartman–Grobman).

The remainder of the introduction will provide an overview of *PDE*s, an account of the notation used, and some guidelines for using this text.

Ordinary Differential Equations vs. Partial Differential Equations

One way to understand a "general approach" to partial differential equations is to contrast them to ordinary differential equations. A good place to begin is with one of the simplest *ODE*s: $dx/dt = a{\cdot}x$ and $x(t_o) = x_o$. This equation is presented to first-year calculus students in the guise of being "separable." That is, matching variables are "multiplied or divided" through to obtain the form $\frac{dx}{x} = a\,dt$. Integrating both sides produces $\int_{x_o}^{x} \frac{d\xi}{\xi} = \int_{t_o}^{t} a\,d\tau$ and, after some algebra, the solution is formed as $x(t) = e^{a(t-t_o)}x_o$.

An example of a "simple" *PDE* is the equation $a\frac{\partial u}{\partial x} = \frac{\partial u}{\partial t}$ with initial condition $u(x,t_o) = u_o(x)$. One approach to solving this equation is to apply the *Fourier transform*† $\mathfrak{F}$ to both sides of this equation to obtain $(i \cdot a\omega)\mathfrak{F}[u(x,t)](\omega) = \frac{\partial}{\partial t}\mathfrak{F}[u(x,t)](\omega)$. Using the *ODE* example above results in

$$\mathfrak{F}[u(x,t)](\omega) = e^{i \cdot a\omega(t-t_o)} \cdot \mathfrak{F}[u(x,t_o)](\omega) = \mathfrak{F}[\delta(x - a(t-t_o))](\omega) \cdot \mathfrak{F}[u_o(x)](\omega) =$$
$$\mathfrak{F}[(\delta \odot u_o)(x - a(t-t_o))](\omega) = \mathfrak{F}[u_o(x - a(t-t_o))](\omega) \text{ or } u(x,t) = u_o(x + a(t-t_o)).$$

Here the symbol $\odot$ represents *convolution* which is detailed in Chapter 4, §4.1.5. Indeed, a direct computation shows that $\frac{\partial u}{\partial x} = u_o'(x + a(t-t_o))$ while $\frac{\partial u}{\partial t} = a \cdot u_o'(x + a(t-t_o))$. Hence, the function $u(x,t) = u_o(x + a(t-t_o))$ satisfies $a\frac{\partial u}{\partial x} = \frac{\partial u}{\partial t}$, $u(x,t_o) = u_o(x + a(t_o - t_o)) = u_o(x)$, and thereby solves the *PDE*.

It is evident that the solution of the "simple" *PDE* requires more sophisticated mathematics than does the solution of the "simple" *ODE*. In fact, the solution of the *PDE* requires the solution of

† Chapter 4 will go into considerable detail about the Fourier transform and its application to *PDE*s.

the *ODE*. This is then the first lesson in *PDE* theory: To solve *PDE*s, find a "transformation" which reduces it to a *simpler* equation such as an *ODE*.

In a parallel manner, examine the n–dimensional *ODE* $\frac{dx}{dt} = A\boldsymbol{x}$ and $\boldsymbol{x}(t_o) = \boldsymbol{x}_o$ where $\boldsymbol{x} \in \mathbb{R}^n$ and $A \in \mathcal{M}at_{n\mathrm{x}n}(\mathbb{R})$. The solution of this equation is $\boldsymbol{x}(t) = e^{A(t-t_o)}\boldsymbol{x}_o$. How is $e^{A(t-t_o)}$ calculated? Assume that A has n independent eigenvalues $\lambda_1, \lambda_2, \ldots, \lambda_n$ and associated eigenvectors $\boldsymbol{v}_1, \boldsymbol{v}_2, \ldots, \boldsymbol{v}_n$. Then $A = V \cdot \Lambda \cdot V^T$ where $V = \begin{bmatrix} \boldsymbol{v}_1 & \boldsymbol{v}_2 & \cdots & \boldsymbol{v}_n \end{bmatrix}$ is the matrix whose columns consist of the eigenvectors of A and $\Lambda = \begin{bmatrix} \lambda_1 & 0 & \cdots & 0 \\ 0 & \lambda_2 & \cdots & 0 \\ \vdots & \vdots & \ddots & \vdots \\ 0 & 0 & \cdots & \lambda_n \end{bmatrix}$ is the diagonal matrix of eigenvalues of A. It is a direct result of linear algebra that

$$e^{A(t-t_o)} = V \cdot \begin{bmatrix} e^{\lambda_1(t-t_o)} & 0 & \cdots & 0 \\ 0 & e^{\lambda_2(t-t_o)} & \cdots & 0 \\ \vdots & \vdots & \ddots & \vdots \\ 0 & 0 & \cdots & e^{\lambda_n(t-t_o)} \end{bmatrix} \cdot V^T.$$

What is the comparable *PDE*? This is where linear algebra gives way to functional analysis and the idea of linear operators is introduced. Matrices are *finite dimensional* linear operators with a finite collection of eigenvalues $\{\lambda_1, \lambda_2, \ldots, \lambda_n\}$ and eigenvectors $\{\boldsymbol{v}_1, \boldsymbol{v}_2, \ldots, \boldsymbol{v}_n\}$. The eigenvalues and eigenvectors, in turn, contain the geometric and analytic information of the matrix. Partial differential equations encounter *infinite dimensional* linear operators L. These operators have a *spectrum* $\sigma(L) = \{\lambda: Lu = \lambda u$ for each u in a particular function space$\}$. These spectral values λ are the eigenvalues of L but are infinite in number. The corresponding functions u which satisfy $Lu = \lambda u$ for $\lambda \in \sigma(L)$ are the *eigenfunctions* of the operator L. Within the scope of *PDE*s, a *function space* is a set of functions that *span* or *form a basis* for the solutions of a particular equation. The solution of the *PDE* $\frac{\partial u}{\partial t}(\boldsymbol{x},t) = Lu(\boldsymbol{x},t)$, $\boldsymbol{x} \in \Omega \subset \mathbb{R}^n$, $t > t_o$ and $u(\boldsymbol{x},t_o) = u_o(\boldsymbol{x})$ for $\boldsymbol{x} \in \partial\Omega$ is a function of the eigenvalues and eigenfunctions of the linear operator L. If $\Omega = \mathbb{R}^n$, then $u(\boldsymbol{x}, t) = e^{L(t-t_o)}u_o(\boldsymbol{x})$ and the solution operator $e^{L(t-t_o)}$ can be derived via the Fourier transform. This is the focus of Chapters 4 and 5. If, conversely, Ω is a bounded subset of $\mathbb{R}^n$, then $e^{L(t-t_o)}$ is determined by way of the separation of variables transform. When $\frac{\partial u}{\partial t} = 0$, then the equation $Lu = 0$ can also be solved via separation of variables. If, however, the right-hand side replaces 0 with the Dirac delta function, then $Lu = \delta(\boldsymbol{x}_o)$. This gives rise to Green's functions as solutions and necessitates multivariable integration theory.

What happens when the equations become *nonlinear*? For *ODE*s, there are several approaches. If the equation is separable, then direct integration can be applied. While this approach can be successful in a wide variety of cases, even the relatively simple equation $dx/dt = \cos(t)$, $x(t_o) = x_o$ results in the implicit solution $(\sin(x) + 1)/\cos(x) = C_o e^{(t-t_o)}$ and $C_o = (\sin(x_o) + 1)/\cos(x_o)$. In other instances, a power series solution can be achieved. One of the most famous examples is the Bessel function of order ν which satisfies the *ODE* $x^2 \frac{d^2 y}{dx^2} + x \frac{dy}{dx} + (x^2 - \nu^2) y = 0$. When analytic techniques fail, then numerical methods are utilized. Solutions to nonlinear *PDE*s present similar challenges. In Chapter 6, the idea of transforming a nonlinear *PDE* to a linear equation is presented via three examples. A nonlinear wave equation (Klein–Gordon) $\frac{\partial^2 u}{\partial t^2} = \sigma^2 \frac{\partial^2 u}{\partial x^2} + \alpha u^n$ is solved analytically by the separation of variables method. Analytic solutions to Burgers' equation $\frac{\partial u}{\partial t} = \sigma^2 \frac{\partial^2 u}{\partial x^2} - \frac{1}{2} \frac{\partial}{\partial x}(u^2)$ are determined through a change of variables $u = H(v)$ which transforms the nonlinear equation into the linear heat equation $\frac{\partial v}{\partial t} = \sigma^2 \frac{\partial^2 v}{\partial x^2}$. The famous *inverse scattering transform* (*IST*) is applied to the Korteweg de Vries equation $\frac{\partial u}{\partial t} - 3 \frac{\partial}{\partial x}(u^2) + \frac{\partial^3 u}{\partial x^3} = 0$. As part of the transform process, the linear Airy's *PDE* $\frac{\partial v}{\partial t} + \frac{\partial^3 v}{\partial x^3} = 0$ must be solved. While a robust method, the *IST* cannot be universally applied to the general nonlinear problem. Even the specific problem of a nonlinear evolution equation $\frac{\partial u}{\partial t} = Lu + N(u)$ [1] resists an analytic recipe for solution.

Ultimately, as with *ODE*s, there are *PDE*s that cannot be solved by analytic methods. Chapter 7 demonstrates how (1) a numeric method can be used to approximate the solution of a nonlinear system of *PDE*s, and (2) functional analysis is employed to prove that the numerical method is stable and convergent.

The first five chapters of this work develop Maxwell's equations and the analytic methods required to solve them. Along the way, four of the most fundamental *linear* partial differential equations of mathematical physics are examined and solved. These are *Laplace's equation*: $\Delta u = 0$, *Poisson's equation*: $\Delta u = \rho(\mathbf{x})$, *the heat equation*: $\frac{\partial u}{\partial t} = \Delta u$, and *the wave equation*: $\frac{\partial^2 u}{\partial t^2} = \Delta u$. Chapter 6 provides examples of a change of variables/transformation strategy that are effective in solving certain nonlinear *PDE*s. Chapter 7 shows how to manage the numerical solution of a system of nonlinear *PDE*s. Is there a general theory as to when certain classes of nonlinear *PDE*s can be solved analytically? The answer is provided in Chapter 8 by way of a generalization of the

[1] Here L is a linear and N a nonlinear operator.

Hartman–Grobman Theorem. This theorem states that, for nonlinear evolution equations $\frac{\partial u}{\partial t} = Lu + N(u)$ with very general conditions required of the linear operator L and nonlinear operator N, there exists a change of variables $u = H(v)$ so that v satisfies the corresponding linear equation $\frac{\partial v}{\partial t} = Lv$.

Notation, Nomenclature, and List of Symbols

The following is a list of symbols used throughout this text.

*ODE*s = ordinary differential equations
*PDE*s = partial differential equations

Numbers and Sets

$\mathbb{R}$ = the real line $(-\infty,\infty)$
$\mathbb{R}^+$ = the positive real line $[0,\infty)$
$\mathbb{R}^n$ = n–dimensional Euclidean space
$\boldsymbol{x} = (x_1, x_2, \ldots, x_n)$ = an n–dimensional vector
$\|\boldsymbol{x}\| = \sqrt{x_1^2 + x_2^2 + \cdots + x_n^2}$ = the Euclidean norm of a vector
$\mathbb{C}$ = the complex plane = $\{z = x + i\cdot y: x, y \in \mathbb{R}, i = \sqrt{-1}\}$
$\mathbb{Z} = \{\ldots, -3, -2, -1, 0, 1, 2, 3, \ldots\}$ = the set of integers
$\mathbb{Z}^+ = \{1, 2, 3, \ldots\}$ = the set of all positive integers
$\mathbb{Z}_o^+ = \{0, 1, 2, 3, \ldots\}$ = the set of all non–negative integers
$\mathbb{Z}^- = \mathbb{Z} \setminus \mathbb{Z}_o^+ = \{-1, -2, -3, \ldots\}$ = the set of all negative integers
$\mathcal{A} \setminus \mathcal{B} = \{x \in \mathcal{A} \text{ and } x \notin \mathcal{B}\}$ = the *complement* of the set $\mathcal{A}$ with respect to the set $\mathcal{B}$
$\mathbb{B}_n(\boldsymbol{x}_o, r) = \left\{\boldsymbol{x} \in \mathbb{R}^n : \|\boldsymbol{x} - \boldsymbol{x}_o\|^2 \le r\right\}$ is the n–dimensional ball of radius r centered at $\boldsymbol{x}_o$
$\mathbb{S}_{n-1}(\boldsymbol{x}_o, r) = \left\{\boldsymbol{x} \in \mathbb{R}^n : \|\boldsymbol{x} - \boldsymbol{x}_o\|^2 = r\right\}$ is the $(n-1)$–dimensional sphere of radius r centered at $\boldsymbol{x}_o$
When the dimension n is clear, then the notation $\mathbb{B}(\boldsymbol{x}_o, r)$ and $\mathbb{S}(\boldsymbol{x}_o, r)$ is also used
$\mathbb{S}(\boldsymbol{0}, r) = \left\{\boldsymbol{x} \in \mathbb{R}^n : \|\boldsymbol{x}\|^2 = r\right\}$ is the $(n-1)$–dimensional sphere of radius r centered at the origin $\boldsymbol{0}$
$\mathbb{S}_1 = \left\{\boldsymbol{x} \in \mathbb{R}^n : \|\boldsymbol{x}\|^2 = 1\right\}$ is the unit $(n-1)$–dimensional sphere
$\mathcal{C}_n(r) = [0, r] \times [0, r] \times \ldots \times [0, r] \subset \mathbb{R}^n$ is the n–dimensional hypercube with sides of length r
$\mathcal{C}_n(1)$ is the n–dimensional unit hypercube and
$\mathcal{C}_3(1)$ is the unit cube in $\mathbb{R}^3$

Generalized Functions and Distributions

$$\mathcal{I}_\Omega(\boldsymbol{x}) = \begin{cases} 1 & \text{for } \boldsymbol{x} \in \Omega \\ 0 & \text{for } \boldsymbol{x} \notin \Omega \end{cases} = \text{the } \textit{indicator function on the set } \Omega$$

Ω is a domain in $\mathbb{R}^n$ and $\partial\Omega$ is the *boundary* of the domain Ω

$$\delta(x) = \lim_{n\to\infty}\left[n \cdot \mathcal{I}_{\left[-\frac{1}{2n}, \frac{1}{2n}\right]}(x) \right] = \text{the } \textit{Dirac delta function}$$

$$H(x) = \begin{cases} 0 & \text{for } x < 0 \\ \frac{1}{2} & \text{for } x = 0 \\ 1 & \text{for } x > 0 \end{cases} = \text{the } \textit{Heaviside function}$$

The Dirac delta function can also be defined as $\delta(x) = \lim\limits_{\varepsilon\to 0^+} \frac{1}{\varepsilon\sqrt{\pi}} e^{-\frac{x^2}{\varepsilon^2}}$. The Heaviside and Dirac delta functions are related via $H(x) = \int_{-\infty}^{x} \delta(z)\,dz$ and $H'(x) = \delta(x)$.

Linear Algebra, Operators, and Function Spaces

$$A = \begin{bmatrix} a_{1,1} & a_{1,2} & \cdots & a_{1,m} \\ a_{2,1} & a_{2,2} & \cdots & a_{2,m} \\ \vdots & \vdots & \ddots & \vdots \\ a_{n,1} & a_{n,2} & \cdots & a_{n,m} \end{bmatrix} = \text{an } n \times m \text{ matrix}$$

$$\mathcal{M}at_{n\times m}(\mathbb{R}) = \left\{ \left[a_{j,k} \right]_{j=1,k=1}^{n,m} : a_{j,k} \in \mathbb{R} \right\} = \text{the set of all } n \times m \text{ matrices with elements in } \mathbb{R}$$

$$I_{n\times n} = \begin{bmatrix} 1 & 0 & \cdots & 0 \\ 0 & 1 & \cdots & 0 \\ \vdots & \vdots & \ddots & \vdots \\ 0 & 0 & \cdots & 1 \end{bmatrix} = \text{the } n \times n \text{ identity matrix}$$

$\langle c_j \rangle_{j=1}^{\infty} = \{c_1, c_2, \ldots, c_n, \ldots\}$ is an infinite sequence

$\langle c_j \rangle_{j=1}^{n} = \{c_1, c_2, \ldots, c_n\}$ is a finite sequence

$L_p(\Omega) = \{f\colon \int_\Omega |f|^p \, d\omega < \infty\}$ = the *Lebesgue space of order p*

$$\|f\|_p = \left(\int_\Omega |f|^p \, d\omega \right)^{\frac{1}{p}} = \text{the norm on } L_p(\Omega)$$

A *Banach space* is a complete normed space $\left(X, \|\bullet\|_X\right)$ where X is a vector space and $\|\bullet\|_X$ is the norm on X. If $\langle \boldsymbol{x}_n \rangle_{n=1}^{\infty} \subset X$ is a convergent sequence, then *X is complete* provided there exists an element $\xi \in X$ so that $\lim_{n\to\infty} \|\boldsymbol{x}_n - \xi\| = 0$. The Lebesgue spaces are Banach spaces.

$\| \bullet \|_X$ is the norm on the space X.

I = the *identity operator* on a function space. That is, for $\boldsymbol{x} \in X$, $I(\boldsymbol{x}) = \boldsymbol{x}$.

L is a *linear operator* on a function space X provided $L(\alpha_1 \cdot \boldsymbol{x}_1 + \alpha_2 \cdot \boldsymbol{x}_2) = \alpha_1 \cdot L(\boldsymbol{x}_1) + \alpha_2 \cdot L(\boldsymbol{x}_2)$ for any $\boldsymbol{x}_1, \boldsymbol{x}_2 \in X$, and all scalars α_1, α_2.

A linear operator $L: X \to Y$ is *bounded* provided there exists a constant c_L so that $\|Lx\|_Y \le c_L \cdot \|x\|_X$ for any $x \in X$.

An operator $F: X \to Y$ is a *contraction* if $\|Fx\|_Y < \|x\|_X$ for any $x \in X$.

An operator $F: X \to Y$ is an *expansion* if $\|Fx\|_Y > \|x\|_X$ for any $x \in X$.

If $S(t): X \to X$ is a family of bounded linear operators mapping the Banach space X into itself for $t \in [0, \infty)$ so that (*i*) $T(0) = I$ and (*ii*) $S(t+\tau) = S(t)S(\tau)$ for each t, $\tau \ge 0$, then S is a *semigroup*.

If L is a bounded linear operator on the Banach space X so that $Au = \left.\frac{d}{dt} S(t)u\right|_{t=0}$, then L is the *infinitesimal generator of the semigroup* $S(t)$. If $S(t)$ is uniformly continuous, then L is the *infinitesimal generator of a C_o–semigroup* $S(t)$ in which case $\lim_{t\to\tau} \|S(t) - S(\tau)\|_X = 0$ and S is written as $S(t) = e^{tL}$.

$\sigma(L) = \{\lambda: Lu = \lambda u\}$ = the *spectrum* of a linear operator L. When $L \in \mathcal{M}at_{n\mathrm{x}n}(\mathbb{R})$ is an n x n matrix, then $\sigma(L)$ is the set of *eigenvalues*.

Initial and Boundary Conditions

If L is a linear and N a nonlinear operator on the space X, then $\frac{\partial u}{\partial t} = Lu + N(u)$ is referred to as a *nonlinear evolution equation*. This is also called a *quasi–linear equation*. The initial condition $u(\boldsymbol{x}, t_o) = f(\boldsymbol{x})$ is called the *Cauchy data* of the equation. For *PDE*s that are second order in time, $\frac{\partial^2 u}{\partial t^2} = Lu + N(u)$, the Cauchy data take the form $u(\boldsymbol{x},t_o) = f(\boldsymbol{x})$ and $\frac{\partial u}{\partial t}(\boldsymbol{x},t_o) = g(\boldsymbol{x})$.

For a *PDE* of the form $\frac{\partial u}{\partial t}(\boldsymbol{x},t) = Lu(\boldsymbol{x},t) + N\big(u(\boldsymbol{x},t)\big)$, $\boldsymbol{x} \in \Omega \subset \mathbb{R}^n$ with Ω a bounded domain, the initial condition $u(\boldsymbol{x}, t) \equiv 0$ for $\boldsymbol{x} \in \partial\Omega$ is called a *Dirichlet boundary condition*. The notion of Dirichlet boundary conditions and Cauchy data can be extended for more general forms of *PDE*s.

Indeed, the most general form of a partial differential equation is established in (0.1) by the somewhat cumbersome set of symbols below.

$\boldsymbol{\alpha} = (\alpha_1, \alpha_2, \dots, \alpha_n)$, $|\boldsymbol{\alpha}| = \alpha_1 + \alpha_2 + \cdots + \alpha_n$

$\boldsymbol{x} = (x_1, x_2, \dots, x_{n+1}) \in \Omega \times [0,T) \subset \mathbb{R}^{n+1}$

$\widehat{\boldsymbol{x}} = (x_1, x_2, \cdots, x_n) \in \Omega \subseteq \mathbb{R}^n$ = the truncated vector of $\boldsymbol{x} \in \mathbb{R}^{n+1}$

$$D^{\alpha}u(\boldsymbol{x}) = \frac{\partial^{|\alpha|}u(\boldsymbol{x})}{\partial x_1^{\alpha_1}\partial x_2^{\alpha_2}\cdots\partial x_n^{\alpha_n}}, \quad D^k u(\boldsymbol{x}) = \left\{D^{\alpha}u(\boldsymbol{x}) : |\boldsymbol{\alpha}| = k\right\}$$

$$\sum_{|\alpha|=k} f_{\alpha}\left(D^{k-1}u, D^{k-2}u, \cdots, Du, u, \boldsymbol{x}\right)D^{\alpha}u + f_o\left(D^{k-1}u, D^{k-2}u, \cdots, Du, u, \boldsymbol{x}\right) = 0, \; \boldsymbol{x} \in \Omega \times [0,T) \quad (0.1)$$

If (0.1) is a *PDE* of order k in time, then the *Cauchy data* are

$$u\left(\widehat{\boldsymbol{x}}, t_o\right) = \phi_o\left(\widehat{\boldsymbol{x}}\right), \frac{\partial u}{\partial t}\left(\widehat{\boldsymbol{x}}, t_o\right) = \phi_1\left(\widehat{\boldsymbol{x}}\right), \cdots, \frac{\partial^k u}{\partial t^k}\left(\widehat{\boldsymbol{x}}, t_o\right) = \phi_k\left(\widehat{\boldsymbol{x}}\right).$$

If Ω is a bounded set, then $u(\boldsymbol{x}, t) \equiv \psi(\boldsymbol{x})$ for $\boldsymbol{x} \in \partial\Omega$ is the general *Dirichlet boundary condition*. It is often the case that $\psi(\boldsymbol{x}) \equiv 0$.

If the initial conditions for (0.1) are $u\left(\widehat{\boldsymbol{x}}, t_o\right) + \frac{\partial u}{\partial \boldsymbol{n}}\left(\widehat{\boldsymbol{x}}, t_o\right) = \psi\left(\widehat{\boldsymbol{x}}\right)$, this constitutes the *Robin boundary condition*. Here, $\frac{\partial u}{\partial \boldsymbol{n}} = \boldsymbol{n} \bullet \nabla u$ is the derivative with respect to the normal vector $\boldsymbol{n}$.

Differential (Linear) Operators

Perhaps the most famous differential operator examined in the study of partial differential equations is the *Laplacian* $\Delta = \frac{\partial^2 u}{\partial x_1^2} + \frac{\partial^2 u}{\partial x_2^2} + \cdots + \frac{\partial^2 u}{\partial x_n^2}$. The Laplacian is directly related to the *gradient operator* $\nabla = \left(\frac{\partial}{\partial x_1}, \frac{\partial}{\partial x_2}, \cdots, \frac{\partial}{\partial x_n}\right)$ and is often written as $\nabla^2 = \Delta$ where $\nabla^2 = \nabla \bullet \nabla$ is the dot product of the gradient with itself. The partial differential operator $\Delta_t \equiv \frac{\partial}{\partial t} - \Delta$ is called the *diffusion operator* and defines the heat equation $\left(\frac{\partial}{\partial t} - \Delta\right)u = 0$. Just as the solution of the matrix equation $A\boldsymbol{x} = \boldsymbol{b}$ is written as $\boldsymbol{x} = A^{-1}\boldsymbol{b}$, the solution of the *inhomogeneous heat equation* $\Delta_t u = \rho(\boldsymbol{x})$ can be viewed as the inversion of the heat operator $u(\boldsymbol{x}, t) = \Delta_t^{-1}\rho(\boldsymbol{x})$. Chapter 4 shows how the Fourier transform is used to invert Δ_t. The *d'Alembertian* or *d'Alembert's operator* $\square_t \equiv \frac{\partial^2}{\partial t^2} - \Delta$

corresponds to the wave equation as $\left(\frac{\partial^2}{\partial t^2} - \Delta\right)u = 0$, and the solution of the inhomogeneous equation $\square_t u(\boldsymbol{x},t) = \rho(\boldsymbol{x},t)$ is solved via the formalism $u(\boldsymbol{x},t) = \square_t^{-1}\rho(\boldsymbol{x},t)$. Chapter 5 will expand upon these ideas.

Function Spaces

Function spaces are used to determine the form of the solution of a particular *PDE*. Existence and uniqueness results depend heavily upon the function spaces on which the solutions are defined. This relation will be seen throughout the book but especially in Chapters 7 and 8.

$C^0(E) = \{f : f(x) \text{ is continuous for all } x \in E\}$

$C^m(E) = \left\{ f : \frac{d^n f}{dx^n}(x) \in C^0(E) \text{ for all } n \le m,\ x \in E \right\}$

$C^\infty(E) = \left\{ f : \frac{d^n f}{dx^n}(x) \in C^0(E) \text{ for all } n \ge 0,\ x \in E \right\}$

$C^\omega(E) = \left\{ f \in C^\infty(E) : \text{the series } \sum_{n=1}^{\infty} \frac{d^n f}{dx^n}(x_o)\cdot(x - x_o)^n \text{ exists and converges for all } x \in E \right\}$

$L_p(\Omega) = \{f: \int_\Omega |f|^p \, d\omega < \infty\}$

$L_p^{(T)}(\Omega) = \{f \in L_p(\Omega) : f(x+T) = f(x)\} = L_p(\Omega)$ functions which are T–periodic

$\mathscr{C}(X,Y)$ is the set of all continuous maps from X into Y

$\mathscr{C}^m(X,Y)$ is the set of all m–times continuously Fréchet–differentiable maps from X into Y

$f'(u)v$ is the *Fréchet derivative* of the operator $f(u)$ evaluated at v and is computed as

$$f'(u)v = \lim_{h\to 0} \frac{f(u + h\cdot v) - f(u)}{h}$$

$\mathscr{L}(X,Y)$ is the set of all linear maps from X into Y

$\mathscr{C}_*^0(X,Y) = \{f \in \mathscr{C}(X,Y)$: f is uniformly bounded and uniformly continuous$\}$

$\mathscr{M}(X,Y) = \{f \in \mathscr{C}^1(X,Y)$: f is bounded and $f(0) = 0 = f'(0)v$ for all $v \in Dom[f'(0)]\}$

$\mathscr{B}_X(0,r) = \{\boldsymbol{x} \in X: \|\boldsymbol{x}\|_X \le r\} \equiv$ the solid sphere (ball) of radius r in X–space

$H_m(\Omega) = \{f: D^{\boldsymbol{q}}(f) \in L_2(\Omega) \text{ for all } |\boldsymbol{q}| < m\} = W^{m,2}(\Omega) =$ the *Sobolev space* of order m

$D^{\boldsymbol{q}} = \frac{\partial^{|\boldsymbol{q}|}}{\partial x_1^{q_1}\partial x_2^{q_2}\cdots\partial x_n^{q_n}}$, $\boldsymbol{q} = (q_1, q_2, \ldots, q_n)$, $|\boldsymbol{q}| = \sum_{j=1}^{n} q_j$, $q_j \in \mathbb{Z}^+$

$H^1([0,T];Z) = \left\{ f : \int_0^T \left(\|f(t)\|_Z^2 + \left\| \frac{\partial f}{\partial t}(t) \right\|_Z^2 \right) dt < \infty \right\}$

If F: $X \rightarrow Y$ is an operator mapping the function space X into the space Y, then the *domain* of F is X and the *range* of F is Y. This is written as $Dom[F]$ and $Range[F]$, respectively.

MATLAB® Code

A catalog of all MATLAB files (M–files) used in this book is contained in the Appendix. The code blocks used to generate most of the figures and all of the calculations presented in this text are contained on the website below.

http://www.morganclaypoolpublishers.com/Costa/

While most of the computations only require MATLAB, some of the analytic calculations are performed using the Symbolic Math Toolbox (an add-on set of specialized M–files that access a computer algebra package).

Chapter 1

The Equations of Maxwell

The aim and purpose of this text is to show the reader how to find solutions of both linear and nonlinear partial differential equations (*PDE*s). It will focus almost exclusively on equations from mathematical physics. This is not surprising since *PDE*s model phenomena which evolve in both space and time. As remarked in the introduction, ordinary differential equations (*ODE*s) evolve exclusively with respect to a single variable (generally time). *PDEs* evolve with respect to several variables. In mathematical physics, those variables are temporal (t) and spatial ($\boldsymbol{x}$). Among the most famous and meaningful *PDE*s in mathematical physics are Maxwell's equations. These equations, expressed by James Clerk Maxwell (1831–1879), provide a complete model of electromagnetic radiation unifying electricity, magnetism, and light [85] under the umbrella of classical Newtonian physics. They are a crowning achievement not only of $19^{\underline{th}}$ century science but of human history.

To solve Maxwell's equations, a seemingly circuitous route is taken first through the distribution of electric charges over a bounded domain, to heat conduction, and finally to wave motion. These phenomena, in turn, are modeled by four fundamental equations of classical physics. Once solutions of these equations are determined, then solutions of Maxwell's equations can be achieved. In this chapter, Maxwell's equations are derived by utilizing the theory of multivariate integration. As a first step, the aforementioned "big four" *PDE*s of mathematical physics are listed.

Laplace's equation: $\Delta u = 0$

Poisson's equation: $\Delta u = \rho(\boldsymbol{x})$

The heat equation: $\dfrac{\partial u}{\partial t} = \Delta u$

The wave equation: $\dfrac{\partial^2 u}{\partial t^2} = \Delta u$

The symbol Δ is called the *Laplacian* or *Laplace's operator* and corresponds to the n–dimensional analog of the second derivative for $n \geq 2$. More precisely, $\Delta = \sum_{j=1}^{n} \dfrac{\partial^2}{\partial x_j^2}$. The Laplacian can also be viewed as the inner product of the *gradient operator* $\nabla \equiv \left(\frac{\partial}{\partial x_1}, \frac{\partial}{\partial x_2}, \cdots, \frac{\partial}{\partial x_n} \right)$ with itself. Specifically,

$$\nabla \bullet \nabla = \left(\frac{\partial}{\partial x_1}, \frac{\partial}{\partial x_2}, \cdots, \frac{\partial}{\partial x_n} \right) \bullet \left(\frac{\partial}{\partial x_1}, \frac{\partial}{\partial x_2}, \cdots, \frac{\partial}{\partial x_n} \right) = \sum_{j=1}^{n} \frac{\partial^2}{\partial x_j^2} = \Delta .$$

Consequently, the Laplacian is often denoted as $\nabla^2 = \Delta$. The partial differential expression $\frac{\partial}{\partial t} - \Delta$ is called the *diffusion operator*. It can be seen that $\left(\frac{\partial}{\partial t} - \Delta \right) u = 0$ is an alternate form of the heat equation. The *d'Alembertian* or *d'Alembert's operator* $\frac{\partial^2}{\partial t^2} - \Delta$ corresponds to the wave equation as $\left(\frac{\partial^2}{\partial t^2} - \Delta \right) u = 0$. These mathematical expressions are used first to describe the equations under study but later as a shorthand for a general theory of partial differential equations. For example, the matrix equation $A\boldsymbol{x} = \boldsymbol{b}$ is known to have the solution $\boldsymbol{x} = A^{-1}\boldsymbol{b}$ whenever the inverse of the matrix A exists. As will be seen in subsequent chapters, this idea is extended to the partial differential operators $\Delta_t \equiv \frac{\partial}{\partial t} - \Delta$ and $\Box_t \equiv \frac{\partial^2}{\partial t^2} - \Delta$. Namely, a solution of the *inhomogeneous heat equation* $\Delta_t u = \rho(\boldsymbol{x})$ can be viewed as $u(\boldsymbol{x}, t) = \Delta_t^{-1} \rho(\boldsymbol{x})$. The exact meaning of, and formulae corresponding to, this inverse operator are detailed in the later chapters.

For now, a summary of the important vector differential operators acting over three–dimensional Euclidean space $\mathbb{R}^3$ is the next step in this process.

§1.1. grad, div, and curl

Let u: $\mathbb{R}^3 \to \mathbb{R}$ and $\boldsymbol{F}$: $\mathbb{R}^3 \to \mathbb{R}^3$ be *smooth* functions. That is, let u and $\boldsymbol{F}$ be so that all derivatives of a specified order exist and are continuous in the spatial variables $x_1 = x$, $x_2 = y$, and $x_3 = z$. This notion is made more precise later in the section. The function u is called a *scalar–valued function* since it has values on the real line $\mathbb{R}$. Conversely, the function $\boldsymbol{F}$ is a *vector field* since it has values on $\mathbb{R}^3$. The *gradient* of a scalar–valued function u is the vector field

$$grad(u) = \nabla u = \left(\frac{\partial u}{\partial x}, \frac{\partial u}{\partial y}, \frac{\partial u}{\partial z} \right). \tag{1.1.1}$$

The *divergence* of the vector field $\boldsymbol{F} = (F_1, F_2, F_3)$ is the scalar–valued function

$$div(\boldsymbol{F}) = \nabla \bullet \boldsymbol{F} = \frac{\partial F_1}{\partial x} + \frac{\partial F_2}{\partial y} + \frac{\partial F_3}{\partial z}. \tag{1.1.2}$$

The *curl* of a vector field $\boldsymbol{F}$ is the cross product of $\boldsymbol{F}$ with the gradient operator ∇

$$curl(\boldsymbol{F}) = \nabla \times \boldsymbol{F} = \det\begin{bmatrix} \boldsymbol{i} & \boldsymbol{j} & \boldsymbol{k} \\ \dfrac{\partial}{\partial x} & \dfrac{\partial}{\partial y} & \dfrac{\partial}{\partial z} \\ F_1 & F_2 & F_3 \end{bmatrix} = \boldsymbol{i}\left(\frac{\partial F_3}{\partial y} - \frac{\partial F_2}{\partial z}\right) - \boldsymbol{j}\left(\frac{\partial F_3}{\partial x} - \frac{\partial F_1}{\partial z}\right) + \boldsymbol{k}\left(\frac{\partial F_2}{\partial x} - \frac{\partial F_1}{\partial y}\right). \tag{1.1.3}$$

The vectors $\boldsymbol{i}, \boldsymbol{j}$, and $\boldsymbol{k}$ are the fundamental unit vectors in the direction of the x–, y–, and z–axes. The notation $\boldsymbol{e}_1$, $\boldsymbol{e}_2$, and $\boldsymbol{e}_3$ is also used. More precisely,

$$\boldsymbol{i} = \boldsymbol{e}_1 = [1,0,0], \boldsymbol{j} = \boldsymbol{e}_2 = [0,1,0], \boldsymbol{k} = \boldsymbol{e}_3 = [0,0,1]. \tag{1.1.4}$$

Observe that the curl is *only* defined on three–dimensional vector fields. While the gradient and divergence can readily be extended to n–dimensions ($n \geq 3$), the generalized curl requires an exploration into exterior calculus and differential forms. These ideas are beyond the focus of this work. The interested reader, however, is referred to Flanders [25], Bamberg and Sternberg [6], and Spivak [72, 73].

In subsequent chapters, the solutions of the fundamental linear partial differential equations, Laplace's, Poisson's, the heat, and the wave equations are detailed. These equations are associated with the 19th century French mathematicians Pierre–Simon Laplace, Siméon Denis Poisson, Jean le Rond d'Alembert, Peter Gustav Lejeune Dirichlet, and Augustin–Louis Cauchy. While noted to be German [83], Dirichlet was born in Düren in 1805. At that time, Düren was part of the French Empire. Also, Dirichlet attended the University of Paris and therefore is accorded honorary status as one of the influential French mathematicians who advanced the study of partial differential equations.

These historical curiosities aside, before describing the aforementioned mathematicians' contributions to partial differential equations, a short detour into the work of French and British physicists is in order.

One of the primary forces acting on physical systems is the interaction of electric and magnetic fields. By combining the laws of the French physicists Charles–Augustin de Coulomb (1785) and André–Marie Ampère (1822) along with his countryman Michael Faraday (1831), James Clerk Maxwell described the interaction between electric and magnetic fields using four partial differential equations which bear his name. In addition to providing an effective mathematical model to a vital physical system, *Maxwell's equations* represent a seminal scientific achievement. These relations can be derived by using two of the fundamental integral theorems of multivariate calculus: Stokes' Theorem and Gauss's Divergence Theorem. The development begins with *Faraday's law*.

§1.2. Faraday's Law

Let $\mathfrak{S}$ be a surface bounded by a closed *orientable curve* γ. A circle, for example, is an orientable curve in that an object's traverse about the circle does not change the direction or orientation of the object. For example, an observer walking along the interior of the circumference will always have the "up" direction pointing toward the circle's center. As Figure 1.1a illustrates, the direction of the arrows does not change orientation as a particle moves about the circle's circumference.

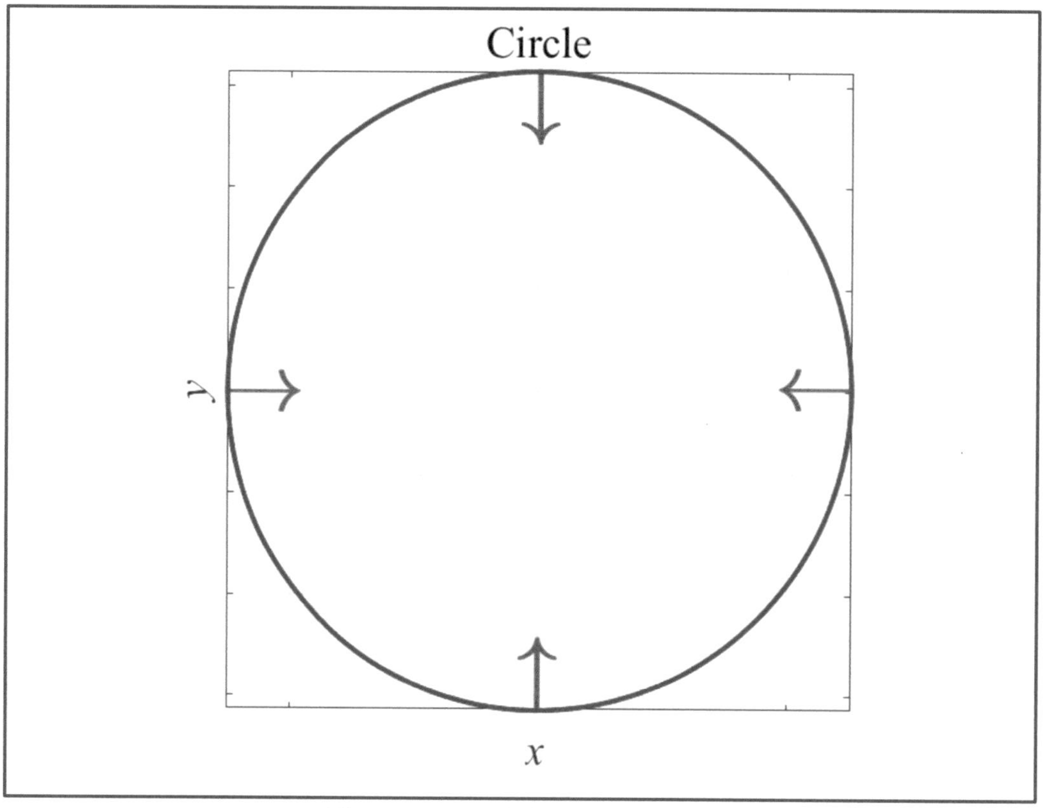

Figure 1.1a. Orientable circle.

Conversely, the direction of a particle's orientation (pointing first positive and then negative along the z–axis) changes on a Möbius band as a particle travels about the edge of the strip. This is depicted in Figure 1.1b. Consequently, the Möbius band is an example of a *non–orientable* surface. A Möbius band of radius R can be generated via the parametric equations

$$f_x(s, t) = \cos(t)\cdot(R + s\cdot\cos(t/2)), f_y(s, t) = \sin(t)\cdot(R + s\cdot\sin(t/2)), f_z(s, t) = s\cdot\sin(t/2)$$

with $s \in [-R, R]$ and $t \in [0, 2\pi]$.

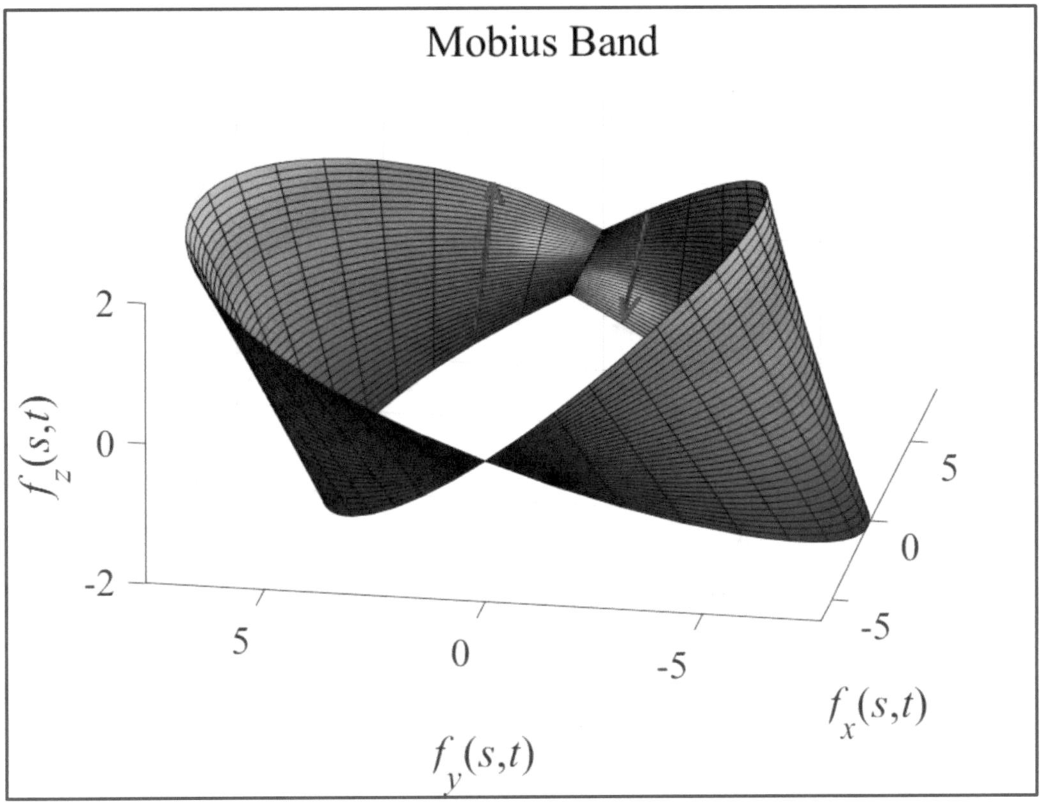

Figure 1.1*b*. Non–orientable Möbius band.

To develop *Faraday's law*, electric and magnetic field flows are restricted to orientable surfaces. Let ***B*** and ***E*** be magnetic and electric fields, respectively, acting over the surface $\mathbb{S}$ with orientable boundary γ. Further, suppose that $\mathbb{E}$ is the electro-motive force and F_B the magnetic flux over $\mathbb{S}$. The flux is a measure of how much of a vector field passes through a surface as per equation (1.2.2). Faraday's law states that the rate of change in the magnetic flux with respect to time is negatively proportional to the electro-motive force. The constant of proportionality is the speed of light c = 299,792,458 meters per second. Hence, Faraday's law is written, mathematically, as

$$\frac{dF_B}{dt} = \frac{d}{dt} Flux(\boldsymbol{B}) = -c\mathbb{E}. \tag{1.2.1}$$

Consider the geometry of the surface depicted in Figure 1.2. Let γ be the orientable curve that bounds the surface $\mathbb{S}$. That is, $\gamma = \partial\mathbb{S}$ is the boundary of the surface $\mathbb{S}$. Further, let V be the volume encapsulated by the surface $\mathbb{S}$. Let $\boldsymbol{n}(\boldsymbol{x}_o)$ and $\boldsymbol{t}(\boldsymbol{x}_o)$ be the (outward) normal and tangent vectors to the surface $\mathbb{S}$ at the point $\boldsymbol{x}_o$. If dγ is the element of arc length along the boundary γ and $\boldsymbol{i}, \boldsymbol{j}$, and $\boldsymbol{k}$ are the elementary unit vectors in $\mathbb{R}^3$ as defined in (1.1.4), then at a given point $\boldsymbol{x} \in \partial\mathbb{S} = \gamma$, $\boldsymbol{t}(\boldsymbol{x}) = \boldsymbol{i}\frac{dx}{d\gamma} + \boldsymbol{j}\frac{dy}{d\gamma} + \boldsymbol{k}\frac{dz}{d\gamma}$.

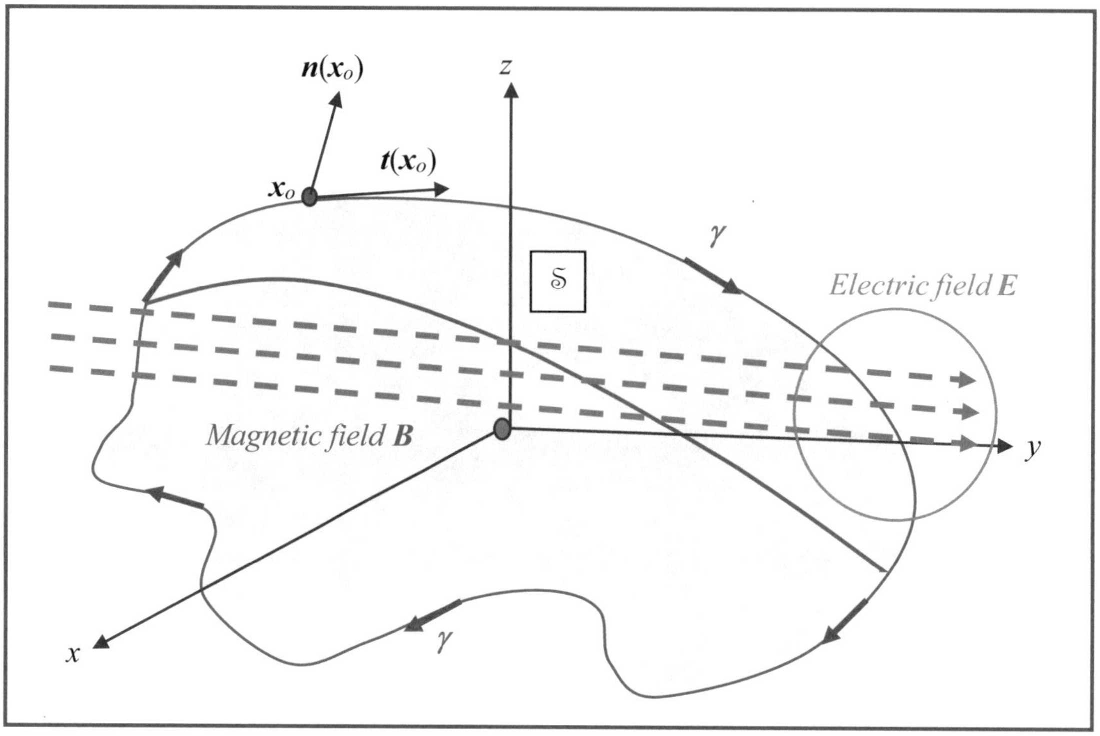

Figure 1.2. The magnetic $\boldsymbol{B}$ and electric $\boldsymbol{E}$ fields acting on a surface $\mathbb{S}$.

If the surface $\mathbb{S}$ is described as $z = f(x, y)$ by a sufficiently smooth function f, then the outward normal has the form $\boldsymbol{n}(\boldsymbol{x}) = \frac{1}{\sqrt{1+\left(\frac{df}{dx}\right)^2+\left(\frac{df}{dy}\right)^2}}\left(-\boldsymbol{i}\frac{df}{dx}-\boldsymbol{j}\frac{df}{dy}+\boldsymbol{k}\right)$. The *flux* of a vector field $\boldsymbol{B}$ is defined with respect to a surface integral over $\mathbb{S}$

$$F_B = Flux(\boldsymbol{B}) = \iint_{\mathbb{S}} \boldsymbol{B}(\boldsymbol{x},t)\bullet\boldsymbol{n}(\boldsymbol{x})\,d\boldsymbol{\sigma} \tag{1.2.2}$$

where $d\boldsymbol{\sigma}$ is the element of surface area. The electro-motive force $\mathbb{E}$, however, is the line integral of the electric field $\boldsymbol{E}$ over the boundary curve γ

$$\mathbb{E} = \oint_{\gamma} \boldsymbol{E}(\boldsymbol{x},t)\bullet\boldsymbol{t}(\boldsymbol{x})\,d\gamma\,. \tag{1.2.3}$$

Assume that $\boldsymbol{B}$ and $\boldsymbol{E}$ are sufficiently smooth vector valued functions so that the first temporal and second spatial derivatives are continuous. That is, $\boldsymbol{B}, \boldsymbol{E} \in C^2(\Omega) \times C^1([0, T])$ where the spaces of continuous functions are defined as follows:

$C^0(\Omega) = \{f: f \text{ is a continuous function on } \Omega\}$
$C^1(\Omega) = \{f: \nabla f \text{ is a continuous function on } \Omega\} = \{f: \nabla f \in C^0(\Omega)\}$
$C^2(\Omega) = \{f: \Delta f \text{ is a continuous function on } \Omega\} = \{f: \Delta f \in C^0(\Omega)\}$ and more generally

$$C^p(\Omega) = \left\{ f : \frac{\partial^{|\boldsymbol{m}|} f}{\partial x_1^{m_1} \partial x_2^{m_2} \cdots \partial x_n^{m_n}} \in C^0(\Omega) \text{ for all } \boldsymbol{x} \in \Omega \text{ and } |\boldsymbol{m}| = m_1 + m_2 + \cdots + m_n \le p \right\}$$

is the set of all functions in which the pth-order derivative is continuous with respect to each variable. The set of all functions that are infinitely differentiable is denoted as $C^\infty(\Omega)$. Those functions which are $C^\infty(\Omega)$ and have convergent Taylor series representations are called *analytic* and designated as $C^\omega(\Omega)$.

The statement $\boldsymbol{E} \in C^2(\Omega) \times C^1([0, T])$ means that the vector field $\boldsymbol{E}(\boldsymbol{x}, t)$, as a function of $\boldsymbol{x} \in \Omega \subset \mathbb{R}^n$, has $\Delta\boldsymbol{E} \in C^0(\Omega)$ while, as a function of $t \in [0, T]$, $\frac{\partial}{\partial t}\boldsymbol{E} \in C^0([0, T])$. That is, $C^2(\Omega) \times C^1([0,T])$ $= \left\{ \boldsymbol{F} : \Delta\boldsymbol{F} \in C^0(\Delta) \text{ and } \frac{\partial \boldsymbol{F}}{\partial t} \in C^0([0,T]) \right\}$.

A *domain* $\mathbb{S} \subset \mathbb{R}^n$ is an open, connected set. The surface $\mathbb{S}$ is a domain whenever it is bounded by a smooth orientable curve γ. By Leibniz's formula, $\frac{d}{dt} Flux(\boldsymbol{B}) = \frac{dF_B}{dt}$ $= \frac{d}{dt}\iint_{\mathbb{S}} \boldsymbol{B}(\boldsymbol{x},t)\bullet\boldsymbol{n}(\boldsymbol{x})\, d\sigma = \iint_{\mathbb{S}} \frac{\partial \boldsymbol{B}}{\partial t}(\boldsymbol{x},t)\bullet\boldsymbol{n}(\boldsymbol{x})\, d\sigma$. The famous result of vector calculus, attributed to the Irish mathematician George Stokes, is critical to the simplification of Faraday's equation (1.2.1).

Stokes' Theorem: *If, as a function of the spatial variable* $\boldsymbol{x}$, $\boldsymbol{E} \in C^1(\mathbb{S}) \cap C^0(\gamma)$, *then* $\oint_\gamma \boldsymbol{E}(\boldsymbol{x},t)\bullet\boldsymbol{t}(\boldsymbol{x})\, d\gamma = \iint_{\mathbb{S}} \nabla\times\boldsymbol{E}(\boldsymbol{x},t)\bullet\boldsymbol{n}(\boldsymbol{x})\, d\sigma$ *for any domain* $\mathbb{S}$ *with boundary* $\partial\mathbb{S} = \gamma$.

Now by (1.2.3) and Stokes' Theorem, $\mathcal{E} = \oint_\gamma \boldsymbol{E}(\boldsymbol{x},t)\bullet\boldsymbol{t}(\boldsymbol{x})\, d\gamma = \iint_{\mathbb{S}} \nabla\times\boldsymbol{E}(\boldsymbol{x},t)\bullet\boldsymbol{n}(\boldsymbol{x})\, d\sigma$.

Consequently, (1.2.1) becomes $\iint_{\mathbb{S}} \frac{\partial \boldsymbol{B}}{\partial t}(\boldsymbol{x},t)\bullet\boldsymbol{n}(\boldsymbol{x})\, d\sigma = -c\iint_{\mathbb{S}} \nabla\times\boldsymbol{E}(\boldsymbol{x},t)\bullet\boldsymbol{n}(\boldsymbol{x})\, d\sigma$ or

$$\nabla\times\boldsymbol{E} = -\frac{1}{c}\frac{\partial \boldsymbol{B}}{\partial t}. \qquad \textit{(Faraday's Law)} \qquad (1.2.4)$$

§1.3. Coulomb's Law

The next portion of the effort to establish Maxwell's equations is *Coulomb's law*. To that end, let $q(t)$ be the charge of an electric field $\boldsymbol{E}$ over a surface $\mathbb{S}$ bounded by the smooth, closed, orientable curve γ. By Gauss's law, the charge is proportional to the flux of the electric field over the surface. That is,

$$4\pi q(t) = Flux(\boldsymbol{E}) = \iint_{\mathbb{S}} \boldsymbol{E}(\boldsymbol{x},t)\bullet\boldsymbol{n}(\boldsymbol{x})\, d\boldsymbol{\sigma}\,. \tag{1.3.1}$$

Let $\frac{1}{\varepsilon}\rho(\boldsymbol{x}, t)$ be the normalized charge density[2] of the electric field $\boldsymbol{E}$ over the volume V subtended by the surface $\mathbb{S}$. Coulomb's law states that the electric charge over the volume V is equal to the integral of the charge density over that volume. Namely,

$$q(t) = \iiint_{V} \tfrac{1}{\varepsilon}\rho(\boldsymbol{x},t)\, dV\,. \tag{1.3.2}$$

Figure 1.3 illustrates the geometry.

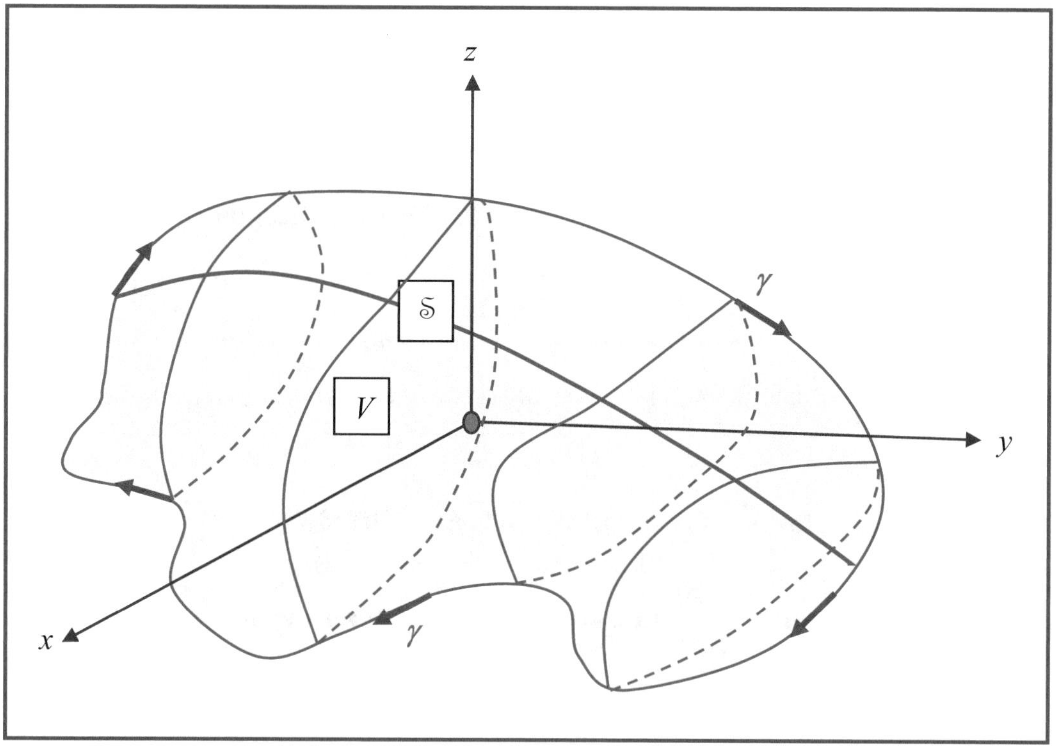

Figure 1.3. The volume V is subtended by the surface $\mathbb{S}$ and bounded by the curve γ.

[2] Traditionally, the charge density $\rho(\boldsymbol{x}, t)$ is taken over the dielectric field $\boldsymbol{D}$ which is related to the electric field $\boldsymbol{E}$ by the dielectric constant $\varepsilon = 8.85419 \times 10^{-12}$ F/m (*Farads/meter*). That is, $\boldsymbol{D} = \varepsilon\boldsymbol{E}$. In this treatment, the normalized density $\rho(\boldsymbol{x}, t)$ is used.

To write the charge as a differential equation, enlist *Gauss's Divergence Theorem*.

Gauss's Divergence Theorem: *If, as a function of the spatial variable* $\boldsymbol{x}$, $\boldsymbol{E} \in C^1(V) \cap C^0(\mathbb{S})$, *then for any volume* V *with boundary the domain* $\mathbb{S}$,

$$\iint_{\mathbb{S}} \boldsymbol{E}(\boldsymbol{x},t) \bullet \boldsymbol{n}(\boldsymbol{x})\, d\boldsymbol{\sigma} = \iiint_V div\big(\boldsymbol{E}(\boldsymbol{x},t)\big)\, dV = \iiint_V \nabla \bullet \boldsymbol{E}(\boldsymbol{x},t)\, dV .$$

Combine equations (1.3.1)–(1.3.2) to obtain $4\pi \iiint_V \frac{1}{\varepsilon}\rho(\boldsymbol{x},t)\, dV = \iint_{\mathbb{S}} \boldsymbol{E}(\boldsymbol{x},t) \bullet \boldsymbol{n}(\boldsymbol{x})\, d\boldsymbol{\sigma}$. By applying the Divergence Theorem, this equation becomes $\iiint_V \frac{4\pi}{\varepsilon}\rho(\boldsymbol{x},t)\, dV = \iiint_V \nabla \bullet \boldsymbol{E}(\boldsymbol{x},t)\, dV$ for an arbitrary volume V. Equating the integrands yields

$$\nabla \bullet \boldsymbol{E}(\boldsymbol{x},t) = \frac{4\pi}{\varepsilon}\rho(\boldsymbol{x},t). \qquad \textit{(Coulomb's Law)} \qquad (1.3.3)$$

The final phase in the construction of Maxwell's equations, is *Ampere's law*. At first it may appear that Maxwell is undeserving as the progenitor of the equations which bear his name. After all, it seems that he only collected and unified Ampere's, Coulomb's, and Faraday's laws. This superficial observation, however, neglects Maxwell's work to modify the mathematical form of Ampere's law; to unify the equations; to connect the magnetic $\boldsymbol{B}$ and electric $\boldsymbol{E}$ fields to the *wave equation*; and to present a coherent model of electro-magnetic theory.

§1.4. Ampere's Law

Ampere's law states that the electric current flux through a surface $\mathbb{S}$ bounded by a closed orientable curve γ is proportional to the magnetic loop tension around γ. The proportionality constant depends on the units of measure imposed upon the respective fields. Let $\boldsymbol{I}(\boldsymbol{x}, t)$ be the electric current, $\boldsymbol{H}(\boldsymbol{x}, t)$ be the magnetic loop tension, and $\mu_o = 4\pi \times 10^{-7}\ H/m$ be the free space permeability constant[3]. If Gaussian units[4] are employed, $\frac{4\pi}{c}\, Flux(\boldsymbol{I}(\boldsymbol{x}, t)) = \oint_\gamma \boldsymbol{H}(\boldsymbol{x},t) \bullet \boldsymbol{t}(\boldsymbol{x})\, d\gamma$.

Observe that the magnetic loop tension is directly proportional to the magnetic field so that $\boldsymbol{H}(\boldsymbol{x}, t) = \frac{1}{\mu_o}\boldsymbol{B}(\boldsymbol{x}, t)$. From the perspective of classical Newtonian physics, there are no free magnetic poles. Thus, the magnetic field and consequently magnetic loop tension have a divergence of 0.

[3] This constant is the measure of the amount of resistance produced when forming a magnetic field in a vacuum. It is measured in *Henries* per meter (*H*/*m*). For more details, see Assis and Chaib [5].

[4] For details concerning the difference between Gaussian vs. *MKS* units, see Marion [50]. For example, the *MKS* unit 1 *ampere* = 3 x 10^9 *statamperes* (a Gaussian unit); 1 volt (*MKS* unit) = $\frac{1}{300}$ *statvolt* (Gaussian unit).

$$\nabla \bullet \boldsymbol{B} = 0 = \nabla \bullet \boldsymbol{H} \qquad \text{(no magnetic charges)} \qquad (1.4.1)$$

By following the first sentence of this section, Ampere's law is written as (1.4.2)

$$\frac{4\pi}{c} Flux(\boldsymbol{I}(\boldsymbol{x}, t)) = \frac{4\pi}{c} \iint_{\mathfrak{S}} \boldsymbol{I}(\boldsymbol{x},t) \bullet \boldsymbol{n}(\boldsymbol{x})\, d\sigma = \oint_{\gamma} \boldsymbol{H}(\boldsymbol{x},t) \bullet \boldsymbol{t}(\boldsymbol{x})\, d\gamma = \oint_{\gamma} \frac{1}{\mu_o} \boldsymbol{B}(\boldsymbol{x},t) \bullet \boldsymbol{t}(\boldsymbol{x})\, d\gamma\,. \qquad (1.4.2)$$

Maxwell's rule states that the electric current is the sum of the dielectric displacement[5] and the current density of moving charges. That is, $\boldsymbol{I}(\boldsymbol{x}, t) = \frac{1}{4\pi} \frac{\partial \boldsymbol{D}}{\partial t}(\boldsymbol{x},t) + \boldsymbol{J}(\boldsymbol{x},t)$. The dielectric and electric fields are directly proportional via the dielectric constant ε (see Footnote 2 above): $\boldsymbol{D}(\boldsymbol{x}, t) = \varepsilon \cdot \boldsymbol{E}(\boldsymbol{x}, t)$. By Ampere's law and Stokes' Theorem, Maxwell's law can be implemented so that

$$\iint_{\mathfrak{S}} \frac{1}{\mu_o} \nabla \times \boldsymbol{B}(\boldsymbol{x},t) \bullet \boldsymbol{n}(\boldsymbol{x})\, d\sigma = \oint_{\gamma} \frac{1}{\mu_o} \boldsymbol{B}(\boldsymbol{x},t) \bullet \boldsymbol{t}(\boldsymbol{x})\, d\gamma = \iint_{\mathfrak{S}} \frac{4\pi}{c} \boldsymbol{I}(\boldsymbol{x},t) \bullet \boldsymbol{n}(\boldsymbol{x})\, d\sigma$$

$$= \iint_{\mathfrak{S}} \frac{4\pi}{c} \left(\frac{1}{4\pi} \frac{\partial \boldsymbol{D}}{\partial t}(\boldsymbol{x},t) + \boldsymbol{J}(\boldsymbol{x},t) \right) \bullet \boldsymbol{n}(\boldsymbol{x})\, d\sigma = \iint_{\mathfrak{S}} \left(\frac{\varepsilon}{c} \frac{\partial \boldsymbol{E}}{\partial t}(\boldsymbol{x},t) + \frac{4\pi}{c} \boldsymbol{J}(\boldsymbol{x},t) \right) \bullet \boldsymbol{n}(\boldsymbol{x})\, d\sigma\,.$$

By equating the integrands, Ampere's law can be written as

$$\frac{1}{\mu_o} \nabla \times \boldsymbol{B}(\boldsymbol{x},t) = \frac{\varepsilon}{c} \frac{\partial \boldsymbol{E}}{\partial t}(\boldsymbol{x},t) + \frac{4\pi}{c} \boldsymbol{J}(\boldsymbol{x},t)\,. \qquad \textit{(Ampere's Law)} \qquad (1.4.3)$$

Equations (1.2.4), (1.3.3), (1.4.1), and (1.4.3) combine to produce *Maxwell's equations* (1.4.4).

$$\begin{array}{|ll|} \hline \nabla \bullet \boldsymbol{E}(\boldsymbol{x},t) = \frac{4\pi}{\varepsilon} \rho(\boldsymbol{x},t) & \text{Coulomb's law} \\ \nabla \times \boldsymbol{E}(\boldsymbol{x},t) = -\frac{1}{c} \frac{\partial \boldsymbol{B}}{\partial t}(\boldsymbol{x},t) & \text{Faraday's law} \\ \frac{1}{\mu_o} \nabla \times \boldsymbol{B}(\boldsymbol{x},t) = \frac{\varepsilon}{c} \frac{\partial \boldsymbol{E}}{\partial t}(\boldsymbol{x},t) + \frac{4\pi}{c} \boldsymbol{J}(\boldsymbol{x},t) & \text{Ampere's law} \\ \nabla \bullet \boldsymbol{B}(\boldsymbol{x},t) = 0 & \text{no magnetic charges} \\ \hline \end{array} \qquad (1.4.4)$$

Maxwell's equations are often written in the "4–field form" with electric $\boldsymbol{E}$, dielectric $\boldsymbol{D}$, magnetic $\boldsymbol{B}$, and magnetic loop tension $\boldsymbol{H}$ fields as per (1.4.5).

[5] The *dielectric displacement* is the derivative of the dielectric field with respect to time. A *dielectric field* is one that does not conduct direct electric current.

$$\begin{array}{|ll|}\hline \nabla\bullet\boldsymbol{D}(\boldsymbol{x},t)=4\pi\,\rho(\boldsymbol{x},t) & \text{Coulomb's law}\\ \nabla\times\boldsymbol{E}(\boldsymbol{x},t)=-\frac{1}{c}\frac{\partial\boldsymbol{B}}{\partial t}(\boldsymbol{x},t) & \text{Faraday's law}\\ \nabla\times\boldsymbol{H}(\boldsymbol{x},t)=\frac{1}{c}\frac{\partial\boldsymbol{D}}{\partial t}(\boldsymbol{x},t)+\frac{4\pi}{c}\boldsymbol{J}(\boldsymbol{x},t) & \text{Ampere's law}\\ \nabla\bullet\boldsymbol{B}(\boldsymbol{x},t)=0 & \text{no magnetic charges}\\ \hline\end{array} \tag{1.4.5}$$

Maxwell's equations provide a complete mathematical description of the evolution[6] and interaction of the electric and magnetic fields acting through a surface with a closed, orientable curve γ. Since the electric $\boldsymbol{E}$ and magnetic $\boldsymbol{B}$ fields are intertwined via Maxwell's equations, can the (non–normalized) charge density ρ from Coulomb's law be related to the current field density of the charges $\boldsymbol{J}$? As one might expect, the answer is *yes*. Suppose that the electric charge density ρ has a velocity $\boldsymbol{v}(\boldsymbol{x}, t)$. Moreover, suppose the charge through any surface $\mathbb{S}$ with boundary γ is *conserved*. A charge is conserved provided that no new charge is created or destroyed as the electric field $\boldsymbol{E}$ passes through $\mathbb{S}$. This geometry is depicted in Figure 1.4.

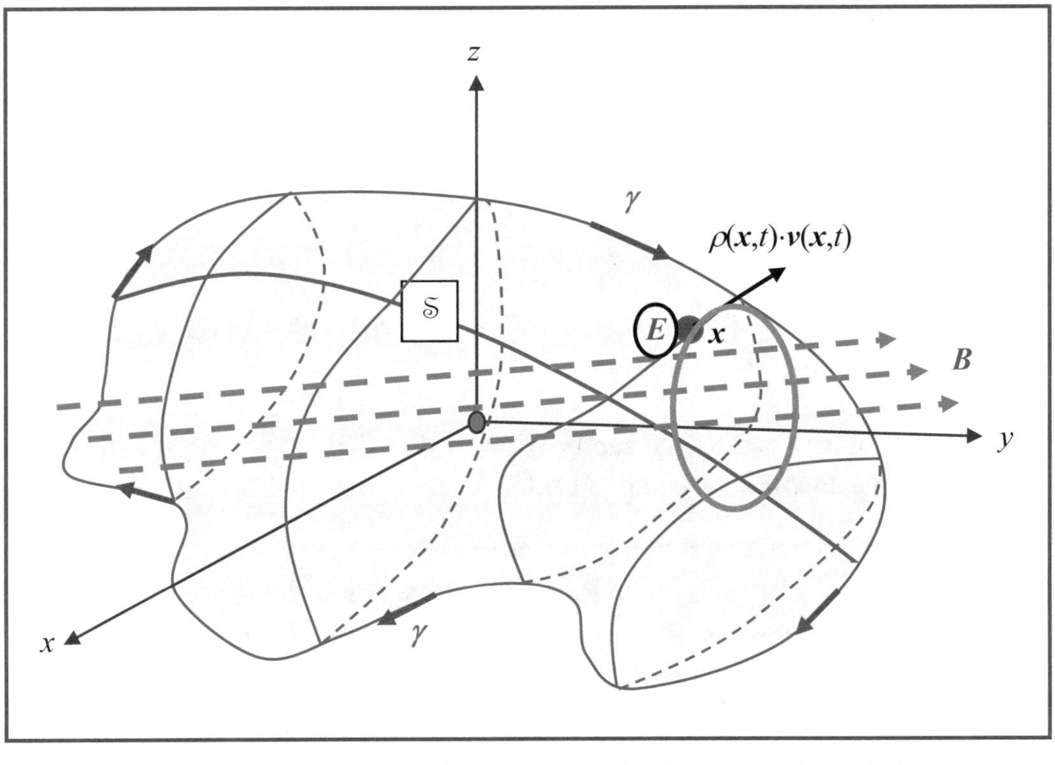

Figure 1.4. Electric $\boldsymbol{E}$ and magnetic $\boldsymbol{B}$ fields passing through $\mathbb{S}$.

As a first step in finding a relation between ρ and $\boldsymbol{J}$, determine the rate at which the density is changing within the volume V subtended by $\mathbb{S}$ (review Figure 1.3). The concentration of the charge

[6] That is, derivative with respect to time t.

in V is $q(t) = \iiint\limits_V \frac{1}{\varepsilon}\rho(\boldsymbol{x},t)\,dV$. Therefore, the rate of change of the charge flowing *into* the finite volume V is

$$\frac{dq}{dt}(t) = -\frac{d}{dt}\iiint\limits_V \frac{1}{\varepsilon}\rho(\boldsymbol{x},t)\,dV = -\iiint\limits_V \frac{1}{\varepsilon}\frac{\partial\rho}{\partial t}(\boldsymbol{x},t)\,dV. \qquad \textit{(flow in)}$$

Note the negative sign on the right-hand side since the *flow into* the volume is, by convention, negative. Here it is assumed that, as a function of t, $\rho \in C^1([0, T])$. The rate at which the charge q *flows out* of the surface $\mathbb{S}$ is, however, defined to be

$$\frac{dq}{dt}(t) = \iint\limits_{\mathbb{S}} \left(\rho(\boldsymbol{x},t)\,\boldsymbol{v}(\boldsymbol{x},t)\right)\bullet\boldsymbol{n}(\boldsymbol{x})\,d\sigma.$$

As q is assumed to be conserved, any extra terms required for charges created or destroyed within the volume V are taken to be 0. By Gauss's Divergence Theorem,

$$\frac{dq}{dt}(t) = \iint\limits_{\mathbb{S}} \left(\rho(\boldsymbol{x},t)\,\boldsymbol{v}(\boldsymbol{x},t)\right)\bullet\boldsymbol{n}(\boldsymbol{x})\,d\sigma = \iiint\limits_V \nabla\bullet\left(\rho(\boldsymbol{x},t)\,\boldsymbol{v}(\boldsymbol{x},t)\right)dV. \qquad \textit{(flow out)}$$

Equating the *flows into* and *out of* $\mathbb{S}$ yields

$$-\iiint\limits_V \frac{1}{\varepsilon}\frac{\partial\rho}{\partial t}(\boldsymbol{x},t)\,dV = \iiint\limits_V \nabla\bullet\left(\rho(\boldsymbol{x},t)\,\boldsymbol{v}(\boldsymbol{x},t)\right)dV \quad \text{or} \quad \frac{\partial\rho}{\partial t}(\boldsymbol{x},t) + \varepsilon\nabla\bullet\left(\rho(\boldsymbol{x},t)\,\boldsymbol{v}(\boldsymbol{x},t)\right) = 0.$$

The current density of an electric field can be expressed as $\boldsymbol{J}(\boldsymbol{x}, t) = \rho(\boldsymbol{x}, t)\cdot\boldsymbol{v}(\boldsymbol{x}, t)$. These results are summarized as the *continuity equation* (1.4.6).

$$\boxed{\frac{\partial\rho}{\partial t}(\boldsymbol{x},t) + \varepsilon\nabla\bullet\boldsymbol{J}(\boldsymbol{x},t) = 0 \qquad \text{(continuity equation)}} \qquad (1.4.6)$$

This relation is sometimes referred to as the "fifth Maxwell equation" or, more properly, the *conservation of charge law*. It establishes how the charge ρ and current density $\boldsymbol{J}$ evolve together. The continuity equation can be derived directly from Maxwell's equation (1.4.5) by using Ampere's and Coulomb's laws.

At the beginning of this chapter, four fundamental linear partial differential equations are identified. This begs the question: *Why mention them before the development of Maxwell's equations*? The answer is that the magnetic $\boldsymbol{B}$ and electric $\boldsymbol{E}$ fields are the derivatives of functions that obey the inhomogeneous *wave equation* (1.4.7)

$$\square_t u \equiv \left(\frac{\partial^2}{\partial t^2} - \Delta \right) u = f \, . \tag{1.4.7}$$

To show that the solution of Maxwell's equations leads to a solution of (1.4.7), recall that the fourth equation in (1.4.5) indicates that the divergence of the magnetic field is zero: $\nabla \bullet \boldsymbol{B} = 0$. Suppose that, with respect to the spatial coordinates $\boldsymbol{x}$, the magnetic field is continuously differentiable. That is, $\boldsymbol{B} \in C^1(\Omega)$, $\Omega \subset \mathbb{R}^3$ a domain and $\nabla \bullet \boldsymbol{B} = 0$. In this case, $\boldsymbol{B}$ is called a *solenoidal vector field* on Ω. Such fields are of considerable importance in partial differential equations and differential manifolds, among other mathematical topics. The proof of the theorem below can be found in Zachmanoglou and Thoe [87] as well as any number of other texts. It will be stated without proof.

Solenoidal Vector Field Theorem: *Let* $\boldsymbol{F}$ *be a non–vanishing, solenoidal vector field defined on the domain* $\Omega \subset \mathbb{R}^3$. *Then, for any point* $\boldsymbol{x}_o = (x_o, y_o, z_o) \in \Omega$, *there exists a neighborhood* $\Omega_o \subseteq \Omega$ *and a field* $\boldsymbol{A} \in C^1(\Omega_o)$ *so that* $\boldsymbol{F}(\boldsymbol{x}) = curl(\boldsymbol{A}(\boldsymbol{x})) = \nabla \times \boldsymbol{A}(\boldsymbol{x})$ *for all* $\boldsymbol{x} \in \Omega_o$.

Since the magnetic field $\boldsymbol{B}$ is solenoidal, take $\boldsymbol{A}$ to be the field of the Solenoidal Vector Field Theorem so that $\boldsymbol{B} = \nabla \times \boldsymbol{A}(\boldsymbol{x})$. In a comparable fashion, write the electric field $\boldsymbol{E}$ as the sum of a potential function u and the time derivative of $\boldsymbol{A}$: $\boldsymbol{E}(\boldsymbol{x}, t) = -\left(\nabla u(\boldsymbol{x},t) + c \frac{\partial \boldsymbol{A}}{\partial t}(\boldsymbol{x},t) \right)$. Inserting these alternate forms of $\boldsymbol{B}$ and $\boldsymbol{E}$ into Ampere's law produces

$$\begin{aligned} \tfrac{1}{\mu_o} \nabla \times \boldsymbol{B}(\boldsymbol{x},t) &= \tfrac{1}{\mu_o} \nabla \times \left(\nabla \times \boldsymbol{A}(\boldsymbol{x},t) \right) \\ &= \frac{\varepsilon}{c} \frac{\partial \boldsymbol{E}}{\partial t}(\boldsymbol{x},t) + \frac{4\pi}{c} \boldsymbol{J}(\boldsymbol{x},t) \\ &= -\frac{\varepsilon}{c} \frac{\partial}{\partial t} \left(\nabla u(x,t) + c \frac{\partial \boldsymbol{A}}{\partial t}(\boldsymbol{x},t) \right) + \frac{4\pi}{c} \boldsymbol{J}(\boldsymbol{x},t) \end{aligned} .$$

After some simplification and the observation that $\frac{\partial}{\partial t}$ and ∇ commute, the equation above leads to

$$\nabla \times \left(\nabla \times \boldsymbol{A} \right) = -\frac{\varepsilon \mu_o}{c} \left(\nabla \frac{\partial u}{\partial t} + c \frac{\partial^2 \boldsymbol{A}}{\partial t^2} \right) + \frac{4\pi \mu_o}{c} \boldsymbol{J} \, .$$

It is a straightforward exercise in vector calculus to show that

$$\Delta \mathbf{A} = grad\left(div(\boldsymbol{A}) \right) - curl\left(curl(\boldsymbol{A}) \right) = \nabla \left(\nabla \bullet \boldsymbol{A} \right) - \nabla \times \left(\nabla \times \boldsymbol{A} \right) . \tag{1.4.8}$$

Therefore,

$$\Delta \mathbf{A} = \nabla\left(\nabla\bullet \boldsymbol{A} + \frac{\varepsilon\mu_o}{c}\frac{\partial u}{\partial t}\right) + \varepsilon\mu_o\frac{\partial^2 \boldsymbol{A}}{\partial t^2} - \frac{4\pi\mu_o}{c}\boldsymbol{J}. \qquad (1.4.9)$$

The dielectric constant ε and permeability constant μ_o are inversely proportional to the square of the speed of light c^2, see e.g., Stratton [76]. That is, $\varepsilon\mu_o = \frac{1}{c^2}$. If the vector field $\boldsymbol{A}$ and potential u satisfy a continuity equation $\boxed{\nabla\bullet \boldsymbol{A} + \frac{\varepsilon\mu_o}{c}\frac{\partial u}{\partial t} = 0}$, then (1.4.9) becomes

$$\frac{\partial^2 \boldsymbol{A}}{\partial t^2} - c^2\Delta \boldsymbol{A} = 4\pi c\mu_o \boldsymbol{J}. \qquad (1.4.10)$$

But (1.4.10) is precisely a vector version of the inhomogeneous wave equation.

In a similar fashion, assuming the electric field can be written as $\boldsymbol{E} = -\left(\nabla u + c\frac{\partial \boldsymbol{A}}{\partial t}\right)$, then by Coulomb's law,

$$\frac{4\pi}{\varepsilon}\rho = \nabla\bullet \boldsymbol{E} = -\nabla\bullet\left(\nabla u + c\frac{\partial \boldsymbol{A}}{\partial t}\right) = \frac{\partial^2 u}{\partial t^2} - (\nabla\bullet\nabla)u - \frac{\partial^2 u}{\partial t^2} - c\nabla\bullet\frac{\partial \boldsymbol{A}}{\partial t} = \frac{\partial^2 u}{\partial t^2} - \Delta u - \frac{\partial}{\partial t}\left(\frac{\partial u}{\partial t} - c\nabla\bullet \boldsymbol{A}\right).$$

If the continuity equation $\boxed{\frac{\partial u}{\partial t} - c\nabla\bullet \boldsymbol{A} = 0}$ is imposed, then the equation above becomes the well-recognized inhomogeneous wave equation (1.4.11)

$$\frac{\partial^2 u}{\partial t^2}(\boldsymbol{x},t) - \Delta u(\boldsymbol{x},t) = \frac{4\pi}{\varepsilon}\rho(\boldsymbol{x},t). \qquad (1.4.11)$$

It is seen that Maxwell's equations are intimately associated with solutions of the (inhomogeneous) wave equation. Indeed, the solutions of (1.4.10) and (1.4.11) for the vector field $\boldsymbol{A}$ and potential u result in the immediate recovery of the magnetic $\boldsymbol{B}$ and electric $\boldsymbol{E}$ fields. Therefore, further analysis of Maxwell's equations is postponed so that the four fundamental linear partial differential equations can be examined. Namely, Laplace's equation $\Delta u = 0$, Poisson's equation $\Delta u = \rho(\boldsymbol{x})$, the (inhomogeneous) heat equation $\frac{\partial u}{\partial t} + \Delta u = \rho(\boldsymbol{x}, t)$, and the (inhomogeneous) wave equation $\frac{\partial^2 u}{\partial t^2} + \Delta u = \rho(\boldsymbol{x}, t)$. Once these equations and their solutions are better understood, then so will be the solutions of Maxwell's equations.

In summary, this chapter provides a development of Maxwell's equations by way of multivariate differential operators and multivariate integration theory. In the next chapter, two time-independent *PDE*s associated with the Laplacian operator are examined and their solutions described.

Exercises

1.1. Derive the continuity equation (1.4.6) from Maxwell's equations (1.4.7).
1.2. Validate equation (1.4.8).

Chapter 2

Laplace's Equation: Classical and Modern Theory

Laplace's, Poisson's, the heat, and the wave equations are often referred to as the fundamental linear partial differential equations. They are prominent in mathematical physics. Laplace's equation $\Delta u = 0$ describes the steady–state temperature distribution over a region $\Omega \subset \mathbb{R}^n$. This equation is also used to model the distribution of electric charges over Ω. In a comparable manner, Poisson's equation $\Delta u = \rho(\boldsymbol{x})$ represents the temperature or electric charge distribution given an external heat/field source ρ acting on Ω. Myint–U with Debnath [52] provide more detailed descriptions of the role of these equations in engineering and mathematical physics.

Functions that satisfy Laplace's equation are called *harmonic*. As presented in Examples 2.1–2.2, the solution of Laplace's equation, subject to particular boundary conditions, involve the sine and cosine functions. This can be interpreted as the motion of a taut surface or (in one–dimension) a string undergoing vibration ("harmonic motion").

The heat equation $\dfrac{\partial u}{\partial t} = \Delta u$ models the conduction and dispersion of heat within a physical medium. The wave equation $\dfrac{\partial^2 u}{\partial t^2} = \Delta u$ describes the propagation of sound, electromagnetic waves,[7] and vibrations. Since these equations are so fundamental to modeling natural phenomena, the natural question to ask is: *How are these equations solved*? The answer is by a combination of *classical* and *modern* mathematical methods. Before precisely describing these techniques, some informal descriptions are in order.

Heuristically, the classical approach uses well established analytic methods to determine a solution (that is, a formula) of the equation. One portion of the classical approach is to reduce the solution of a *PDE* to solutions of ordinary differential equations (*ODE*s) or even algebraic equations. The

[7] As is the case of Maxwell's equations in Chapter 1.

most famous of these classical methods is *separation of variables* in which the function described by the *PDE* is written as the product of functions of each spatial and temporal variable. The separation of variables method, in turn, is frequently associated with *Fourier series*.

The so-called "modern" approach employs the power of vector calculus and differential geometry by way of Stokes' and Green's Theorems, *Green's functions*, generalized functions or *distributions*, and *Fourier transforms*. Modern methods use functional analysis to obtain representations of the solutions. Such representations are concise though not always reducible to analytic form.

These as yet unexplained techniques are the focus of this chapter.

§2.1. Laplace's Equation and the Classical Approach (Part 1)

Let $\Omega \subset \mathbb{R}^n$ be a *domain* (that is, an open, connected set) with smooth boundary $\partial\Omega$. The *closure* $\overline{\Omega}$ of a domain Ω is the closed set which is a union of the set with its boundary $\overline{\Omega} = \Omega \cup \partial\Omega$. If Ω is bounded, then the boundary $\partial\Omega$ is a curve (in $n = 2$ spatial dimensions) or surface[8] ($n \geq 3$). As mentioned above, a function $u \in C^2(\Omega)$ that satisfies Laplace's equation (2.1.1) is called harmonic

$$\Delta u = \nabla^2 u = \sum_{j=1}^{n} \frac{\partial^2 u}{\partial x_j^2} = 0. \tag{2.1.1}$$

How can harmonic functions be constructed? That is, how can (2.1.1) be solved? For the one–dimensional problem (i.e., $n = 1$), it is evident that $u(x) = x + 1$ satisfies $\dfrac{d^2u}{dx^2} = 0$. The same is true for any scalar multiple of $u(x)$. What happens when $n > 1$? The development of harmonic functions by John [43] and Evans [23] leads to the consideration of the general case. Since the one–dimensional harmonic function is along the line $y = x + 1$, it appears reasonable to guess that the n–dimensions harmonic functions are along the radial length $r = \|\boldsymbol{x}\| = \sqrt{x_1^2 + x_2^2 + \cdots + x_n^2}$. Observing that $\partial r/\partial x_j = x_j/r = x_j \cdot r^{-1}$ and $\partial^2 r/\partial x_j^2 = 1/r - x_j^2/r^3$, the Laplacian of r takes the simple form $\Delta r = (n-1)/r$. Plainly, the Laplacian of the radial length is *not* equal to 0 and therefore $r = \|\boldsymbol{x}\|$ is *not* a harmonic function. Rather than a linear function of r, assume that $u(\boldsymbol{x}) = \psi(r)$ solves (2.1.1) for a sufficiently smooth function ψ with non-zero derivatives. In this case,

$$\frac{\partial u}{\partial x_j} = \psi'(r)\frac{\partial r}{\partial x_j} \text{ and } \frac{\partial^2 u}{\partial x_j^2} = \psi'(r)\frac{\partial^2 r}{\partial x_j^2} + \psi''(r)\left(\frac{\partial r}{\partial x_j}\right)^2 = \left(\frac{1}{r} - \frac{x_j^2}{r^3}\right)\psi'(r) + \frac{x_j^2}{r^2}\psi''(r).$$

[8] For $n \geq 4$, $\partial\Omega$ is a *hypersurface*.

Combining terms, Laplace's equation (2.1.1) yields

$$0 = \Delta u = \left(\frac{n}{r} - \frac{x_1^2 + x_2^2 + \cdots + x_n^2}{r^3}\right)\psi'(r) + \frac{x_1^2 + x_2^2 + \cdots + x_n^2}{r^2}\psi''(r) = \left(\frac{n-1}{r}\right)\psi'(r) + \psi''(r).$$

This is a separable *ODE* which reduces to $\dfrac{d}{d\zeta}\ln(\zeta) = \dfrac{1-n}{r}$ for $\zeta = \psi'$. Integrating produces $\psi'(r)$ $= \zeta = \dfrac{c_o}{r^{n-1}}$ for some constant c_o. Assuming $r \neq 0$ (that is, $r > 0$), a second integration gives the solution

$$\psi(r) = \begin{cases} b_2 \ln(r) + b_1 & \text{for } n = 2 \\ \dfrac{b_n}{r^{n-2}} + b_1 & \text{for } n \geq 3 \end{cases}. \tag{2.1.2}$$

Without loss of generality, set $b_1 = 0$ in (2.1.2). To determine the values of b_n in (2.1.2), an excursion into "modern" methods is required.

§2.2. Green's Theorem and the Modern Approach

Plainly, the radial variable $r = \|\boldsymbol{x}\|$ can be replaced by $r - r_o = \|\, \boldsymbol{x} - \boldsymbol{x}_o \,\|$ in the family of harmonic functions (2.1.2) above. Hence, the function $\psi(r - r_o)$ is harmonic everywhere on $\Omega \subset \mathbb{R}^n$ except $\boldsymbol{x} = \boldsymbol{x}_o$. That is, $\psi(r - r_o)$ is harmonic for all $\boldsymbol{x} \in \Omega \setminus \{\boldsymbol{x}_o\}$. As noted in §2.1, the constants b_n in (2.1.2) require *Green's Theorem* to determine their value.

Green's Theorem: *Let* $u, v \in C^2(\overline{\Omega})$ *on the domain* $\Omega \subset \mathbb{R}^n$. *Then*

(i) $$\int_\Omega v\,\Delta u\, d\omega = \int_{\partial\Omega}\left(v\frac{\partial u}{\partial \boldsymbol{n}} - u\frac{\partial v}{\partial \boldsymbol{n}}\right)d\sigma - \int_\Omega \nabla u \bullet \nabla v\, d\omega$$

(ii) $$\int_\Omega v\,\Delta u\, d\omega = \int_{\partial\Omega}\left(v\frac{\partial u}{\partial \boldsymbol{n}} - u\frac{\partial v}{\partial \boldsymbol{n}}\right)d\sigma + \int_\Omega u\,\Delta v\, d\omega \text{ and}$$

(iii) $$\int_\Omega \nabla u \bullet \nabla u\, d\omega = \int_{\partial\Omega} u\frac{\partial u}{\partial \boldsymbol{n}}d\sigma - \int_\Omega u\,\Delta u\, d\omega. \qquad (\textit{Energy Identity})$$

The symbol $\int_\Omega f\, d\omega$ means the multi–dimensional integral of the function f over the domain Ω with respect to the n–dimensional volume element $d\boldsymbol{\omega}$. Similarly, $\int_{\partial\Omega} f\, d\sigma$ is the integral over the domain boundary $\partial\Omega$ with respect to the n–dimensional surface element $d\boldsymbol{\sigma}$. Finally, $\boldsymbol{n}$ is the (outward) normal vector. To prove Green's Theorem, recall the single field version of the Divergence Theorem (which can be found in any standard calculus text book). That is,

Divergence (Gauss–Green) Theorem: *If* $\psi \in C^1(\overline{\Omega})$ *on the domain* $\Omega \subset \mathbb{R}^n$, *then* $\int\limits_{\Omega} \nabla \psi \, d\omega = \int\limits_{\partial\Omega} \psi \bullet \boldsymbol{n} \, d\sigma$.

Proof of Green's Theorem: Suppose $u, v \in C^2(\overline{\Omega})$ and $\psi = u \cdot v$. Then $\psi \in C^2(\overline{\Omega}) \subset C^1(\overline{\Omega})$ and $\nabla \psi = \nabla(u \cdot v) = (\nabla u) \cdot v + (\nabla v) \cdot u$. Therefore, by the Divergence Theorem, $\int\limits_{\Omega} (\nabla u \cdot v + \nabla v \cdot u) \, d\omega = \int\limits_{\partial\Omega} (u \cdot v) \bullet \boldsymbol{n} \, d\sigma$ or $\boxed{\int\limits_{\Omega} \nabla u \cdot v \, d\omega = -\int\limits_{\Omega} \nabla v \cdot u \, d\omega + \int\limits_{\partial\Omega} (u \cdot v) \bullet \boldsymbol{n} \, d\sigma}$. This is the n–dimensional version of integration by parts. Using this relation and substituting ∇v for v, then

$$\int\limits_{\Omega} \nabla u \cdot \nabla v \, d\omega = -\int\limits_{\Omega} \nabla(\nabla v) \cdot u \, d\omega + \int\limits_{\partial\Omega} (u \cdot \nabla v) \bullet \boldsymbol{n} \, d\sigma \equiv -\int\limits_{\Omega} v \cdot \Delta u \, d\omega + \int\limits_{\partial\Omega} u \frac{\partial v}{\partial \boldsymbol{n}} \, d\sigma$$

which is precisely relation (*i*).

When $u = v$, then (*i*) becomes (*iii*).

Now interchange u and v in (*i*) to obtain the relation $\int\limits_{\Omega} \nabla v \cdot \nabla u \, d\omega = -\int\limits_{\Omega} u \cdot \Delta v \, d\omega + \int\limits_{\partial\Omega} v \frac{\partial u}{\partial \boldsymbol{n}} \, d\sigma$.

Subtracting this equation from (*iii*) yields $\int\limits_{\Omega} (u \Delta v - v \Delta u) \, d\omega = \int\limits_{\partial\Omega} \left(u \frac{\partial v}{\partial \boldsymbol{n}} - v \frac{\partial u}{\partial \boldsymbol{n}} \right) d\sigma$ which establishes (*ii*). ■

As stated at the beginning of this section, the function $\psi(r - r_o)$ defined in equation (2.1.2) obeys Laplace's equation $\Delta \psi = 0$ everywhere except at $r = r_o$ or, equivalently, at $\boldsymbol{x} = \boldsymbol{x}_o$. In order to properly account for this exception and determine the values of the constants b_n in (2.1.2), the notion of a *distribution* or *generalized function* is required. In particular, the *Dirac delta function* $\delta(\boldsymbol{x})$ is among the most well-known distributions. The Dirac delta function is treated as a *point source* in that all of its "energy" is concentrated at a single point.

The *identifier* or *indicator function* is defined as $\mathcal{S}_{\Omega}(\boldsymbol{x}) = \begin{cases} 1 & \text{for } \boldsymbol{x} \in \Omega \\ 0 & \text{otherwise} \end{cases}$. In one–spatial dimension, the sequence $\delta_m(x) = m \cdot \mathcal{S}_{[-1/2m,\ 1/2m]}(x)$ defines the Dirac delta function as $\delta(x) = \lim\limits_{m \to \infty} \delta_m(x)$. This is illustrated in Figure 2.1. The more general n–dimensional version of the Dirac delta function can be defined as the limit on $\delta_m(\boldsymbol{x}) = m^n \cdot \mathcal{S}_{C_n(1/2m)}(\boldsymbol{x})$ where $C_n(1/2m) =$

$\left[\frac{-1}{2m},\frac{1}{2m}\right]\times\left[\frac{-1}{2m},\frac{1}{2m}\right]\times\cdots\times\left[\frac{-1}{2m},\frac{1}{2m}\right]$ is the n–dimensional hypercube with volume m^{-n}. If $\boldsymbol{x} = [x_1, x_2, \ldots, x_n]$, then the n–dimensional version of the Dirac delta function is simply a product of its one–dimensional components: $\boldsymbol{\delta}(\boldsymbol{x}) = \delta(x_1)\cdot\delta(x_2)\cdot \cdots \cdot\delta(x_n)$. Informally, $\boldsymbol{\delta}(\boldsymbol{x})$ is 0 everywhere except $\boldsymbol{x} = \boldsymbol{0}$ where it takes infinite value. The most prominent property of the Dirac delta function is its behavior within integrals. That is, for any integrable function f: $\mathbb{R}\to\mathbb{R}$, $\int_{-\infty}^{\infty} f(x)\delta(x)\,dx = f(0)$ and more generally, for $\boldsymbol{f}$: $\mathbb{R}^n\to\mathbb{R}^n$, $\int_{\mathbb{R}^n} \boldsymbol{f}(\boldsymbol{x})\delta(\boldsymbol{x})\,d\boldsymbol{x} = \boldsymbol{f}(\boldsymbol{0})$. A direct extension of this result is the relation

$$\int_{\mathbb{R}^n} \boldsymbol{f}(\boldsymbol{x})\delta(\boldsymbol{x}-\boldsymbol{x}_o)\,d\boldsymbol{x} = \boldsymbol{f}(\boldsymbol{x}_o). \tag{2.2.1}$$

As the Dirac delta function can be written as the product of one–dimensional delta functions, all with compact support, then $\boldsymbol{\delta}(\boldsymbol{x}) = \lim_{m\to\infty}\prod_{j=1}^{n}\delta_m(x_j)$ and $\int_{\mathbb{R}^n}\boldsymbol{\delta}(\boldsymbol{x})\,d\boldsymbol{x} = \lim_{m\to\infty}\int_{-\infty}^{\infty}\delta_m(x_1)\,dx_1\cdots\int_{-\infty}^{\infty}\delta_m(x_n)\,dx_n$

$$= \lim_{m\to\infty}\int_{-1/2m}^{1/2m} m\,dx_1\cdots\int_{-1/2m}^{1/2m} m\,dx_n = \lim_{m\to\infty}\left(\left[m\cdot\frac{1}{m}\right]\times\left[m\cdot\frac{1}{m}\right]\times\cdots\times\left[m\cdot\frac{1}{m}\right]\right) = 1.$$

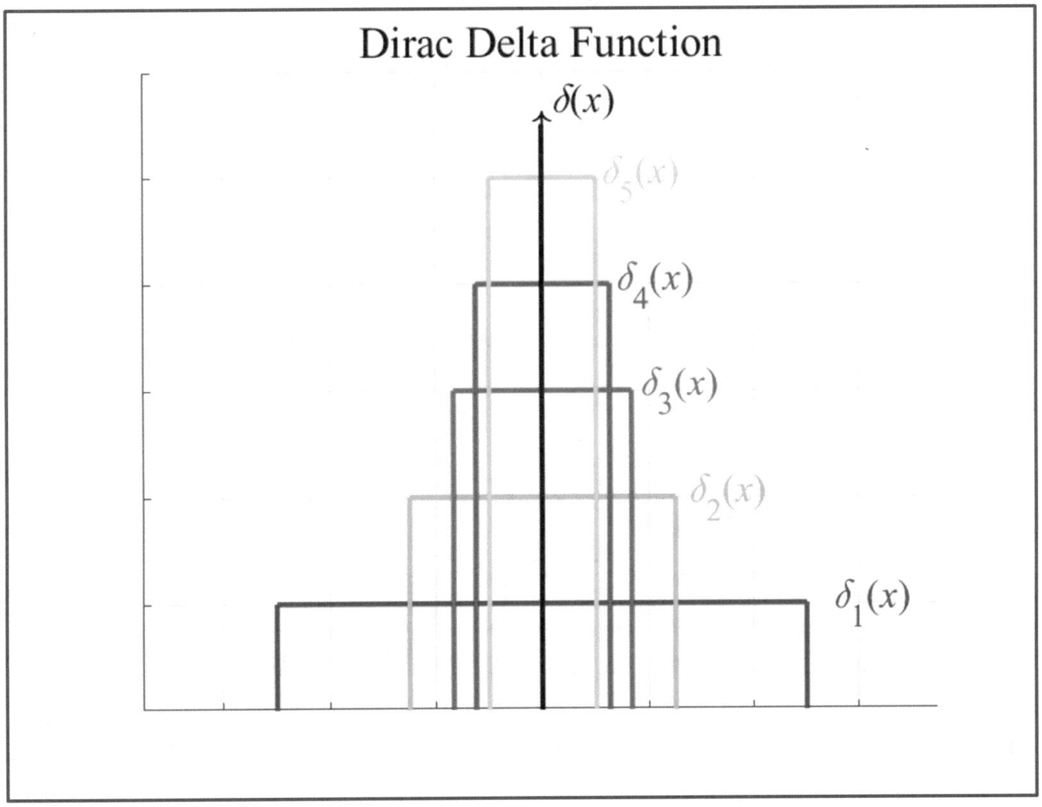

Figure 2.1. The sequence $\delta_n(x)$ converging to the Dirac delta function.

Equivalently, the Dirac delta function can simply be defined as the function that satisfies (2.2.1). The family of harmonic functions defined in (2.1.2) satisfy the following variation of Laplace's equation

$$\Delta\psi = \delta(\boldsymbol{x}-\boldsymbol{x}_o), \boldsymbol{x} \in \Omega \subset \mathbb{R}^n. \tag{2.2.2}$$

Now Green's second integral identity is used to determine the coefficients b_n of (2.1.2). Let $u \in C^2(\overline{\Omega})$ and $\boldsymbol{x}_o \in \Omega \subset \mathbb{R}^n$, Ω a bounded domain. Define $\mathbb{B}_n(\boldsymbol{x}_o,\varepsilon) = \{\boldsymbol{x} \in \mathbb{R}^n : \|\boldsymbol{x}-\boldsymbol{x}_o\| < \varepsilon\}$ as the n–dimensional ball of radius ε about the point $\boldsymbol{x}_o$. Similarly, let $\mathbb{S}_{n-1}(\boldsymbol{x}_o,\varepsilon) = \{\boldsymbol{x} \in \mathbb{R}^n : \|\boldsymbol{x}-\boldsymbol{x}_o\| = \varepsilon\}$ be the $(n-1)$–dimensional sphere of radius ε and center $\boldsymbol{x}_o$. Let Ω_ε be the domain Ω minus the ball of radius ε: $\Omega_\varepsilon = \Omega \setminus \mathbb{B}_n(\boldsymbol{x}_o,\varepsilon)$. See Figure 2.2.

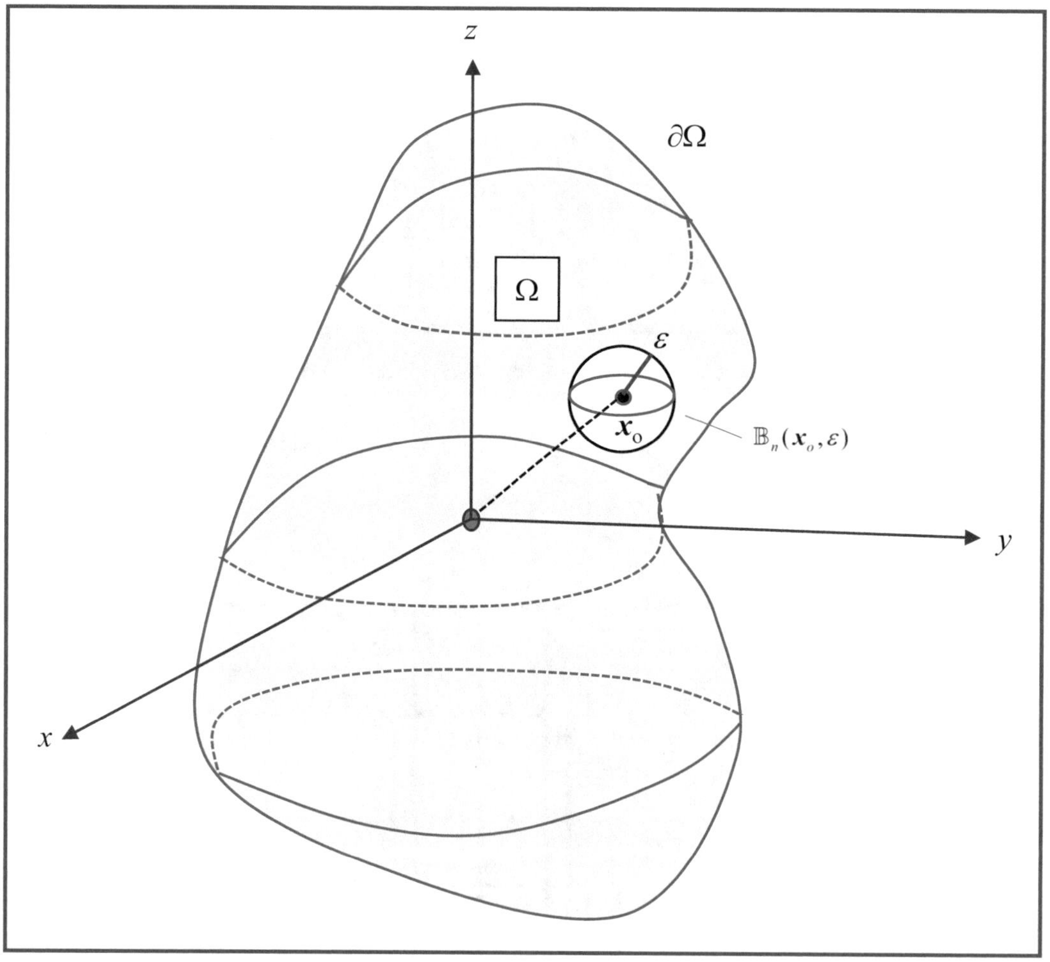

Figure 2.2. Removing a ball of radius ε from the domain Ω.

In the modified domain Ω_ε, $\psi(r-r_o)$ is harmonic. Also, $\Omega_\varepsilon = \Omega \cup \mathbb{S}_{n-1}(\boldsymbol{x}_o,\varepsilon)$. Therefore, by the second integral identity of Green's Theorem,

$$\int_{\Omega_\varepsilon} \psi\, \Delta u\, d\omega = \int_{\partial\Omega_\varepsilon} \left(\psi \frac{\partial u}{\partial \boldsymbol{n}} - u \frac{\partial \psi}{\partial \boldsymbol{n}} \right) d\sigma + \int_{\Omega_\varepsilon} u\, \Delta \psi\, d\omega$$

$$= \int_{\partial\Omega} \left(\psi \frac{\partial u}{\partial \boldsymbol{n}} - u \frac{\partial \psi}{\partial \boldsymbol{n}} \right) d\sigma + \int_{\mathbb{S}_{n-1}(\boldsymbol{x}_o,\varepsilon)} \left(\psi \frac{\partial u}{\partial \boldsymbol{n}} - u \frac{\partial \psi}{\partial \boldsymbol{n}} \right) d\sigma.$$

This holds since ψ is harmonic on Ω_ε and consequently $\Delta\psi(r-r_o) \equiv 0$ on Ω_ε. The integral over $\partial\Omega_\varepsilon$ becomes the integral over the component boundaries $\partial\Omega$ and $\mathbb{S}(\boldsymbol{x}_o,\varepsilon)$. On the sphere, the exterior normal points *inward toward* the center $\boldsymbol{x}_o$ so that the sign is negative. Thus, $\frac{\partial \psi}{\partial \boldsymbol{n}} = -\frac{d\psi}{dr} = -\psi'(r-r_o) = -\psi'(\varepsilon)$ and $\psi(r-r_o) = \psi(\varepsilon)$ for all $\boldsymbol{x} \in \mathbb{S}_{n-1}(\boldsymbol{x}_o,\varepsilon)$ when ε is sufficiently small. Therefore,

$$\int_{\mathbb{S}_{n-1}(\boldsymbol{x}_o,\varepsilon)} \left(\psi \frac{\partial u}{\partial \boldsymbol{n}} - u \frac{\partial \psi}{\partial \boldsymbol{n}} \right) d\sigma = \int_{\mathbb{S}_{n-1}(\boldsymbol{x}_o,\varepsilon)} \left(\psi(\varepsilon) \frac{\partial u}{\partial \boldsymbol{n}} - u\,\psi'(\varepsilon) \right) d\sigma = \psi(\varepsilon) \int_{\mathbb{B}_n(\boldsymbol{x}_o,\varepsilon)} \Delta u\, d\omega + b_n\, \varepsilon^{1-n} \int_{\mathbb{S}_{n-1}(\boldsymbol{x}_o,\varepsilon)} u\, d\sigma$$

by Green's identity (*ii*) and the definition of ψ in (2.1.2). Since $u \in C^2(\overline{\Omega})$, then $u(\boldsymbol{x}) \to u(\boldsymbol{x}_o)$ as $\varepsilon \to 0$ for $\boldsymbol{x} \in \mathbb{S}_{n-1}(\boldsymbol{x}_o,\varepsilon)$. Therefore, $\int_{\mathbb{S}_{n-1}(\boldsymbol{x}_o,\varepsilon)} u\, d\sigma = u(\boldsymbol{x}_o)\, \omega_n\, \varepsilon^{n-1}$ where $\omega_n = \dfrac{2\left(\sqrt{\pi}\right)^n}{\Gamma\left(\frac{n}{2}\right)}$ is the surface area of the n–dimensional unit sphere $\mathbb{S}_{n-1}(\boldsymbol{0},1)$. Furthermore, the assumption $u \in C^2(\overline{\Omega})$ means that the integral $\psi(\varepsilon) \int_{\mathbb{B}_n(\boldsymbol{x}_o,\varepsilon)} \Delta u\, d\omega = \psi(\varepsilon)\, \Delta u(\boldsymbol{x}_o)\, \frac{\omega_n}{n} \varepsilon^n$ tends to 0 as $\varepsilon \to 0$ for $\boldsymbol{x} \in \mathbb{B}_n(\boldsymbol{x}_o,\varepsilon)$.

Hence, $\lim_{\varepsilon\to 0} \left(\psi(\varepsilon) \int_{\mathbb{B}_n(\boldsymbol{x}_o,\varepsilon)} \Delta u\, d\omega \right) = 0$ so that $\lim_{\varepsilon\to 0} \left(\int_{\mathbb{S}_{n-1}(\boldsymbol{x}_o,\varepsilon)} \left(\psi \frac{\partial u}{\partial \boldsymbol{n}} - u \frac{\partial \psi}{\partial \boldsymbol{n}} \right) d\sigma \right) = b_n\, \omega_n\, u(\boldsymbol{x}_o)$. Select $b_n = 1/\omega_n$ so that

$$\int_{\Omega} \psi\, \Delta u\, d\omega = \lim_{\varepsilon\to 0} \int_{\Omega_\varepsilon} \psi\, \Delta u\, d\omega$$

$$= \lim_{\varepsilon\to 0} \left(\int_{\partial\Omega_\varepsilon} \left(\psi \frac{\partial u}{\partial \boldsymbol{n}} - u \frac{\partial \psi}{\partial \boldsymbol{n}} \right) d\sigma + b_n \varepsilon^{1-n} \int_{\mathbb{S}_{n-1}(\boldsymbol{x}_o,\varepsilon)} u\, d\sigma \right)$$

$$= \int_{\partial\Omega} \left(\psi \frac{\partial u}{\partial \boldsymbol{n}} - u \frac{\partial \psi}{\partial \boldsymbol{n}} \right) d\sigma + u(\boldsymbol{x}_o)$$

or $$u(\boldsymbol{x}_o) = \int\limits_{\Omega} \psi \, \Delta u \, d\omega + \int\limits_{\partial\Omega} \left(u \frac{\partial \psi}{\partial \boldsymbol{n}} - \psi \frac{\partial u}{\partial \boldsymbol{n}} \right) d\sigma .$$

A harmonic function which satisfies Laplace's equation (2.2.2) has the form (2.2.3) with $r = \|\boldsymbol{x}\|$ and $r_o = \|\boldsymbol{x}_o\|$.

$$\psi(r - r_o) = \begin{cases} \dfrac{1}{2\pi} \ln(r - r_o) & \text{for } n = 2 \\ \dfrac{1}{\omega_n} \dfrac{1}{(r - r_o)^{n-3}} & \text{for } n \geq 3 \\ \omega_n = \dfrac{2\left(\sqrt{\pi}\right)^n}{\Gamma\left(\dfrac{n}{2}\right)} & \end{cases} \tag{2.2.3}$$

The family of radial functions $\psi(r - r_o) \equiv G(\boldsymbol{x}, \boldsymbol{x}_o)$ are the *Green's functions* for Laplace's equation. These results are summarized below.

Green's Representation Theorem: *Let $u \in C^2(\overline{\Omega})$ for a domain $\Omega \subset \mathbb{R}^n$. For any $\boldsymbol{x}_o \in \Omega$ and the corresponding Green's function $G(\boldsymbol{x}, \boldsymbol{x}_o)$ for Laplace's equation, then $u(\boldsymbol{x}_o)$ has the integral representation*

$$u(\boldsymbol{x}_o) = \int\limits_{\Omega} G(\boldsymbol{x}, \boldsymbol{x}_o) \Delta u(\boldsymbol{x}) \, d\omega + \int\limits_{\partial\Omega} \left(u(\boldsymbol{x}) \frac{\partial G}{\partial \boldsymbol{n}}(\boldsymbol{x}, \boldsymbol{x}_o) - G(\boldsymbol{x}, \boldsymbol{x}_o) \frac{\partial u}{\partial \boldsymbol{n}}(\boldsymbol{x}) \right) d\sigma . \tag{2.2.4}$$

An alternate proof of Green's Representation Theorem can be derived using property (2.2.1) of the Dirac delta function and the Green's function (2.2.3). Take $u \in C^2(\overline{\Omega})$ and let $G(\boldsymbol{x}, \boldsymbol{x}_o)$ be the Green's function for Laplace's equation. Then G satisfies $\Delta G = \delta(\boldsymbol{x} - \boldsymbol{x}_o)$. Multiplying both sides of this equation by $u(\boldsymbol{x})$ and integrating over Ω produces

$$\begin{aligned} u(\boldsymbol{x}_o) &= \int\limits_{\Omega} u(\boldsymbol{x}) \delta(\boldsymbol{x} - \boldsymbol{x}_o) \, d\omega = \int\limits_{\Omega} u(\boldsymbol{x}) \Delta G(\boldsymbol{x}, \boldsymbol{x}_o) \, d\omega \\ &= \int\limits_{\Omega} G(\boldsymbol{x}, \boldsymbol{x}_o) \Delta u(\boldsymbol{x}) \, d\omega + \int\limits_{\partial\Omega} \left(u(\boldsymbol{x}) \frac{\partial G}{\partial \boldsymbol{n}}(\boldsymbol{x}, \boldsymbol{x}_o) - G(\boldsymbol{x}, \boldsymbol{x}_o) \frac{\partial u}{\partial \boldsymbol{n}}(\boldsymbol{x}) \right) d\sigma \end{aligned}$$

which is precisely (2.2.4). ■

For more details on, and a deeper development of, Green's functions see Roach [67].

§2.3. The Classical Approach (Part 2)

The harmonic Green's function of (2.2.3) and the integral representation (2.2.4) are derived using modern methods. While the Green's functions in (2.2.3) are of analytic form, they were produced via classical methods (i.e., *ODE*s). The *coefficients* of the baseline solutions, however, required the integral methods of Green's and Gauss's Divergence Theorems, so too with the integral representation (2.2.4). The Green's functions explicit formulation permit a return to the classical approach.

As mentioned in the introduction of this chapter, just prior to §2.1, the classical method is used to reduce Laplace's *PDE* to an *ODE*. Also, the fundamental solutions to Laplace's equation (2.2.3) are radially symmetric and specialized. They do not specify behavior at the boundary. Consider the basic first order *ODE* $\frac{dx}{dt} = k\,x$ which has solution $x_c(t) = c\cdot e^{kt}$ for arbitrary constant c. Plainly, this solution is not unique and gives rise to a family of solutions $\{x_c(t)\colon c \in \mathbb{R}\}$. How can c be specified so that the solution is unique? That answer is to require an additional condition on the system. In this case, if an *initial condition* in time is specified, say $x(t = 0) = x(0) = x_o$, then it must be the case that $c = x_o$. Therefore, $x(t) = x_o\cdot e^{kt}$ is a unique solution of the *initial value problem* $\frac{dx}{dt} = k\,x$ and $x(0) = x_o$.

For second order *ODE*s, two conditions are required to ensure uniqueness. For example, $x(t) = c_1\cdot\cos(kt) + c_2\cdot\sin(kt)$ is a family of solutions to the equation $\frac{d^2x}{dt^2} = -k^2x$. To obtain a unique solution, that is, to find unique values of c_1 and c_2, additional conditions must be prescribed. If $x(0) = x_o$ and $x(\pi/2k) = x_1$, then $c_1 = x_o$ and $c_2 = x_1$ so that $x(t) = x_o\cdot\cos(kt) + x_1\cdot\sin(kt)$ is the unique solution to the *boundary value problem* $\frac{d^2x}{dt^2} = -k^2x$, $x(0) = x_o$, and $x(\pi/2k) = x_1$. The solution $x(t)$ has prescribed values on the boundary of the time interval $[0, \pi/2k]$.

If, instead of specifying points on the boundary of the time interval $[t_o, T]$, conditions on the function $x(t)$ and its derivative $x'(t)$ at time $t = t_o$ are preset, then a unique solution to the *mixed value problem* $\frac{d^2x}{dt^2} = -k^2x$, $x(t_o) = x_o$, and $x'(t_o) = x'_o$ can be found: $x(t) = x_o\cdot\cos(k[t - t_o]) + \frac{x'_o}{k}\cdot\sin(k[t - t_o])$.

In partial differential equations, the conditions corresponding to initial value, boundary value, and mixed value problems are imposed on the boundary of a domain $\Omega \subset \mathbb{R}^n$. For a known function f, Laplace's equation has the variations listed below. See Gustafson [31] for more details.

The Dirichlet initial value problem

$$\left.\begin{array}{ll}\Delta u(\boldsymbol{x})=0, & \boldsymbol{x}\in\Omega\\ u(\boldsymbol{x})=f(\boldsymbol{x}), & \boldsymbol{x}\in\partial\Omega\end{array}\right\} \tag{2.3.1}$$

The Neumann boundary value problem

$$\left.\begin{array}{ll}\Delta u(\boldsymbol{x})=0, & \boldsymbol{x}\in\Omega\\ \dfrac{\partial u}{\partial \boldsymbol{n}}(\boldsymbol{x})=f(\boldsymbol{x}), & \boldsymbol{x}\in\partial\Omega\end{array}\right\} \tag{2.3.2}$$

Here, $\dfrac{\partial}{\partial \boldsymbol{n}}$ is the exterior normal[9] derivative of u on Ω: $\dfrac{\partial u}{\partial \boldsymbol{n}}(\boldsymbol{x})=\nabla u(\boldsymbol{x})\bullet\boldsymbol{n}(\boldsymbol{x})$.

The Robin mixed value problem

$$\left.\begin{array}{ll}\Delta u(\boldsymbol{x})=0, & \boldsymbol{x}\in\Omega\\ \dfrac{\partial u}{\partial \boldsymbol{n}}(\boldsymbol{x})+u(\boldsymbol{x})=f(\boldsymbol{x}), & \boldsymbol{x}\in\partial\Omega\end{array}\right\} \tag{2.3.3}$$

Finally, there is an inhomogeneous version of (2.3.1) for a known function $\rho(\boldsymbol{x})$.

Dirichlet–Poisson problem

$$\left.\begin{array}{ll}\Delta u(\boldsymbol{x})=\rho(\boldsymbol{x}), & \boldsymbol{x}\in\Omega\\ u(\boldsymbol{x})=f(\boldsymbol{x}), & \boldsymbol{x}\in\partial\Omega\end{array}\right\} \tag{2.3.4}$$

Clearly, the Neumann and Robin problems can be extended by the introduction of an inhomogeneous term $\rho(\boldsymbol{x})$ to produce *Neumann–Poisson* and *Robin–Poisson* problems. Before solving some examples of these problems, a brief discussion of one of the principal classical techniques in the solution of *PDE*s is presented, namely *separation of variables*.

The general idea behind the separation of variables is to write a proposed solution as a product of functions of the independent variables. That is, if $u(\boldsymbol{x}) = u(x_1, x_2)$ is the desired solution, then assume $u(\boldsymbol{x}) = X_1(x_1)\cdot X_2(x_2)$ where X_1 and X_2 are independent functions. For linear *PDE*s with constant coefficients, this generally leads to two separate *ODE*s in the functions X_1 and X_2. As a refresher, the basic approach to solving linear *ODE*s is presented. Consider the general n^{th}–order

[9] For a definition of the normal vector, see Chapter 1. $D_n = \frac{\partial}{\partial n}$ is also called the *directional derivative* or the gradient in the direction of the normal vector $\boldsymbol{n}$.

ODE with constant coefficients as defined via the linear differential operator $L[x] = \sum_{j=0}^{n} \alpha_j \frac{d^j x}{dt^j} = \alpha_o x + \alpha_1 \frac{dx}{dt} + \alpha_2 \frac{d^2 x}{dt^2} + \cdots + \alpha_n \frac{d^n x}{dt^n}$. Assume that $x(t) = e^{\lambda t}$. A direct computation shows that $L[e^{\lambda t}] = e^{\lambda t} \sum_{j=0}^{n} \alpha_j \lambda^j$. Therefore, to solve the *ODE* $L[x] = 0$, it is sufficient to solve the algebraic equation $\chi_L(\lambda) \equiv \sum_{j=0}^{n} \alpha_j \lambda^j = 0$ for λ. The symbol $\chi_L(\lambda)$ is called the *characteristic function* of the differential operator L and $\chi_L(\lambda) = 0$ is the *characteristic equation* for the *ODE* $L[x] = 0$. If λ_1, λ_2, … , λ_n are the roots of the characteristic equation $\chi_L(\lambda) = 0$, then $e^{\lambda_1 t}$, $e^{\lambda_2 t}$, … , $e^{\lambda_n t}$ are n–linearly independent solutions of $L[x] = 0$. Thus, $x(t) = \sum_{j=1}^{n} c_j e^{\lambda_j t}$ is a general and non-unique solution of the *ODE* $L[x] = 0$. The coefficients c_j are determined when boundary and/or initial conditions are prescribed. A readable and thorough exposition on ordinary differential equations can be found in Boyce and DiPrima [12].

Now employ the method of separation of variables to solve Laplace's equation with Dirichlet conditions.

§2.4. Some Examples

To get an understanding of how the classical technique of separation of variables is implemented, two examples are undertaken. The first example details the method of separation of variables in standard two–dimensional Euclidean coordinates. The second example requires a shift to polar coordinates.

Example 2.1. Solve Laplace's equation in $\mathbb{R}^2$ on the square of length L with Dirichlet boundary conditions. More precisely, find a harmonic function satisfying (2.4.1).

$$\left.\begin{aligned}
&\Delta u(x,y) = 0 \quad \text{for} \quad (x,y) \in \Omega = \left\{(x,y) \in \mathbb{R}^2 : 0 \le x, y \le L\right\} \\
&u(x,-L/2) = -\sinh\left(\frac{2\pi}{L} x\right) = u(x, L/2) \\
&u(-L/2, y) = -\sinh(\pi)\left(\cos\left(\frac{2\pi}{L} y\right) + \sin\left(\frac{2\pi}{L} y\right)\right) \\
&u(L/2, y) = \sinh(\pi)\left(\cos\left(\frac{2\pi}{L} y\right) + \sin\left(\frac{2\pi}{L} y\right)\right)
\end{aligned}\right\} \tag{2.4.1}$$

Solution: To solve (2.4.1), assume that the solution $u(x, y) = X(x)\cdot Y(y)$ where X and Y are functions of the independent variables x and y. Since the initial conditions in (2.4.1) are not identically zero, then it must be the case that $X(x) \neq 0$ for all x and $Y(y) \neq 0$ for all y. Then $\frac{\partial u}{\partial x} = X'(x)\cdot Y(y)$ and $\frac{\partial^2 u}{\partial x^2} = X''(x)\cdot Y(y)$. Similarly, $\frac{\partial^2 u}{\partial y^2} = Y(x)\cdot Y''(y)$. Thus, u is harmonic provided $0 = \Delta u = \frac{\partial^2 u}{\partial x^2} + \frac{\partial^2 u}{\partial y^2} = X''(x)\cdot Y(y) + X(x)\cdot Y''(y) \Leftrightarrow X''(x)\cdot Y(y) = -X(x)\cdot Y''(y)$ or $\frac{X''(x)}{X(x)} = -\frac{Y''(y)}{Y(y)}$. But the right-hand side of this equation is purely a function of y while the left-hand side is purely of function of x. Since the variables x and y are independent, this means that these ratios are both equal to a constant C_o. The left-hand side therefore becomes $\frac{X''(x)}{X(x)} = C_o$ or $X''(x) - C_o X(x) = 0$. Using the idea of the characteristic equation as discussed at the end of §2.3, the *ODE* in x becomes equivalent to solving the algebraic equation $\lambda^2 - C_o = 0$ or $\lambda = \pm\sqrt{C_o}$. This means that $X(x) = c_1 e^{\sqrt{C_o}x} + c_2 e^{-\sqrt{C_o}x}$. The Dirichlet boundary conditions $u(x, -L/2) = u(x, L/2) = -\sinh(2\pi x/L)$ mean that $u(x,-L/2) = X(x)\cdot Y(-L/2) = -\sinh(2\pi x/L) = u(x, L/2) = X(x)\cdot Y(-L/2)$ or $Y(-L/2) = Y(L/2) = \pm 1$. Without loss of generality, take $Y(-L/2) = Y(L/2) = -1$. Then $X(x) = \sinh(2\pi x/L)$. This can happen provided $c_1 = c_2 = ½$ and $C_o = (2\pi/L)^2$. Using this information, the second *ODE* $\frac{Y''(y)}{Y(y)} = -C_o = -(2\pi/L)^2$ becomes $Y''(y) + (2\pi/L)^2 Y(y) = 0$. The corresponding characteristic equation for this *ODE* is $\lambda^2 + (2\pi/L)^2 = 0 \Rightarrow \lambda = \pm i(2\pi/L)$. Hence, $Y(y) = b_1 e^{i\frac{2\pi}{L}y} + b_2 e^{-i\frac{2\pi}{L}y} = b_1(\cos(2\pi y/L) + i\sin(2\pi y/L)) + b_2(\cos(2\pi y/L) - i\sin(2\pi y/L)) = (b_1 + b_2)\cos(2\pi y/L) + i(b_1 - b_2)\sin(2\pi y/L)$. Now $X(x) = \sinh(2\pi x/L)$ means that $X(L/2) = \sinh(\pi)$ and $X(L/2) = -\sinh(\pi)$. By the Dirichlet boundary conditions

$$u(-L/2, y) = -\sinh(\pi)\left(\cos\left(\frac{2\pi}{L}y\right) + \sin\left(\frac{2\pi}{L}y\right)\right) = X\left(-\frac{L}{2}\right)\left(\cos\left(\frac{2\pi}{L}y\right) + \sin\left(\frac{2\pi}{L}y\right)\right)$$
$$= X\left(-\frac{L}{2}\right)Y(y)$$

and

$$u(L/2, y) = \sinh(\pi)\left(\cos\left(\frac{2\pi}{L}y\right) + \sin\left(\frac{2\pi}{L}y\right)\right) = X\left(\frac{L}{2}\right)\left(\cos\left(\frac{2\pi}{L}y\right) + \sin\left(\frac{2\pi}{L}y\right)\right)$$
$$= X\left(\frac{L}{2}\right)Y(y).$$

This can be achieved provided $(b_1 + b_2) = 1$ and $(b_1 - b_2) = -i$ or $b_1 = ½(1-i)$ and $b_2 = ½(1+i)$.

Therefore,

$$\boxed{u(x,y) = \sinh\left(\frac{2\pi x}{L}\right)\cdot\left(\cos\left(\frac{2\pi y}{L}\right)+\sin\left(\frac{2\pi y}{L}\right)\right)} \tag{2.4.2}$$

is the unique solution to (2.4.1). Figure 2.3 illustrates (2.4.2) with $L = 2$. ■

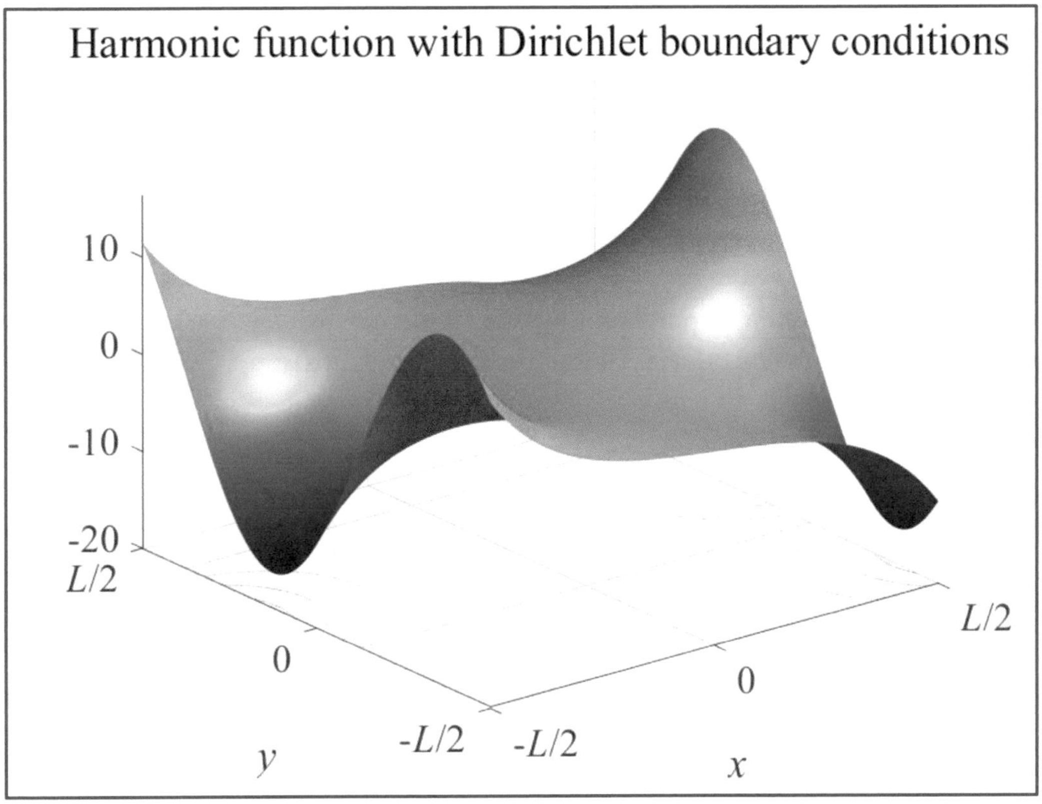

Figure 2.3. Harmonic function satisfying Dirichlet boundary conditions.

Rather than determining the steady–state distribution of an electric field over the square of length L, instead consider the domain to be the open unit disk centered at the origin. This geometry requires a change in coordinates which will require a change in the form of the Laplacian Δ.

Example 2.2. Let the domain on which the Laplacian is defined be the two–dimensional unit disk with center at the origin; $\Omega = \mathbb{B}_2(\boldsymbol{0},1) = \{\boldsymbol{x} \in \mathbb{R}^2 : \|\boldsymbol{x}\| < 1\}$. Then the boundary of Ω is the unit circle $\partial\Omega = \mathbb{S}_1(\boldsymbol{0},1) = \{\boldsymbol{x} \in \mathbb{R}^2 : \|\boldsymbol{x}\| = 1\}$ and the *Dirichlet* problem is to solve

$$\left.\begin{array}{ll} \Delta u(\boldsymbol{x}) = 0 & \text{for } \boldsymbol{x} \in \mathbb{B}_2(\boldsymbol{0},1) \\ u(\boldsymbol{x}) = f(\boldsymbol{x}) & \text{for } \boldsymbol{x} \in \mathbb{S}_1(\boldsymbol{0},1) \end{array}\right\}. \tag{2.4.3}$$

Solution: Since the domain is the unit disk, a change to polar coordinates is required. The change of variables $x = r\cos(\theta)$, $y = r\sin(\theta)$ results in a new form of the Laplacian. That is, $u(x, y) \mapsto u(r, \theta)$ and $\Delta u = \frac{\partial^2 u}{\partial r^2} + \frac{1}{r}\frac{\partial u}{\partial r} + \frac{1}{r^2}\frac{\partial^2 u}{\partial \theta^2}$. Now implementing the separation of variables approach, assume that u is a product of functions of the independent variables: $u(r, \theta) = R(r)\cdot\Theta(\theta)$. For non-zero initial condition $f(\boldsymbol{x})$, it must be the case that $R(r)$ and $\Theta(\theta)$ are not identically zero. Then $0 = \Delta u = \frac{\partial^2}{\partial r^2}\big(R(r)\cdot\Theta(\theta)\big) + \frac{1}{r}\frac{\partial}{\partial r}\big(R(r)\cdot\Theta(\theta)\big) + \frac{1}{r^2}\frac{\partial^2}{\partial \theta^2}\big(R(r)\cdot\Theta(\theta)\big)$. Proceeding as per Example 2.1, it is seen that $\frac{\partial}{\partial r}\big(R(r)\cdot\Theta(\theta)\big) = R'(r)\cdot\Theta(\theta)$, $\frac{\partial^2}{\partial r^2}\big(R(r)\cdot\Theta(\theta)\big) = R''(r)\cdot\Theta(\theta)$, and $\frac{\partial^2}{\partial \theta^2}\big(R(r)\cdot\Theta(\theta)\big) = R(r)\cdot\Theta''(\theta)$.

Hence, $0 = R''(r)\cdot\Theta(\theta) + \frac{1}{r}R'(r)\cdot\Theta(\theta) + \frac{1}{r^2}R(r)\cdot\Theta''(\theta)$ or $\frac{r^2 R''(r) + r\,R'(r)}{R(r)} = -\frac{\Theta''(\theta)}{\Theta(\theta)}$. The right-hand side of this equation is purely a function of the independent variable θ while the left-hand side is purely a function of r. Therefore, the two ratios must be equal to a constant, $\frac{r^2 R''(r) + r\,R'(r)}{R(r)} = -\frac{\Theta''(\theta)}{\Theta(\theta)} = \lambda$ (constant). Further, on $\partial\mathbb{B}_2(\boldsymbol{0},1) = \mathbb{S}_1(\boldsymbol{0},1)$, $r = 1$ so that $f(\boldsymbol{x}) = u(x, y) = u(r, \theta) = u(1, \theta) = R(1)\cdot\Theta(\theta) \equiv \phi(\theta)$. That is, $f(\boldsymbol{x}) = \phi(\theta)$ on the domain boundary $\mathbb{S}_1(\boldsymbol{0},1)$. Consequently, the Dirichlet problem becomes the set of ordinary differential equations (2.4.4)

$$\left.\begin{aligned} r^2 R''(r) + rR'(r) - \lambda R(r) = 0 \quad & \text{for } 0 \le r \le 1 \\ \Theta''(\theta) + \lambda\Theta(\theta) = 0 \quad & \text{for } -\pi \le \theta \le \pi \\ u(1,\theta) = \phi(\theta) \quad & \end{aligned}\right\}. \tag{2.4.4}$$

To find a representation for the function R, assume that it is an analytic function. That is, assume R has a convergent Taylor series $R(r) = \sum_{n=0}^{\infty} a_n r^n$. Then $R'(r) = \sum_{n=1}^{\infty} n\cdot a_n r^{n-1}$ and $R''(r) = \sum_{n=2}^{\infty} n\cdot(n-1)\cdot a_n r^{n-2}$. Using the first *ODE* of (2.4.4), these computations imply that

$$0 = r^2\sum_{n=2}^{\infty} n\cdot(n-1)\cdot a_n r^{n-2} + r\sum_{n=1}^{\infty} n\cdot a_n r^{n-1} - \lambda\sum_{n=0}^{\infty} a_n r^n = \sum_{n=2}^{\infty} n\cdot(n-1)\cdot a_n r^n + \sum_{n=1}^{\infty} n\cdot a_n r^n - \lambda\sum_{n=0}^{\infty} a_n r^n$$

$$= \sum_{n=2}^{\infty}\big[n\cdot(n-1) + n - \lambda\big]a_n r^n + a_1 r - (a_1 r + a_o)\lambda = \sum_{n=2}^{\infty}\big[n^2 - \lambda\big]a_n r^n + (1-\lambda)a_1 r - \lambda a_o.$$

Now $a_n = 0$ for all $n \in \mathbb{Z}_o^+$ solves the equation above, but such a set of coefficients produces the trivial solution $R(r) \equiv 0$. If $\phi(\theta) \neq 0$, then the boundary condition is not satisfied by $R = 0$.

Therefore, take $\lambda_n = n^2$ for $n \in \mathbb{Z}^+$. Notice that this includes the values $\lambda_1 = 1$ and $\lambda_o = 0$ which cancel the coefficients a_1 and a_o. These values for λ permit R to be a non-trivial analytic function with primary solutions $R_n(r) = a_n \cdot r^n$. In addition, $\Theta(\theta)$ now satisfies the equation $\Theta''(\theta) + n^2\Theta(\theta) = 0$. The associated characteristic equation for this *ODE* is $\chi(\rho) = \rho^2 + n^2 = 0$ which means that $\rho_n = \pm i\, n$ for $n \in \mathbb{Z}^+$ are the characteristic roots. Therefore, the primary solutions for Θ are $\Theta_n(\theta) = b_n e^{in\theta} + c_n e^{-in\theta} = (b_n + c_n)\cdot\cos(n\theta) + i(b_n - c_n)\cdot\sin(n\theta)$ so that $\Theta_n(\theta) = \beta_n \cdot \cos(n\theta) + \beta_n \cdot \sin(n\theta)$ for $b_n = ½\,(1-i)\cdot\beta_n$ and $c_n = ½\,(1+i)\cdot\beta_n$. Note that $\gamma_n \cdot \sin(n\theta)$ can be substituted for $\beta_n \cdot \sin(n\theta)$ as any multiple of $\sin(n\theta)$ satisfies $\Theta''(\theta) + n^2\Theta(\theta) = 0$. Thus for constants A_n and B_n, $u_n(r,\theta) = R_n(r)\cdot \Theta_n(\theta) = A_n \cdot r^n \cos(n\theta) + B_n \cdot r^n \sin(n\theta)$ comprise a family of solutions to $\Delta u = 0$ for any $n \in \mathbb{Z}^+$. Since the Laplacian is a linear operator, then the sum of any two solutions of $\Delta u = 0$ is also a solution. Subsequently,

$$u(r,\theta) = \sum_{n\in\mathbb{Z}^+} u_n(r,\theta) = A_o + \sum_{n=1}^{\infty} A_n \cdot r^n \cos(n\theta) + B_n \cdot r^n \sin(n\theta) \tag{2.4.5}$$

is a solution of Laplace's equation $\Delta u = 0$. This idea that the sum of solutions is also a solution is called the *superposition principle*. The function $u(r,\theta)$ of (2.4.5) solves (2.4.4), provided $f(\mathbf{x}) = u(1,\theta) = \phi(\theta)$ satisfies

$$\phi(\theta) = A_o + \sum_{n=1}^{\infty} A_n \cdot \cos(n\theta) + B_n \cdot \sin(n\theta). \tag{2.4.6}$$

As will be seen in the next chapter, this representation of the function $\phi(\theta)$ is possible with respect to relatively mild conditions on ϕ. In fact, (2.4.6) is the *Fourier series* for the function $\phi(\theta)$. Thus, for all functions ϕ which have a Fourier series, (2.4.5)–(2.4.6) solves (2.4.4). ■

In this chapter, two distinct methodologies to the solution of Laplace's and Poisson's equations are described. The classical approach uses the separation of variables transformation to obtain analytic, series solutions of Laplace's equation. The modern approach utilizes Green's Representation Theorem, Green's functions, and the generalized function $\delta(\mathbf{x})$ to solve Laplace's and Poisson's equations. The notions of harmonic functions (i.e., functions $u(\mathbf{x})$ satisfying $\Delta u(\mathbf{x}) = 0$), Dirichlet, Neumann, and Robin boundary conditions are presented.

Exercises

2.1 Under the change of variables $x = r\cos(\theta)$, $y = r\sin(\theta)$, show that the Laplacian operator $\Delta = \frac{\partial^2}{\partial x^2} + \frac{\partial^2}{\partial y^2}$ becomes $\frac{\partial^2}{\partial r^2} + \frac{1}{r}\frac{\partial}{\partial r} + \frac{1}{r^2}\frac{\partial^2}{\partial \theta^2}$. This is the *polar coordinate form of the Laplacian.*

2.2 Show that, under the change of variables $x = r\sin(\varphi)\cos(\psi)$, $y = r\sin(\varphi)\sin(\psi)$, $z = r\cos(\varphi)$, the three–dimensional Laplacian $\Delta = \frac{\partial^2}{\partial x^2} + \frac{\partial^2}{\partial y^2} + \frac{\partial^2}{\partial z^2}$ becomes the *spherical coordinate form of the Laplacian* $\frac{\partial^2}{\partial r^2} + \frac{2}{r}\frac{\partial}{\partial r} + \frac{1}{r^2}\frac{\partial^2}{\partial \varphi^2} + \frac{1}{r^2}\cot(\varphi)\frac{\partial}{\partial \varphi} + \frac{1}{r^2\sin^2(\varphi)}\frac{\partial^2}{\partial \psi^2}$.

2.3 Solve the Neumann problem below on the unit square.

$$\left.\begin{aligned} &\Delta u = 0 \quad \text{for} \quad 0 < x < 1, 0 < y < 1 \\ &\frac{\partial u}{\partial x}(0, y) = y, \frac{\partial u}{\partial x}(1, y) = y, \frac{\partial u}{\partial y}(x, 0) = x, \frac{\partial u}{\partial y}(x, 1) = x \end{aligned}\right\} \tag{2.4.7}$$

2.4 Solve the Robin problem in polar coordinates on the unit disk.

$$\left.\begin{aligned} &\Delta u = 0 \quad \text{for} \quad r < 1, 0 < \theta < 2\pi \\ &\frac{\partial u}{\partial r}(1, \theta) = \cos(\theta), u(1, 0) = 0, u(1, 2\pi) = 0 \end{aligned}\right\} \tag{2.4.8}$$

Chapter 3

Fourier Series, Bessel Functions, and Mathematical Physics

If asked to write a computer program or design a central processing unit to compute the sine or cosine of any given angle x, the programmer or chip designer would not necessarily reference a table of trigonometric values. Rather, the person charged with such a task would return to calculus, Taylor's Theorem, and the associated series. In particular, the Taylor series for the sine and cosine functions are $\sin(x) = x - \frac{1}{3!}x^3 + \frac{1}{5!}x^5 - \frac{1}{7!}x^7 + \frac{1}{9!}x^9 \cdots$ and $\cos(x) = 1 - \frac{1}{2!}x^2 + \frac{1}{4!}x^4 - \frac{1}{6!}x^6 + \frac{1}{8!}x^8 \cdots$. These series would be expanded until a sufficiently small error between known values of the sine and cosine functions and their respective Taylor approximations are achieved.

One way, then, to represent a smooth function is to write it as a linear combination of the elementary monomials $1, x, x^2, x^3, x^4, \ldots$ via Taylor's Theorem. That is,

$$f(x) = f(x_o) + \sum_{n=1}^{\infty} \frac{d^n f}{dx^n}(x_o) \cdot (x - x_o)^n . \tag{3.0.1}$$

The implicit requirement in writing a Taylor series for a function is that the function has an infinite number of derivatives which can be evaluated at a point x_o. This requires that $f \in C^\infty([a,b]) = \left\{ f : \frac{d^n f}{dx^n} \in C^0([a,b]) \text{ for all } n \geq 0 \right\}$ where $C^0([a, b])$ is the collection of all functions that are continuous on the interval $[a, b]$. Taylor's Theorem also requires that the series (3.0.1) converges. Such functions in which Taylor series exist and converge are called *analytic* and identified as being part of the set of functions $C^\omega([a, b])$.

$$C^\omega([a,b]) = \left\{ f \in C^\infty([a,b]) : \text{the Taylor series (3.0.1) exists and converges} \right\}$$

Not all functions, however, are continuous or continuously differentiable. By recalling from Chapter 2 the notation $\mathcal{S}_{[a,b]}(x) = 1$ for $x \in [a, b]$ and 0 otherwise, it is not difficult to see that the function $f(x, x_o) = (x - x_o)\cdot \mathcal{S}_{[x_o, 2x_o)}(x) + (x - 2x_o)\cdot \mathcal{S}_{[2x_o, 3x_o)}(x) + (x - 3x_o)\cdot \mathcal{S}_{[3x_o, 4x_o)}(x)$ is discontinuous at $x = x_o$, $2x_o$, and $3x_o$. Similarly, $\phi(x, x_o) = |x - x_o|$ does not have a well–defined derivative at $x = x_o$. Such functions are not analytic and are referred to as *non–smooth*. Examples of non–smooth functions are illustrated in Figure 3.1.

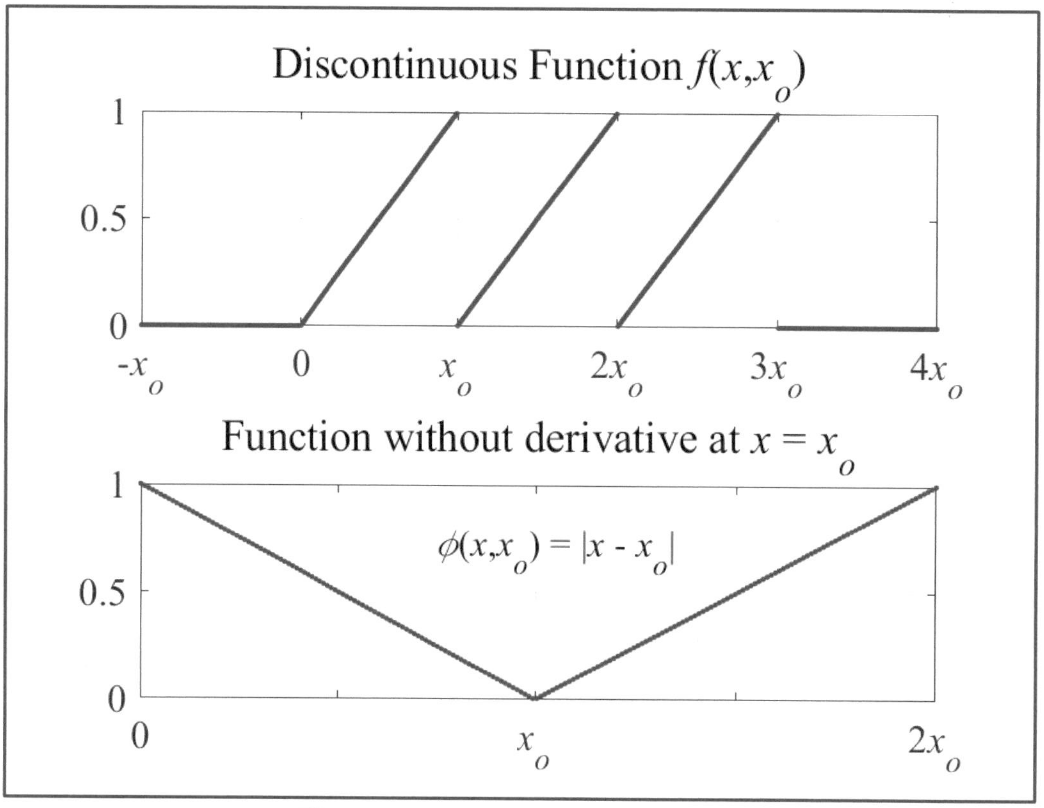

Figure 3.1. Non–smooth functions.

Since only analytic functions can be represented by Taylor series, how can non–smooth functions be approximated? One approach is *Fourier series*.

§3.1. Fourier Series

Any element of a vector space $\mathcal{V}$ can be written as a linear combination of the basis vectors. More precisely, if $\{\boldsymbol{b}_1, \boldsymbol{b}_2, \ldots, \boldsymbol{b}_n\}$ are the basis vectors of $\mathcal{V}$, then for any $\boldsymbol{v} \in \mathcal{V}$, $\boldsymbol{v} = a_1\boldsymbol{b}_1 + a_2\boldsymbol{b}_2 + \cdots + a_n\boldsymbol{b}_n$ for scalars $a_1, a_2, \ldots, a_n$. In this case, $\mathcal{V}$ is an n–dimensional vector space spanned by the basis vectors $\boldsymbol{b}_1, \boldsymbol{b}_2, \ldots, \boldsymbol{b}_n$: $\mathcal{V} = \text{span}\{\boldsymbol{b}_1, \boldsymbol{b}_2, \ldots, \boldsymbol{b}_n\} = \{\boldsymbol{v}\colon \boldsymbol{v} = \sum_{j=1}^{n} a_j\boldsymbol{b}_j,\ a_j \text{ scalars}\}$. When the scalars a_j are all real numbers then $\mathcal{V}$ is a real vector space. If the scalars a_j are complex numbers, then $\mathcal{V}$ is called a complex vector space. Assume further that the basis vectors are *orthonormal*;

that is, with respect to the inner product of the vector space, the basis vectors are perpendicular and have unit length. The inner product on $\mathcal{V}$ is represented by the symbol $<\boldsymbol{u}, \boldsymbol{v}>$ for any two vectors $\boldsymbol{u}, \boldsymbol{v} \in \mathcal{V}$. The length of a vector $\boldsymbol{v}$ is the square root of the inner product; $<\boldsymbol{v}, \boldsymbol{v}> = \| \boldsymbol{v} \|^2$. Moreover, the vectors $\boldsymbol{u}$ and $\boldsymbol{v}$ are orthogonal provided the inner product is zero; $<\boldsymbol{u}, \boldsymbol{v}> = 0$. Thus, the basis vectors $\{\boldsymbol{b}_1, \boldsymbol{b}_2, \dots, \boldsymbol{b}_n\}$ are orthonormal provided $<\boldsymbol{b}_j, \boldsymbol{b}_k> = 0$ for all $j \neq k$ and $\| \boldsymbol{b}_j \|^2 = <\boldsymbol{b}_j, \boldsymbol{b}_j> = 1$ for all $j = 1, 2, \dots, n$. It is a straightforward exercise to show that if $\{\boldsymbol{b}_1, \boldsymbol{b}_2, \dots, \boldsymbol{b}_n\}$ is an orthonormal basis for the vector space $\mathcal{V}$, then any vector $\boldsymbol{v} \in \mathcal{V}$ is written as $\boldsymbol{v} = \sum_{j=1}^{n} a_j \boldsymbol{b}_j$ and $a_j = <\boldsymbol{v}, \boldsymbol{b}_j>$.

Why take this detour into linear algebra? It is to give context to the contention that the monomials $\mathfrak{X} = \{1, x, x^2, x^3, \dots \}$ form a basis for the space of analytic functions $C^{\omega}([a, b])$. In many ways, the space $C^{\omega}([a, b])$ can be treated as an infinite–dimensional vector space with basis $\mathfrak{X}$. In fact, $C^{\omega}([a, b])$ is a *Banach*[10] *space* which is a complete linear space with a norm. For more information on Banach spaces, see Bartle [7] or Kantorovich and Akilov [45].

Now suppose f is a bounded, 2π–periodic function that is not continuously differentiable. Rather than writing f as the linear combination of the monomials $\mathfrak{X}$, consider representing f as the sum of other 2π–periodic functions. The most familiar of these functions are the sine and cosine. Define the *Lebesgue*[11] *space* $L_2^{(2\pi)}([-\pi,\pi]) = \left\{ f : \int_{-\pi}^{\pi} |f(x)|^2\, dx < \infty \text{ and } f(x+2\pi) = f(x) \right\}$ which is the set of all square integrable, 2π–periodic functions over the interval $[-\pi, \pi]$. This collection of functions is a Banach space with norm $\|f\|_2^2 = \int_{-\pi}^{\pi} |f(x)|^2\, dx$ and inner product $<f, g> = \int_{-\pi}^{\pi} f(x) \cdot \overline{g}(x)\, dx$, $\overline{g}(x)$ is the complex conjugate of the function $g(x)$. The most general Lebesgue space is $L_p(\Omega) = \left\{ f : \int_{\Omega} |f(\boldsymbol{x})|^p\, d\boldsymbol{\omega} < \infty \right\}$. The Lebesgue space $L_2^{(2\pi)}([-\pi,\pi])$ has an orthonormal basis $\mathfrak{B}_2[-\pi, \pi] = \{1, \cos(x), \sin(x), \cos(2x), \sin(2x), \cos(3x), \sin(3x), \dots\}$ with respect to the weighted L_2–inner product $<f, g> = \frac{1}{\pi}\int_{-\pi}^{\pi} f(x) \cdot g(x)\, dx$. Note the functions in $L_2^{(2\pi)}([-\pi,\pi])$ are all real valued. Therefore, it is unnecessary to use the complex conjugate $\overline{g}(x)$ within the inner product.

---

[10] Named for the Polish mathematician Stefan Banach (1892–1945) whose 1920 dissertation and 1931 monograph *The Theory of Linear Operators* formally defined the notion of a complete normed linear space.

[11] Named for the French mathematician Henri Lebesgue (1875–1941) whose 1918 treatise *Remarques sur les théories de la mesure et de l'intégration* devised a formal theory for integrals and integration.

A function $f \in L_2^{(2\pi)}([-\pi,\pi])$ can be written as a sum of elements from the basis set $\mathcal{B}_2[-\pi, \pi]$. Such a representation is called a *Fourier series* which, for $f \in L_2^{(2\pi)}([-\pi,\pi])$, yields the formulae (3.1.1).

$$\boxed{\begin{aligned} f(x) &= \tfrac{1}{2}a_o + \sum_{n=1}^{\infty} a_n \cos(nx) + b_n \sin(nx) \\ a_o &= < f,1> = \frac{1}{\pi}\int_{-\pi}^{\pi} f(x)\,dx \\ a_n &= < f,\cos(nx)> = \frac{1}{\pi}\int_{-\pi}^{\pi} f(x)\cdot\cos(nx)\,dx \\ b_n &= < f,\sin(nx)> = \frac{1}{\pi}\int_{-\pi}^{\pi} f(x)\cdot\sin(nx)\,dx \end{aligned}} \tag{3.1.1}$$

Observe that if $f(x) \equiv 1$, then by the second equation in (3.1.1), $a_o = 2$. Hence, the first term in the Fourier series is normalized by ½.

More generally, if $f \in L_2^{(\ell)}([0,\ell])$, then the Fourier coefficients are written as below (see, e.g., Andrews and Shivamoggi [4] or Zachmanoglou and Thoe [87]).

$$\boxed{\begin{aligned} a_n &= \frac{2}{\ell}\int_0^{\ell} f(x)\cdot\cos\left(\tfrac{\pi n}{\ell}x\right)dx,\ n \in \mathbb{Z}_o^+ \\ b_n &= \frac{2}{\ell}\int_0^{\ell} f(x)\cdot\sin\left(\tfrac{\pi n}{\ell}x\right)dx,\ n \in \mathbb{Z}^+ \end{aligned}} \tag{3.1.1ℓ}$$

Consider the example of the periodic extension of $x + x^2$.

Example 3.1: Let $f(x) = x + x^2$ for $-\pi \le x \le \pi$ be extended periodically as per Figure 3.2. Calculate the Fourier series for f.

Solution: Using formulae (3.1.1), it is seen that $a_o = \frac{1}{\pi}\int_{-\pi}^{\pi}(x+x^2)\,dx = \frac{1}{\pi}\left(\tfrac{1}{2}x^2 + \tfrac{1}{3}x^3\right)\Big|_{-\pi}^{\pi} = \tfrac{2}{3}\pi^2$.

Further, direct computations produce the formulae $a_n = \frac{1}{\pi}\int_{-\pi}^{\pi}(x+x^2)\cdot\cos(nx)\,dx = \frac{4}{n^2}(-1)^n$ and $b_n = \frac{1}{\pi}\int_{-\pi}^{\pi}(x+x^2)\cdot\sin(nx)\,dx = \frac{2}{n}(-1)^{n+1}$. Therefore, the Fourier series for f is

$$f(x) = \tfrac{1}{3}\pi^2 + \sum_{n=1}^{\infty}(-1)^n\frac{4}{n^2}\cos(nx) + (-1)^{n+1}\frac{2}{n}\sin(nx).$$

The approximation $f_N(x) = \frac{1}{3}\pi^2 + \sum_{n=1}^{N} (-1)^n \frac{4}{n^2}\cos(nx) + (-1)^{n+1}\frac{2}{n}\sin(nx)$ is illustrated for $N = 3$, 9, 16, and 24 in Figure 3.3. As N increases, the truncated Fourier series $f_N(x)$ better approximates $f(x)$.

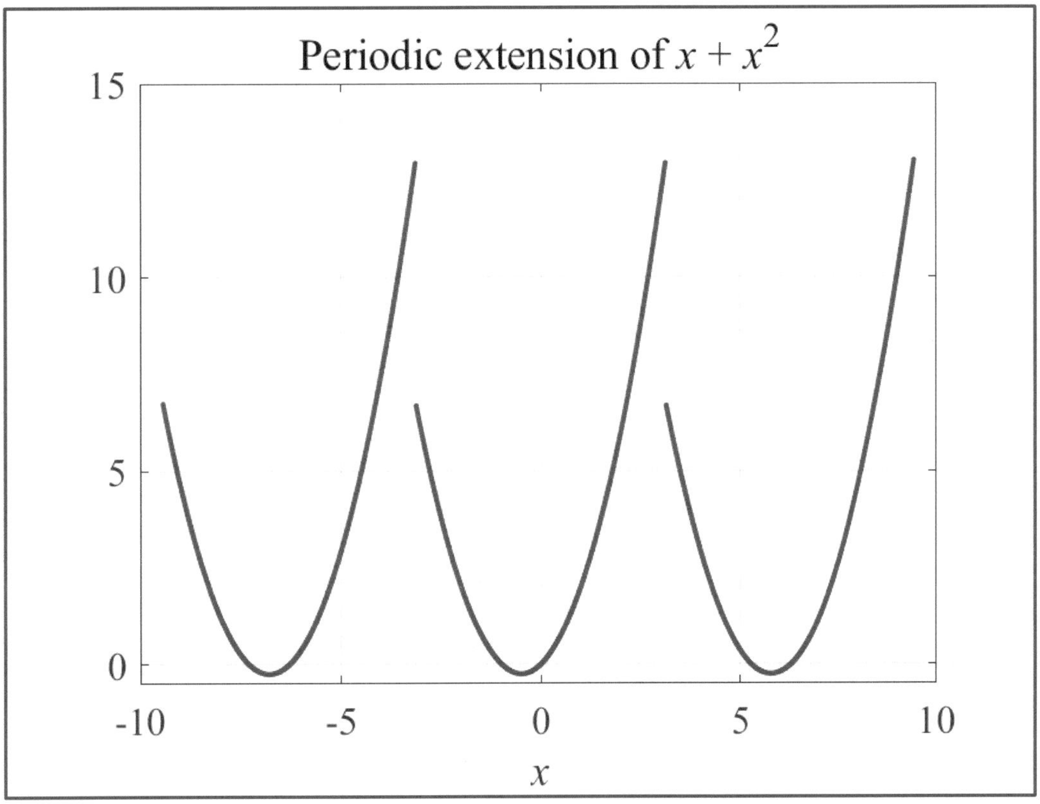

Figure 3.2. Periodic extension.

Now return to Example 2.2 of Chapter 2. In that case, the solution of Laplace's equation $\Delta u(r,\theta) = 0$ for $(r,\theta) \in \mathbb{B}_2(\boldsymbol{0},1)$ with Dirichlet boundary condition $u(1,\theta) = f(\theta)$ on $\partial\mathbb{B}_2(\boldsymbol{0},1) = \mathbb{S}_1(\boldsymbol{0},1)$ is $u(r,\theta) = A_o + \sum_{n=1}^{\infty} A_n r^n \cos(nx) + B_n r^n \sin(nx)$. The boundary condition $u(1,\theta) = f(\theta)$ requires that $f(\theta) = A_o + \sum_{n=1}^{\infty} A_n \cos(nx) + B_n \sin(nx)$. Therefore, the constants A_o, A_n, and B_n are the Fourier coefficients for the function f as expressed by (3.1.1). Since f is defined on the boundary of the unit disk; namely, the unit circle $\mathbb{S}_1(\boldsymbol{0},1)$, it is sensible to require that f be 2π-periodic and integrable over $[-\pi, \pi]$. That is, $f \in L_2^{(2\pi)}([-\pi,\pi])$.

The curious reader may now ask, "Are Fourier series restricted to one dimension?" As the next section indicates, the answer is "no."

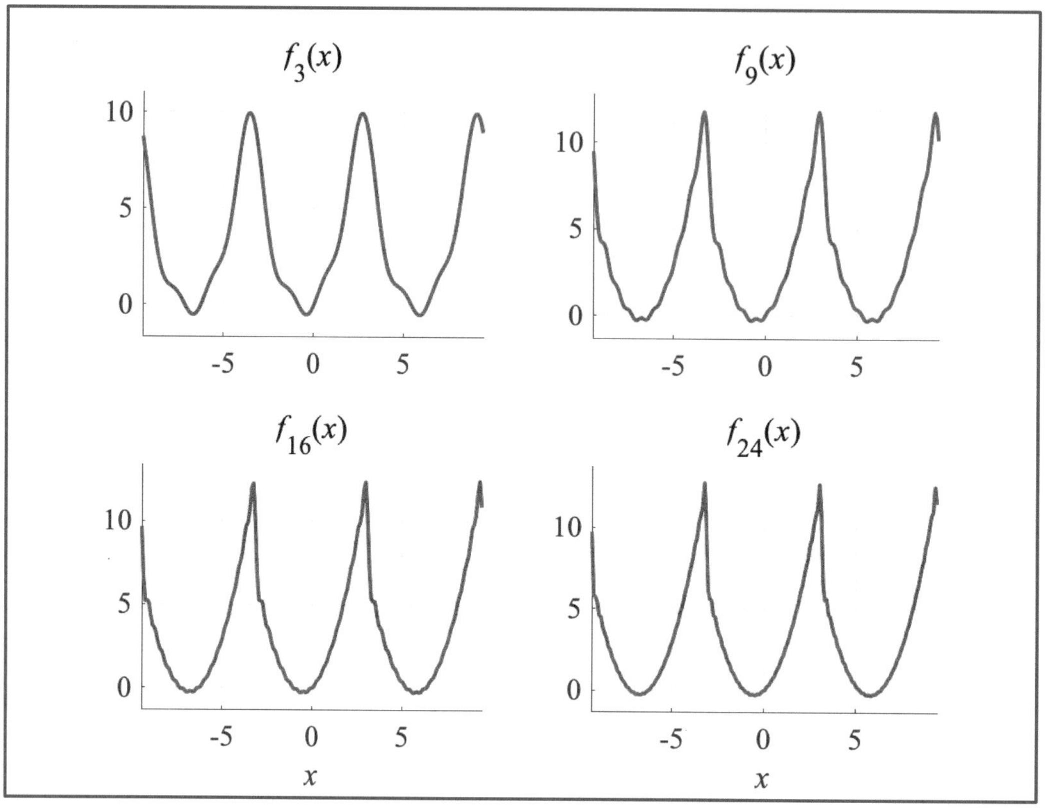

Figure 3.3. Fourier series approximations.

§3.2. Fourier Series in Two Dimensions

The section above developed the definition and mechanism to expand a one–dimensional periodic function which does not need to be either everywhere differentiable or even continuous. Now these ideas are extended to two dimensions. It is left to the reader to generalize the development to n–dimensions. The discussion below follows Myint–U with Debnath [52].

Suppose $f \in L_2^{(2\pi)}([-\pi,\pi]\times[-\pi,\pi])$ which means that $f(x, y)$ is 2π–periodic in both x and y *and* is square–integrable over the domain $[-\pi, \pi] \times [-\pi, \pi]$. That is, $f(x+2\pi, y) = f(x, y)$, $f(x, y+2\pi) = f(x, y)$, and $\int_{-\pi}^{\pi}\int_{-\pi}^{\pi}|f(x,y)|^2\,dx\,dy < \infty$.

Since f is 2π–periodic in x, then the one–dimensional Fourier series for f can be expressed as

$$f(x, y) = \tfrac{1}{2}\, a_o(y) + \sum_{m=1}^{\infty} a_m(y)\cos(mx) + b_m(y)\sin(mx) \text{ with}$$

$$a_o(y) = \frac{1}{\pi}\int_{-\pi}^{\pi} f(x,y)\,dx,\; a_m(y) = \frac{1}{\pi}\int_{-\pi}^{\pi} f(x,y)\cos(mx)\,dx,\; b_m(y) = \frac{1}{\pi}\int_{-\pi}^{\pi} f(x,y)\sin(mx)\,dx.$$

But f is also 2π–periodic in y so that $a_o(y)$, $a_m(y)$, and $b_m(y)$ must also be 2π–periodic. Consequently, each of these functions has a Fourier series. In particular,

$$a_m(y) = \tfrac{1}{2}\, a_{m,o} + \sum_{n=1}^{\infty} a_{m,n} \cos(ny) + b_{m,n} \sin(ny),\ m = 0, 1, 2, \ldots$$

and

$$a_{m,n} = \frac{1}{\pi}\int_{-\pi}^{\pi} a_m(y)\cos(ny)\,dy = \frac{1}{\pi^2}\int_{-\pi}^{\pi}\int_{-\pi}^{\pi} f(x,y)\cos(mx)\cos(ny)\,dx\,dy$$

$$b_{m,n} = \frac{1}{\pi}\int_{-\pi}^{\pi} a_m(y)\sin(ny)\,dy = \frac{1}{\pi^2}\int_{-\pi}^{\pi}\int_{-\pi}^{\pi} f(x,y)\cos(mx)\sin(ny)\,dx\,dy$$

for $n \in \mathbb{Z}^+$.

In a parallel manner, the functions $b_m(y)$ for $m \in \mathbb{Z}^+$ are formulated.

$$b_m(y) = \tfrac{1}{2}\, c_{m,o} + \sum_{n=1}^{\infty} c_{m,n} \cos(ny) + d_{m,n} \sin(ny)$$

$$c_{m,n} = \frac{1}{\pi}\int_{-\pi}^{\pi} b_m(y)\cos(ny)\,dy = \frac{1}{\pi^2}\int_{-\pi}^{\pi}\int_{-\pi}^{\pi} f(x,y)\sin(mx)\cos(ny)\,dx\,dy$$

$$d_{m,n} = \frac{1}{\pi}\int_{-\pi}^{\pi} b_m(y)\sin(ny)\,dy = \frac{1}{\pi^2}\int_{-\pi}^{\pi}\int_{-\pi}^{\pi} f(x,y)\sin(mx)\sin(ny)\,dx\,dy$$

Thus, the *double Fourier series* for the doubly 2π–periodic function $f(x, y)$ is established via (3.2.1).

$$\left.\begin{aligned} f(x,y) &= \tfrac{1}{4}a_{0,0} + \tfrac{1}{2}\sum_{n=1}^{\infty}\left(a_{0,n}\cos(ny) + b_{0,n}\sin(ny)\right) \\ &\quad + \tfrac{1}{2}\sum_{m=1}^{\infty}\left(a_{m,0}\cos(mx) + c_{m,0}\sin(mx)\right) \\ &\quad + \sum_{m=1}^{\infty}\sum_{n=1}^{\infty}\left(a_{m,n}\cos(mx)\cos(ny) + b_{m,n}\cos(mx)\sin(ny)\right) \\ &\quad + \sum_{m=1}^{\infty}\sum_{n=1}^{\infty}\left(c_{m,n}\sin(mx)\cos(ny) + d_{m,n}\sin(mx)\sin(ny)\right) \end{aligned}\right\} \tag{3.2.1}$$

The corresponding coefficients are defined as below.

$$\left.\begin{aligned}
a_{0,0} &= \frac{1}{\pi^2}\int_{-\pi}^{\pi}\int_{-\pi}^{\pi} f(x,y)\,dx\,dy \\
a_{0,n} &= \frac{1}{\pi^2}\int_{-\pi}^{\pi}\int_{-\pi}^{\pi} f(x,y)\cos(ny)\,dx\,dy \\
b_{0,n} &= \frac{1}{\pi^2}\int_{-\pi}^{\pi}\int_{-\pi}^{\pi} f(x,y)\sin(ny)\,dx\,dy \\
a_{m,n} &= \frac{1}{\pi^2}\int_{-\pi}^{\pi}\int_{-\pi}^{\pi} f(x,y)\cos(mx)\cos(ny)\,dx\,dy,\ m,n\ge 1 \\
b_{m,n} &= \frac{1}{\pi^2}\int_{-\pi}^{\pi}\int_{-\pi}^{\pi} f(x,y)\cos(mx)\sin(ny)\,dx\,dy,\ m,n\ge 1 \\
c_{m,n} &= \frac{1}{\pi^2}\int_{-\pi}^{\pi}\int_{-\pi}^{\pi} f(x,y)\sin(mx)\cos(ny)\,dx\,dy,\ m,n\ge 1 \\
d_{m,n} &= \frac{1}{\pi^2}\int_{-\pi}^{\pi}\int_{-\pi}^{\pi} f(x,y)\sin(mx)\sin(ny)\,dx\,dy,\ m,n\ge 1
\end{aligned}\right\} \tag{3.2.2}$$

If f obeys certain symmetry conditions, then (3.2.1) can be considerably simplified. These cases are presented directly.

§3.2.1. f is doubly even

In this case $f(-x, y) = f(x, y)$ and $f(x, -y) = f(x, y)$. Since the sine is an odd function, this means that any coefficients which multiply $\sin(mx)$ or $\sin(ny)$ must be zero. Therefore, the double Fourier series for f is

$$\boxed{f(x,y) = \tfrac{1}{4}a_{0,0} + \tfrac{1}{2}\sum_{n=1}^{\infty} a_{0,n}\cos(ny) + \sum_{m=1}^{\infty}\sum_{n=1}^{\infty} a_{m,n}\cos(mx)\cos(ny)}.$$

Moreover, since the cosine function and f are both even, then the integration can be performed over the interval $[0, \pi]$ and then doubled. This leads to the condensed formula

$$\boxed{a_{m,n} = \left(\frac{2}{\pi}\right)^2\int_0^{\pi}\int_0^{\pi} f(x,y)\cos(mx)\cos(ny)\,dx\,dy}.$$

§3.2.2. f is even in x and odd in y

In this case, $f(-x, y) = f(x, y)$ while $f(x,-y) = -f(x, y)$. As the cosine is even and the sine is odd, then only the terms $\cos(mx)$, $\sin(ny)$, and $\cos(mx)\cdot\sin(ny)$ are maintained. All other coefficients are set to zero. This produces the formulae

$$f(x,y)=\tfrac{1}{4}a_{0,0}+\tfrac{1}{2}\sum_{n=1}^{\infty}b_{0,n}\sin(ny)+\sum_{m=1}^{\infty}\sum_{n=1}^{\infty}b_{m,n}\cos(mx)\sin(ny)$$

with $a_{0,0}$ and $b_{m,n}$ as in (3.2.2). The special case of $b_{0,n}$ is handled in (3.2.2) since cos(0) = 1.

§3.2.3. f is odd in x and even in y

For this scenario, $f(-x, y) = -f(x, y)$ and $f(x,-y) = f(x, y)$ so that only terms involving $\sin(mx)$ and $\cos(ny)$ are retained. This gives

$$f(x,y)=\tfrac{1}{4}a_{0,0}+\tfrac{1}{2}\sum_{n=1}^{\infty}a_{0,n}\cos(ny)+\sum_{m=1}^{\infty}\sum_{n=1}^{\infty}c_{m,n}\sin(mx)\cos(ny)$$

with $a_{0,n}$ and $c_{m,n}$ as in (3.2.2). The special case of $a_{0,0}$ is included in the definition for $a_{0,n}$ as cos(0) = 1.

§3.2.4. f is doubly odd

In this case, $f(-x, y) = -f(x, y)$ and $f(x,-y) = -f(x, y)$. Since the cosine is an odd function, then any coefficients multiplying $\cos(mx)$ or $\cos(ny)$ must be zero. This leaves

$$f(x,y)=\tfrac{1}{4}a_{0,0}+\tfrac{1}{2}\sum_{n=1}^{\infty}b_{0,n}\sin(ny)+\sum_{m=1}^{\infty}\sum_{n=1}^{\infty}d_{m,n}\sin(mx)\sin(ny)$$

with $a_{0,0}$, $b_{0,n}$, and $d_{m,n}$ as in (3.2.2).

After the development of Fourier series for one– and two–spatial dimensions, the mathematical machinery required to solve the Dirichlet (2.3.1), Neumann (2.3.2), and Robin (2.3.3) problems with respect to Laplace's equation is now in place.

Note that, if the interval $[-\pi, \pi]$ in x is replaced by $[0, \ell_1]$ and by $[0, \ell_2]$ in y, then (3.2.2) is replaced by (3.2.2ℓ). These formulae can be derived via the change of variables $\xi = ½\,\ell_1 + (\ell_1/2\pi)\,x$ and $\psi = ½\,\ell_2 + (\ell_2/2\pi)\,y$ and by defining the function $\phi(\xi,\psi) = f(½\,\ell_1 + (\ell_1/2\pi)\,x,\ ½\,\ell_2 + (\ell_2/2\pi)\,y) \equiv F(x, y)$. In this case, $\phi \in L_2^{(\ell_1,\ell_2)}([0,\ell_1]\times[0,\ell_2]) = \left\{f : \int_0^{\ell_2}\int_0^{\ell_1}|f(x,y)|^2\,dx\,dy < \infty, f(x+\ell_1, y+\ell_2) = f(x,y)\right\}$. Indeed, $F(-\pi, -\pi) = f(0, 0)$, $F(-\pi, \pi) = f(0, \ell_2)$, $F(\pi, -\pi) = f(\ell_1, 0)$, and $F(\pi, \pi) = f(\ell_1, \ell_2)$. Hence, if $f \in L_2^{(\ell_1,\ell_2)}([0,\ell_1]\times[0,\ell_2])$, then F is 2π–periodic. The change of variables in ξ and ψ above in the integrals in (3.2.2) produce (3.2.2ℓ). See Myint–U with Debnath [52] for details.

$$\left.\begin{aligned}
a_{0,0} &= \frac{4}{\ell_1\ell_2}\int_0^{\ell_2}\int_0^{\ell_1} f(x,y)\,dx\,dy \\
a_{0,n} &= \frac{4}{\ell_1\ell_2}\int_{-\pi}^{\pi}\int_{-\pi}^{\pi} f(x,y)\cos(\tfrac{n\pi}{\ell_1}y)\,dx\,dy \\
b_{0,n} &= \frac{4}{\ell_1\ell_2}\int_0^{\ell_2}\int_0^{\ell_1} f(x,y)\sin(\tfrac{n\pi}{\ell_2}y)\,dx\,dy \\
a_{m,n} &= \frac{4}{\ell_1\ell_2}\int_0^{\ell_2}\int_0^{\ell_1} f(x,y)\cos(\tfrac{m\pi}{\ell_1}x)\cos(\tfrac{n\pi}{\ell_2}y)\,dx\,dy,\ \ m,n\ge 1 \\
b_{m,n} &= \frac{4}{\ell_1\ell_2}\int_0^{\ell_2}\int_0^{\ell_1} f(x,y)\cos(\tfrac{m\pi}{\ell_1}x)\sin(\tfrac{n\pi}{\ell_2}y)\,dx\,dy,\ \ m,n\ge 1 \\
c_{m,n} &= \frac{4}{\ell_1\ell_2}\int_0^{\ell_2}\int_0^{\ell_1} f(x,y)\sin(\tfrac{m\pi}{\ell_1}x)\cos(\tfrac{n\pi}{\ell_2}y)\,dx\,dy,\ \ m,n\ge 1 \\
d_{m,n} &= \frac{4}{\ell_1\ell_2}\int_0^{\ell_2}\int_0^{\ell_1} f(x,y)\sin(\tfrac{m\pi}{\ell_1}x)\sin(\tfrac{n\pi}{\ell_2}y)\,dx\,dy,\ \ m,n\ge 1
\end{aligned}\right\}\qquad (3.2.2\ell)$$

§3.3. A Return to Laplace's Equation (Part 2)

This section proceeds as a series of examples.

Example 3.3.1. Solve the *Laplace–Dirichlet problem* $\Delta u = 0$ on the cube of length π subject to zero boundary conditions everywhere but the bottom face of the cube. Specifically, solve

$$\left.\begin{aligned}
&\Delta u(\boldsymbol{x}) = 0 && \text{for } \boldsymbol{x}\in C_3(\pi) = \{\boldsymbol{x}\in\mathbb{R}^3 : 0\le x,y,z\le\pi\} \\
&u(0,y,z) = 0 = u(\pi,y,z) && \text{for } 0\le y,z\le\pi \\
&u(x,0,z) = 0 = u(x,\pi,z) && \text{for } 0\le x,z\le\pi \\
&u(x,y,\pi) = 0 && \text{for } 0\le x,y\le\pi \\
&u(x,y,0) = f(x,y)\ne 0 && \text{for } 0\le x,y\le\pi
\end{aligned}\right\}. \qquad (3.3.1)$$

Figure 3.4 illustrates the boundary conditions.

One physical interpretation of (3.3.1) is the steady–state temperature distribution within the cube of length π with initial temperature distribution $f(x, y)$ across the bottom (i.e., the xy–plane) of the cube. A second view is that (3.3.1) models the electrical potential due to the distribution of charges within the cube $C_3(\pi)$ and initial charge $f(x, y)$ across the bottom of the cube.

Solution: Implement the classical method of separation of variables by assuming $u(x, y, z) = X(x)\cdot Y(y)\cdot Z(z)$. In this case, $\Delta u = 0$ implies $0 = X''(x)\cdot Y(y)\cdot Z(z) + X(x)\cdot Y''(y)\cdot Z(z) + X(x)\cdot Y(y)\cdot Z''(z)$. The boundary conditions require that u is not identically zero. Therefore dividing through by

$X(x)\cdot Y(y)\cdot Z(z)$ produces the equation $0 = \frac{X''(x)}{X(x)} + \frac{Y''(y)}{Y(y)} + \frac{Z''(z)}{Z(z)}$ or $-\frac{Z''(z)}{Z(z)} = \frac{X''(x)}{X(x)} + \frac{Y''(y)}{Y(y)}$. Since the left-hand side of this equation is purely a function of z while the right-hand side is a function only of x and y then the right- and left-hand sides of the equations must be constant. That is, $\lambda = -\frac{Z''(z)}{Z(z)} = \frac{X''(x)}{X(x)} + \frac{Y''(y)}{Y(y)}$. Thus, $Z''(z) + \lambda\cdot Z(z) = 0$. Now, $\lambda = \frac{X''(x)}{X(x)} + \frac{Y''(y)}{Y(y)}$ means that $\lambda - \frac{X''(x)}{X(x)} = \frac{Y''(y)}{Y(y)}$ or the left- and right-hand sides are functions solely of the independent variables x and y. This means that the ratios are constant and $\lambda - \frac{X''(x)}{X(x)} = \frac{Y''(y)}{Y(y)} = \mu$. For non-trivial[12] solutions, the boundary conditions $u(0, y, z) = 0 = u(\pi, y, z)$ imply that $X(0) = 0 = X(\pi)$; otherwise either $Y(y) \equiv 0$ or $Z(z) \equiv 0$ which renders $u(x, y, z) \equiv 0$. In a parallel manner, the boundary conditions $u(x, 0, z) = 0 = u(x, \pi, z)$ imply that $Y(0) = 0 = Y(\pi)$ and $u(x, y, \pi) = 0$ means that $Z(\pi) = 0$. This results in three second order ordinary differential equations.

$$(i)\ \begin{cases} X''(x) - \mu X(x) = 0 \\ X(0) = 0 = X(\pi) \end{cases} \qquad (ii)\ \begin{cases} Y''(y) - (\lambda - \mu)Y(y) = 0 \\ Y(0) = 0 = Y(\pi) \end{cases} \qquad (iii)\ \begin{cases} Z''(z) + \lambda Z(z) = 0 \\ Z(\pi) = 0 \end{cases}$$

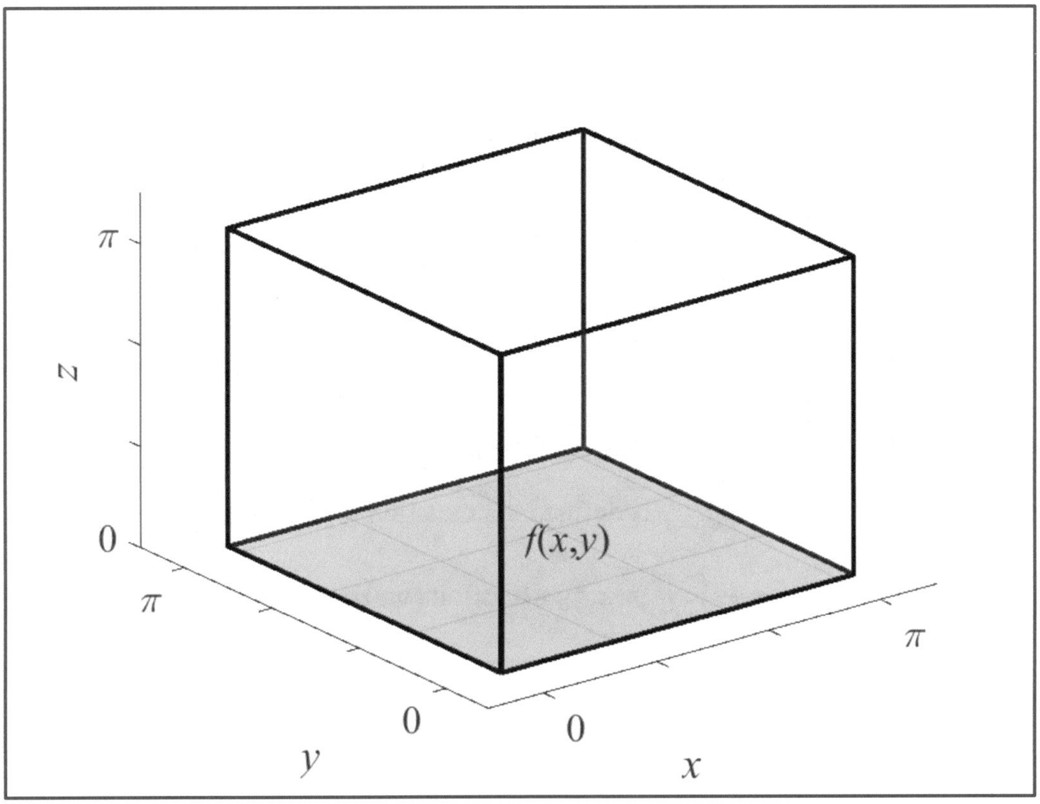

Figure 3.4. The Laplace–Dirichlet problem on $C_3(\pi)$.

[12] The trivial solution to $\Delta u = 0$ is $u(x,y,z) \equiv 0$.

Revisit Chapter 2 and Example §2.1. The first of these equations has characteristic equation $\omega^2 - \mu = 0$ or $X(x) = c_1 e^{\sqrt{\mu}x} + c_2 e^{-\sqrt{\mu}x}$. If $\mu = 0$, then $X(x)$ = constant and the boundary condition $X(0) = 0$ means that $X(x) \equiv 0$ so that $u \equiv 0$ in contradiction to the requirement of a non-trivial solution. Similarly, $\mu > 0$ means that $0 = X(0) = c_1 + c_2$ and $0 = X(\pi) = c_1 e^{\sqrt{\mu}\pi} + c_2 e^{-\sqrt{\mu}\pi} = c_1\left(e^{\sqrt{\mu}\pi} - e^{-\sqrt{\mu}\pi}\right)$ $= c_1 \sinh(\pi\sqrt{\mu}) \Rightarrow c_1 = c_2 = 0$ so that $X(x) \equiv 0$. Again, a contradiction. Therefore, $\mu = -\mu_o^2 < 0$ so that $X(x) = \gamma_1 \cos(\mu_o x) + \gamma_2 \sin(\mu_o x)$. The boundary conditions $0 = X(0) = \gamma_1$ and $0 = X(\pi) = \gamma_2 \sin(\mu_o \pi)$ require $\mu_o = n \in \mathbb{Z}^+$, otherwise $\gamma_2 = 0$ which would compel $X(x)$ to be 0. Consequently, $\mu_n = -n^2$ for $n \in \mathbb{Z}^+$. Hence, $X_n(x) = A_n \sin(nx)$. With this value for $\mu \mapsto \mu_n = -n^2$, now examine equation (*ii*). That is, solving $Y''(y) - (\lambda + n^2)\,Y(y) = 0$ and $Y(0) = 0 = Y(\pi)$ implies that $\lambda + n^2 = -m^2$ and $Y_m(y) = B_m \sin(my)$. Thus, $\lambda = -(m^2 + n^2)$ and, by equation (*iii*), $Z''(z) - (m^2 + n^2)\cdot Z(z) = 0$, $Z(\pi) = 0$. The characteristic equation for this *ODE* is $\omega^2 - (m^2 + n^2) = 0$ resulting in $Z(z) = \zeta_1 \cosh(\omega_{m,n}\,[z-\pi]) + \zeta_2 \sinh(\omega_{m,n}\,[z-\pi])$ which satisfies $Z(\pi) = 0$ when $\zeta_1 = 0$. Subsequently, $Z_{m,n}(z) = C_{m,n} \sinh\left(\sqrt{m^2+n^2}\,[z-\pi]\right)$ and therefore $u_{m,n}(x, y, z) = X_n(x)\cdot Y_m(y)\cdot Z_{m,n}(z) = \sin(nx) \sin(my) \sinh\left(\sqrt{m^2+n^2}\,[z-\pi]\right)$ is harmonic for any $m, n \in \mathbb{Z}^+$. Since the sum of any two harmonic functions is also harmonic, then $u(x, y, z) = \sum_{m=1}^{\infty}\sum_{n=1}^{\infty} U_{m,n}\, u_{m,n}(x,y,z)$ is harmonic for the as yet undetermined constants $U_{m,n}$. This is the *principle of superposition* for the Laplace equation. By the Cauchy condition,

$$f(x, y) = u(x, y, 0) = \sum_{m=1}^{\infty}\sum_{n=1}^{\infty} U_{m,n} \sin(nx)\sin(my)\sinh\left(-\sqrt{m^2+n^2}\,\pi\right)$$

$$= -\sum_{m=1}^{\infty}\sum_{n=1}^{\infty} U_{m,n} \sinh\left(\sqrt{m^2+n^2}\,\pi\right)\sin(nx)\sin(my).$$

This means f is a doubly odd function with no single terms in $\sin(nx)$ or $\sin(my)$. By §3.2.4 the terms $-U_{m,n} \sinh\left(\sqrt{m^2+n^2}\,\pi\right) = d_{m,n}$ as defined via equation (3.2.2). More precisely,

$$U_{m,n} = -\frac{4}{\pi^2 \sinh\left(\pi\sqrt{m^2+n^2}\right)} \int_0^{\pi}\int_0^{\pi} f(x,y)\sin(nx)\sin(my)\,dx\,dy$$

and therefore

$$u(x,y,z) = \sum_{m=1}^{\infty}\sum_{n=1}^{\infty} u_{m,n}(x,y,z)$$

$$= \sum_{m=1}^{\infty}\sum_{n=1}^{\infty} -U_{m,n} \sin(nx)\sin(my)\sinh\left(\sqrt{m^2+n^2}\,[z-\pi]\right).$$

■

Example 3.3.2. Solve the *Laplace–Neumann problem* $\Delta u = 0$ on the cube of length π subject to zero gradient conditions everywhere except the top face of the cube. That is, solve the equation (3.3.2) so that the normal derivative $\frac{\partial u}{\partial \boldsymbol{n}}$ is zero on each face of the cube except on the face in which $z = \pi$. The normal vectors $\boldsymbol{n}$ on the cube faces in which $y = 0$ and $y = \pi$ along with the condition $\frac{\partial u}{\partial \boldsymbol{n}} = f(x, y) \neq 0$ are illustrated in Figure 3.5.

$$\left.\begin{array}{ll} \Delta u(\boldsymbol{x}) = 0 & \text{for } \boldsymbol{x} \in C_3(\pi) = \left\{\boldsymbol{x} \in \mathbb{R}^3 : 0 \le x, y, z \le \pi\right\} \\ \frac{\partial u}{\partial \boldsymbol{n}}(0, y, z) = 0 = \frac{\partial u}{\partial \boldsymbol{n}}(\pi, y, z) & \text{for } 0 \le y, z \le \pi \\ \frac{\partial u}{\partial \boldsymbol{n}}(x, 0, z) = 0 = \frac{\partial u}{\partial \boldsymbol{n}}(x, \pi, z) & \text{for } 0 \le x, z \le \pi \\ \frac{\partial u}{\partial \boldsymbol{n}}(x, y, 0) = 0 & \text{for } 0 \le x, y \le \pi \\ \frac{\partial u}{\partial \boldsymbol{n}}(x, y, \pi) = f(x, y) \neq 0 & \text{for } 0 \le x, y \le \pi \end{array}\right\} \tag{3.3.2}$$

Solution: Again, the separation of variables approach is employed: $u(x, y, z) = X(x)\cdot Y(y)\cdot Z(z)$. The boundary conditions imply that $-X'(0) = 0 = X'(\pi)$, $-Y'(0) = 0 = Y'(\pi)$, and $-Z'(0) = 0$. This is because, on the face of the cube, the normal vectors $\boldsymbol{n}$ are in the direction of the coordinate axes (or their negative direction at 0). For example, at $y = 0$ the normal vector $\boldsymbol{n}$ points *away* from the direction of the positive y–axis. Thereby, on the face of the cube coincident with the plane $y = 0$, the exterior normal is $\boldsymbol{n} = (0, -1, 0)$ so that $0 = \frac{\partial u}{\partial \boldsymbol{n}}(x, 0, z) = \nabla u(x, 0, z)\boldsymbol{\cdot n} = \nabla u(x, 0, z)\bullet(0, -1, 0) = -\frac{\partial u}{\partial y}(x, 0, z) = -X'(x)Y'(0)Z'(z)$. In order to avoid the trivial solution $u(x, y, z) \equiv 0$, then $Y'(0) = 0$. This reasoning is used to confirm that $0 = -X'(0) = X'(\pi) = Y'(\pi) = -Z'(0)$. Therefore, as in Example 3.3.1, the system of three ODEs is obtained below.

$$(i)\ \begin{cases} X''(x) - \mu X(x) = 0 \\ X'(0) = 0 = X'(\pi) \end{cases} \qquad (ii)\ \begin{cases} Y''(y) - (\lambda - \mu)Y(y) = 0 \\ Y'(0) = 0 = Y'(\pi) \end{cases} \qquad (iii)\ \begin{cases} Z''(z) + \lambda Z(z) = 0 \\ Z'(0) = 0 \end{cases}$$

By following the methodology detailed in Example 3.3.1, it is determined that the eigenvalues and eigenfunctions of (*i*)–(*iii*) above are $\mu_n = -n^2$, $X_n(x) = c_x\cdot\cos(nx)$; $\gamma_m = -m^2$, $Y_m(y) = c_y\cdot\cos(my)$; and $\lambda_{m,n} = -(m^2+n^2)$, $Z_{m,n}(z) = \frac{1}{\sqrt{m^2+n^2}}\cosh\left(\sqrt{m^2+n^2}\, z\right)$, respectively, with c_x, $c_y \in \mathbb{R}$ arbitrary constants. The primary solutions of the Laplace–Neumann problem (3.3.2) are therefore

$$u_{m,n}(x,y,z) = \frac{C_{m,n}}{\sqrt{m^2+n^2}} \cos(nx)\cos(my)\cosh\left(\sqrt{m^2+n^2}\, z\right).$$

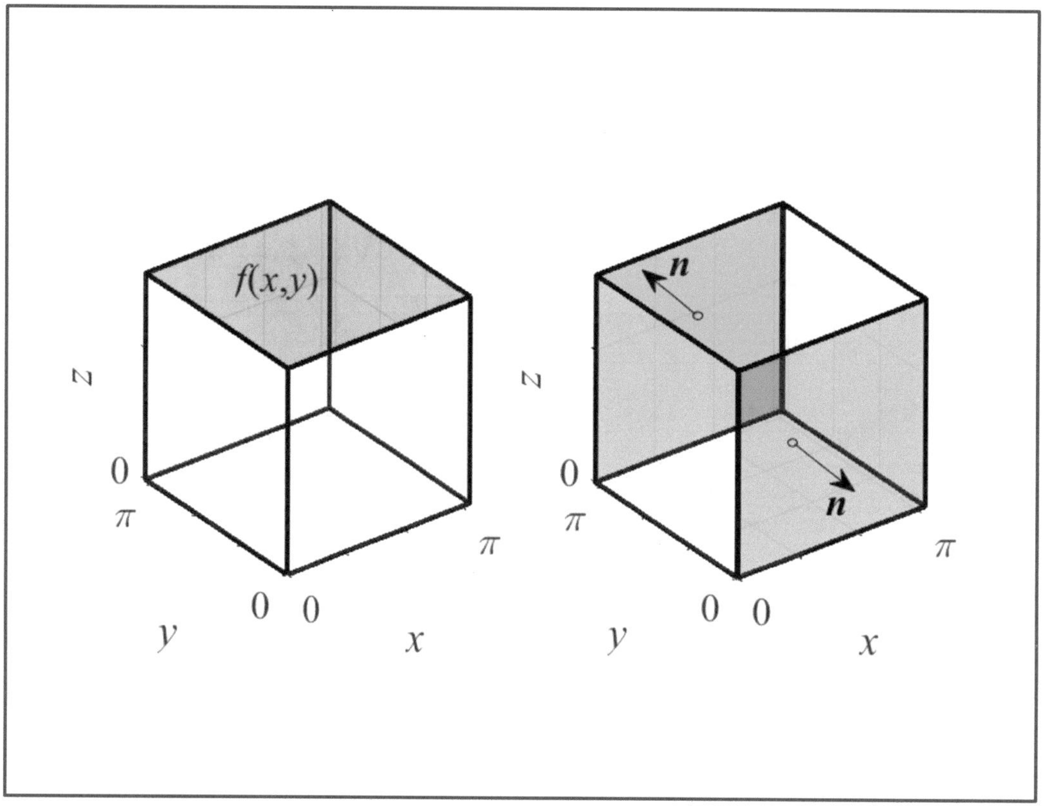

Figure 3.5. The Laplace–Neumann problem on $C_3(\pi)$ with normal vectors $\boldsymbol{n}$ at $y = 0$ and $y = \pi$.

By the superposition principle, the sum of harmonic functions is harmonic and

$$\begin{aligned} u(x,y,z) &= \sum_{m=1}^{\infty}\sum_{n=1}^{\infty} u_{m,n}(x,y,z) \\ &= \sum_{m=1}^{\infty}\sum_{n=1}^{\infty} \frac{C_{m,n}}{\sqrt{m^2+n^2}} \cos(nx)\cos(my)\cosh\left(\sqrt{m^2+n^2}\, z\right). \end{aligned}$$

The Neumann boundary condition $f(x, y) = \dfrac{\partial u}{\partial \boldsymbol{n}}(x,y,\pi) = \dfrac{\partial u}{\partial z}(x,y,\pi)$ means that differentiating the formula above, first with respect to z and then evaluating at $z = \pi$, results in

$$f(x, y) = \sum_{m=1}^{\infty}\sum_{n=1}^{\infty} C_{m,n} \cos(nx)\cos(my)\sinh\left(\sqrt{m^2+n^2}\, \pi\right).$$

This form above indicates that f is doubly even so that by §3.2.1, $C_{m,n} = \dfrac{b_{m,n}}{\sinh\left(\pi\sqrt{m^2+n^2}\right)}$ and the $b_{m,n}$ are the Fourier coefficients of f: $b_{m,n} = \dfrac{4}{\pi^2}\int_0^\pi\int_0^\pi f(x,y)\cos(nx)\cos(my)\,dx\,dy$. ■

The solution of the Robin problem is left as an exercise.

§3.4. Poisson's Equation (Modern Methods Revisited)

The ideas presented thus far in this chapter as well as in Chapter 2 have provided the required mathematical tools to solve Laplace's equation and its variants, namely the Laplace–Dirichlet, Laplace–Neumann, and Laplace–Robin equations. The next step is to solve the inhomogeneous version of Laplace's equation, also called Poisson's equation $\Delta u = \rho(\boldsymbol{x})$. Attempting the classical approach of *separation of variables* in $\mathbb{R}^3$ leads to $u(\boldsymbol{x}) = u(x, y, z) = X(x)\cdot Y(y)\cdot Z(z)$ and $\rho(\boldsymbol{x}) = \rho(x, y, z) = \xi(x)\cdot\psi(y)\cdot\zeta(z)$, so that $\Delta(X(x)\cdot Y(y)\cdot Z(z)) = \xi(x)\cdot\psi(y)\cdot\zeta(z)$ or $X''(x)\cdot Y(y)\cdot Z(z) + X(x)\cdot Y''(y)\cdot Z(z) + X(x)\cdot Y(y)\cdot Z''(z) = \xi(x)\cdot\psi(y)\cdot\zeta(z)$. Assuming that none of $X(x)$, $Y(y)$, or $Z(z)$ is identically zero and then dividing through by $X(x)\cdot Y(y)\cdot Z(z)$ produces

$$\frac{X''(x)}{X(x)} + \frac{Y''(y)}{Y(y)} + \frac{Z''(z)}{Z(z)} = \frac{\xi(x)\cdot\psi(y)\cdot\zeta(z)}{X(x)\cdot Y(y)\cdot Z(z)}. \tag{3.4.1}$$

If the right-hand side of (3.4.1) is *not constant*[13], then the separation of variables method will fail. Therefore, an alternative to the classical approach is required. This invites a return to the modern methods initiated in Chapter 2.

In particular, suppose that G satisfies the Laplace distribution equation

$$\Delta G(\boldsymbol{x}) = \delta(\boldsymbol{x} - \boldsymbol{x}_o),\ \boldsymbol{x} \in \Omega \subset \mathbb{R}^n. \tag{3.4.2}$$

Then G is the *Green's function* for Laplace's equation as per (2.2.3). Next, suppose that $w(\boldsymbol{x})$ satisfies the *Dirichlet problem*

$$\left.\begin{array}{ll} \Delta w(\boldsymbol{x}) = 0, & \text{for } \boldsymbol{x} \in \Omega \subset \mathbb{R}^n \\ w(\boldsymbol{x}) = -G(\boldsymbol{x}, \boldsymbol{x}_o), & \text{for } \boldsymbol{x} \in \partial\Omega \end{array}\right\}. \tag{3.4.3}$$

Set $v(\boldsymbol{x}, \boldsymbol{x}_o) = w(\boldsymbol{x}) + G(\boldsymbol{x}, \boldsymbol{x}_o)$. The solution of the *Poisson–Dirichlet problem* (3.4.4) can subsequently be determined via the function v.

---

[13] If the right–hand side of (3.4.1) *is* constant, then this is precisely the case of Laplace's equation.

$$\left.\begin{aligned} &\Delta u(\boldsymbol{x}) = \rho(\boldsymbol{x}), && \text{for } \boldsymbol{x} \in \Omega \subset \mathbb{R}^n \\ &u(\boldsymbol{x}) = f(\boldsymbol{x}), && \text{for } \boldsymbol{x} \in \partial\Omega \end{aligned}\right\} \tag{3.4.4}$$

Theorem 3.1: *Let G, w* $\in C^2(\Omega)$ *satisfy* (3.4.2) *and* (3.4.3), *respectively. Then for any* $\boldsymbol{x}_o \in \Omega$ *and for* $v(\boldsymbol{x}, \boldsymbol{x}_o) = w(\boldsymbol{x}) + G(\boldsymbol{x}, \boldsymbol{x}_o)$, *the representation in* (3.4.5) *provides a solution of* (3.4.4) at $\boldsymbol{x} = \boldsymbol{x}_o$.

$$\boxed{u(\boldsymbol{x}_o) = \int_{\Omega} v(\boldsymbol{x}, \boldsymbol{x}_o)\rho(\boldsymbol{x})\,d\omega + \int_{\partial\Omega} f(\boldsymbol{x})\frac{\partial v}{\partial \boldsymbol{n}}(\boldsymbol{x}, \boldsymbol{x}_o)\,d\sigma} \tag{3.4.5}$$

Proof: Multiply Poisson's equation $\Delta u = \rho$ by $v = w + G$ and integrate over Ω. Now, using the second identity (*ii*) of Green's Theorem of §2.2, obtain

$$\begin{aligned} \int_{\Omega} v(\boldsymbol{x}, \boldsymbol{x}_o)\,\rho(\boldsymbol{x})\,d\omega &= \int_{\Omega} v(\boldsymbol{x}, \boldsymbol{x}_o)\,\Delta u(\boldsymbol{x})\,d\omega \\ &= \int_{\partial\Omega} \left(v(\boldsymbol{x}, \boldsymbol{x}_o)\frac{\partial u}{\partial \boldsymbol{n}}(\boldsymbol{x}) - u(\boldsymbol{x})\frac{\partial v}{\partial \boldsymbol{n}}(\boldsymbol{x}, \boldsymbol{x}_o) \right) d\sigma + \int_{\Omega} u(\boldsymbol{x})\,\Delta v(\boldsymbol{x}, \boldsymbol{x}_o)\,d\omega. \end{aligned}$$

Since $v(\boldsymbol{x}, \boldsymbol{x}_o) = w(\boldsymbol{x}) + G(\boldsymbol{x}, \boldsymbol{x}_o)$ with w satisfying (3.4.3) and G satisfying (3.4.2), then on $\partial\Omega$, $w(\boldsymbol{x}) = -G(\boldsymbol{x}, \boldsymbol{x}_o)$ and therefore $v(\boldsymbol{x}, \boldsymbol{x}_o) = -G(\boldsymbol{x}, \boldsymbol{x}_o) + G(\boldsymbol{x}, \boldsymbol{x}_o) \equiv 0$ on $\partial\Omega$. By supposition, u satisfies (3.4.4) so that on $\partial\Omega$, $u(\boldsymbol{x}) = f(\boldsymbol{x})$. Thus, the integral on the boundary retains only the term multiplying the normal derivative on v. Similarly, $\Delta v(\boldsymbol{x}, \boldsymbol{x}_o) = \Delta w(\boldsymbol{x}) + \Delta G(\boldsymbol{x}, \boldsymbol{x}_o) = 0 + \delta(\boldsymbol{x} - \boldsymbol{x}_o)$ on Ω by (3.4.2) and (3.4.3). Therefore,

$$\int_{\partial\Omega} \left(v(\boldsymbol{x}, \boldsymbol{x}_o)\frac{\partial u}{\partial \boldsymbol{n}}(\boldsymbol{x}) - u(\boldsymbol{x})\frac{\partial v}{\partial \boldsymbol{n}}(\boldsymbol{x}, \boldsymbol{x}_o) \right) d\sigma = \int_{\partial\Omega} -f(\boldsymbol{x})\frac{\partial v}{\partial \boldsymbol{n}}(\boldsymbol{x}, \boldsymbol{x}_o)\,d\sigma \text{ and}$$

$$\int_{\Omega} u(\boldsymbol{x})\,\Delta v(\boldsymbol{x}, \boldsymbol{x}_o)\,d\omega = \int_{\Omega} u(\boldsymbol{x})\,\delta(\boldsymbol{x} - \boldsymbol{x}_o)\,d\omega = u(\boldsymbol{x}_o)$$

so that the first equation in the proof becomes

$$\int_{\Omega} v(\boldsymbol{x}, \boldsymbol{x}_o)\,\rho(\boldsymbol{x})\,d\omega = -\int_{\partial\Omega} f(\boldsymbol{x})\frac{\partial v}{\partial \boldsymbol{n}}(\boldsymbol{x}, \boldsymbol{x}_o)\,d\sigma + u(\boldsymbol{x}_o). \qquad \blacksquare$$

It is now a straightforward matter to show that, if G satisfies the Laplace distribution equation (3.4.2), w satisfies the *Laplace–Neumann equation* (3.4.6), and $v(\boldsymbol{x}, \boldsymbol{x}_o) = w(\boldsymbol{x}) + G(\boldsymbol{x}, \boldsymbol{x}_o)$, then (3.4.8) provides a solution to the *Poisson–Neumann problem* (3.4.7).

Laplace–Neumann (Special Case)

$$\left.\begin{aligned} &\Delta w(\boldsymbol{x}) = 0 && \text{for } \boldsymbol{x} \in \Omega \\ &\frac{\partial w}{\partial \boldsymbol{n}}(\boldsymbol{x}) = -\frac{\partial G}{\partial \boldsymbol{n}}(\boldsymbol{x}, \boldsymbol{x}_o) && \text{for } \boldsymbol{x} \in \partial\Omega \end{aligned}\right\} \tag{3.4.6}$$

Poisson–Neumann

$$\left.\begin{aligned} \Delta u(\boldsymbol{x}) &= \rho(\boldsymbol{x}) \quad \text{for } \boldsymbol{x} \in \Omega \\ \frac{\partial u}{\partial \boldsymbol{n}}(\boldsymbol{x}) &= f(\boldsymbol{x}) \quad \text{for } \boldsymbol{x} \in \partial\Omega \end{aligned}\right\} \tag{3.4.7}$$

$$\boxed{u(\boldsymbol{x}_o) = \int_{\Omega} v(\boldsymbol{x},\boldsymbol{x}_o)\,\rho(\boldsymbol{x})\,d\omega - \int_{\partial\Omega} f(\boldsymbol{x})\,v(\boldsymbol{x},\boldsymbol{x}_o)\,d\sigma} \tag{3.4.8}$$

Finally suppose, as above, that G satisfies the Laplace distribution equation (3.4.2), w satisfies the *Laplace–Robin problem* (3.4.9), and $v(\boldsymbol{x}, \boldsymbol{x}_o) = w(\boldsymbol{x}) + G(\boldsymbol{x}, \boldsymbol{x}_o)$. Then (3.4.8) *also* solves the *Poisson–Robin problem* (3.4.10).

Laplace–Robin (Special Case)

$$\left.\begin{aligned} &\Delta w(\boldsymbol{x}) = 0 && \text{for } \boldsymbol{x} \in \Omega \\ &\frac{\partial w}{\partial \boldsymbol{n}}(\boldsymbol{x}) + w(\boldsymbol{x}) = -\left[G(\boldsymbol{x},\boldsymbol{x}_o) + \frac{\partial G}{\partial \boldsymbol{n}}(\boldsymbol{x},\boldsymbol{x}_o) \right] && \text{for } \boldsymbol{x} \in \partial\Omega \end{aligned}\right\} \tag{3.4.9}$$

Poisson–Robin

$$\left.\begin{aligned} &\Delta u(\boldsymbol{x}) = \rho(\boldsymbol{x}) && \text{for } \boldsymbol{x} \in \Omega \\ &\frac{\partial u}{\partial \boldsymbol{n}}(\boldsymbol{x}) + u(\boldsymbol{x}) = f(\boldsymbol{x}) && \text{for } \boldsymbol{x} \in \partial\Omega \end{aligned}\right\} \tag{3.4.10}$$

§3.5. The Heat Equation (A Blend of Modern and Classical)

The temperature distribution in a heat-conducting medium is described by the *heat equation* $\frac{\partial u}{\partial t}(\boldsymbol{x},t) = \sigma^2\,\Delta u(\boldsymbol{x},t)$ for $\boldsymbol{x} \in \Omega \subset \mathbb{R}^n$ and $\kappa = \sigma^2$ is called the *conductivity coefficient*. Throughout this section, the conductivity coefficient will be treated as a constant. For $n = 1$, the heat equation models the temperature distribution over an infinitely thin wire of length $L = |\Omega|$. When the wire is considered over an interval $\Omega = [a, b]$, then $L = |\Omega| = b - a$. As is demonstrated in the development of Laplace's and Poisson's equation, the initial and boundary conditions are crucial elements in the construction of solutions to the heat equation. The remainder of this section will focus on solutions of the heat equation when Ω is a compact set in $n = 1$, 2, and 3 spatial dimensions. This will be accomplished by the usual method of separation of variables. In the case in which the domain is unbounded, and especially $\Omega = \mathbb{R}^n$, then a new approach is required: The Fourier transform which will be deferred to Chapter 4.

Example 3.5.1. Solve the heat equation for the wire of length L and conductivity constant $\kappa = \sigma^2$. Assume that the ends of the wire are held at 0° temperature and that temperature distribution $f(x)$ of the wire at time $t = 0$ is known. That is, solve

$$\left.\begin{array}{ll}\frac{\partial u}{\partial t}(x,t)=\sigma^2\frac{\partial^2 u}{\partial x^2}, & x\in\Omega=[0,L] \text{ and } t>0\\ u(0,t)=0=u(L,t) & \\ u(x,0)=f(x) & \text{(Cauchy data)}\end{array}\right\}. \tag{3.5.1}$$

Solution: Employ the separation of variables method and assume $u(x, t) = X(x)\cdot T(t)$ so that the heat equation produces $X(x)\cdot T'(t)=\sigma^2 X''(x)\cdot T(t)$. Dividing through by $\sigma^2 X(x)\cdot T(t)$, which for non–trivial initial function $f(x)$ must be non-zero, yields $\frac{T'(t)}{\sigma^2 T(t)}=\frac{X''(x)}{X(x)}$. As seen in Examples 3.3.1 and 3.3.2, this means that these ratios are equal to a constant λ (since the left- and right-hand sides of the equation above are exclusively functions of the independent variables t and x, respectively). This fact, combined with the initial conditions in (3.5.1), produces the two ordinary differential equations

$$(i)\ \begin{cases} X''(x)-\lambda X(x)=0\\ X(0)=0=X(L)\end{cases} \qquad (ii)\ T'(t)=\lambda\sigma^2 T(t).$$

The characteristic equation for the *ODE* (*i*) is $r^2 - \lambda = 0$ so that the characteristic roots are $r_{1,2} = \pm\sqrt{\lambda}$. If $\lambda = 0$, then $r_{1,2} = 0$ and $X(t) = c_1 + c_2 x$. The boundary condition $X(0) = 0$ implies that $0 = c_1$ and $0 = X(L) = c_2 L$ implies $c_2 = 0$ or $X(x) \equiv 0$. This violates the assumption of the initial distribution $f(x) \neq 0$. If $\lambda = k^2 > 0$, then $r_{1,2} = \pm k$ so that $X(x) = c_1 e^{kx} + c_2 e^{-kx}$. The boundary conditions produce the two equations $0 = c_1 + c_2$ and $0 = c_1 e^{kL} + c_2 e^{-kL}$. Equivalently, these equations can be written as $A\cdot\boldsymbol{c} = \boldsymbol{0}$ where $A = \begin{bmatrix}1 & 1\\ e^{kL} & e^{-kL}\end{bmatrix}$, $\boldsymbol{c} = [c_1, c_2]^T$, and $\boldsymbol{0} = [0, 0]^T$. Observe $\det(A) = e^{-kL} - e^{kL} = -2\sinh(kL) \neq 0$ for $k\cdot L \neq 0$. Therefore, the matrix A is invertible and subsequently, $\boldsymbol{c} = A^{-1}\cdot\boldsymbol{0} = \boldsymbol{0}$. Hence, $X(x) \equiv 0$ which again violates the assumption on $f(x)$. Therefore, it must be the case that $\lambda = -k^2 < 0$ so that $X(x) = c_1 e^{ikx} + c_2 e^{-ikx} = \gamma_1\cdot\cos(kx) + \gamma_2\cdot\sin(kx)$ for suitable constants γ_1 and γ_2. The boundary condition $0 = X(0) = \gamma_1\cdot\cos(0) + \gamma_2\cdot\sin(0) = \gamma_1$, while $0 = X(L) = \gamma_2\cdot\sin(kL)$. To avoid the trivial solution $X(x) = 0$, take $k = \frac{m\pi}{L}$ and $m \in \mathbb{Z}^+$. Then, $X_m(x) = \sin\left(\frac{m\pi}{L}x\right)$ are the eigenfunctions and $\lambda_m = -\left(\frac{m\pi}{L}\right)^2$ are the eigenvalues of the *ODE* (*i*). Since the solution of the first order, separable *ODE* $T'(t) = \lambda_m \sigma^2 T(t)$ is $T_m(t) = e^{\lambda_m\sigma^2 t} = e^{-(m\pi\sigma/L)^2 t}$, then $u_m(x, t) = X_m(x)\cdot T_m(t) = c'_m\cdot\sin\left(\frac{m\pi}{L}x\right)\cdot e^{-(m\pi\sigma/L)^2 t}$ for some series of constants c'_m. By the superposition principle (see page 5 of the *Introduction*), then

$$u(x,t) = \sum_{m=1}^{\infty} c'_m \cdot e^{-(m\pi\sigma/L)^2 t} \cdot \sin\left(\tfrac{m\pi}{L} x\right)$$

is a solution to the heat equation (3.5.1). But the *Cauchy data* $u(x, 0) = f(x)$ means that $f(x) = \sum_{m=1}^{\infty} c'_m \cdot \sin\left(\tfrac{m\pi}{L} x\right)$. Hence, by formula (3.1.1ℓ) for functions periodic over $[0, L]$, the coefficients are computed as $c'_m = \dfrac{2}{L}\int_0^L f(x)\cdot\sin\left(\tfrac{m\pi}{L}x\right)dx$. One example of an initial heat distribution $f(x) = -4x(x-L)$ is illustrated in Figure 3.6. The following set of commands, using the Symbolic Math Toolbox for MATLAB, produces the formula $c'_m = \left(\dfrac{2L}{\pi m}\right)^3 \dfrac{(-1)^{m+1}\left((-1)^m - 1\right)^2}{L}$.

MATLAB Commands	*(Symbolic Math Toolbox)*

```
% Define a variety of symbolic variables with different characteristics
syms x
syms L positive real
syms m positive integer
% Define the Cauchy data f(x) as per Figure 3.6
f = -4*x*(x - L);
% Compute the coefficients c'_m
c = 2*int(f*sin(pi*m*x/L),x,0,L)/L;
% Print the results in a readable format
pretty(simplify(c))

      m+1  2      m     2
8 (-1)    L  ((-1)  - 1)
-------------------------
          3   3
         m  pi
```

The first seven values of c'_m are $c'_1 = \dfrac{2^5 L^2}{\pi^3}$, $c'_2 = 0$, $c'_3 = \dfrac{2^5 L^2}{(3\pi)^3}$, $c'_4 = 0$, $c'_5 = \dfrac{2^5 L^2}{(5\pi)^3}$, $c'_6 = 0$, $c'_7 = \dfrac{2^5 L^2}{(7\pi)^3}$. It is evident from these values that $c'_m = \begin{cases} \dfrac{2^5 L^2}{(m\cdot\pi)^3} & \text{if } m \text{ is odd} \\ 0 & \text{if } m \text{ is even} \end{cases}$.

Therefore, the solution of the heat equation (3.5.1) with Cauchy data $f(x) = -4\cdot x(x-L)$ is

$$\begin{aligned} u(x,t) &= \sum_{\substack{m=1 \\ m\ odd}}^{\infty} \tfrac{2^5 L^2}{(m\cdot\pi)^3} \exp\left(-\left(\tfrac{m\cdot\pi\cdot\sigma}{L}\right)^2 t\right)\sin\left(\tfrac{m\cdot\pi}{L}x\right) \\ &= \sum_{k=1}^{\infty} \tfrac{2^5 L^2}{([2k-1]\cdot\pi)^3} \exp\left(-\left(\tfrac{[2k-1]\cdot\pi\cdot\sigma}{L}\right)^2 t\right)\sin\left(\tfrac{[2k-1]\cdot\pi}{L}x\right) \end{aligned}.$$

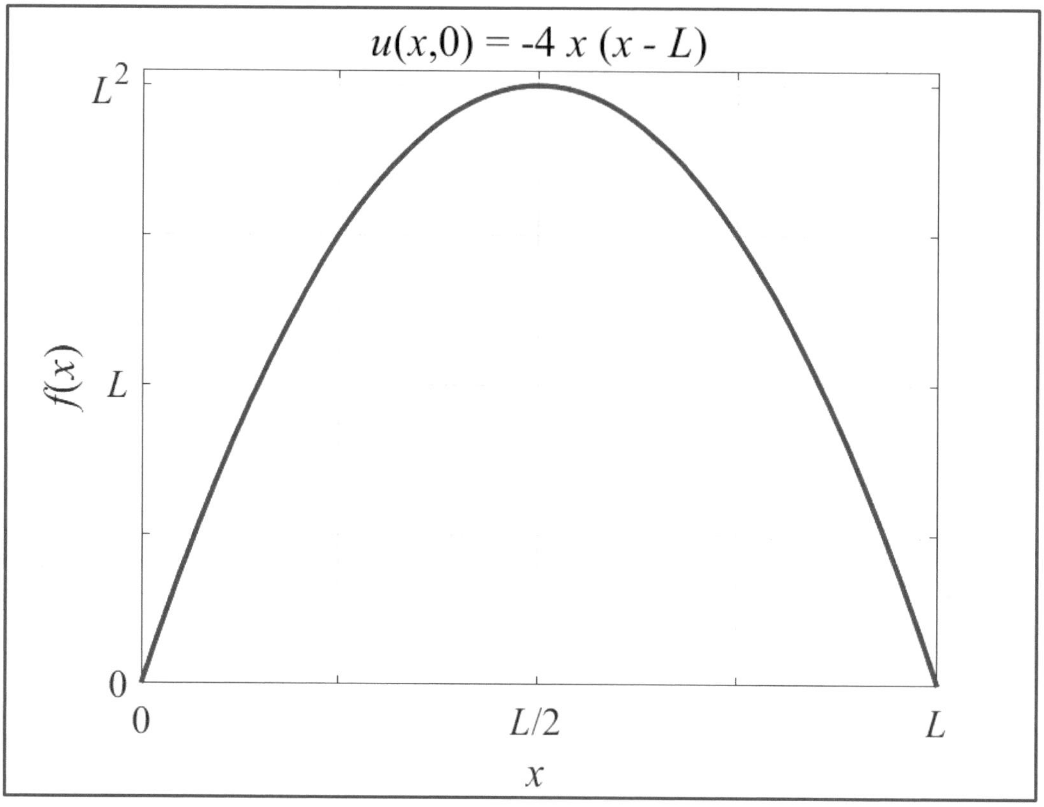

Figure 3.6. Cauchy data for the heat equation.

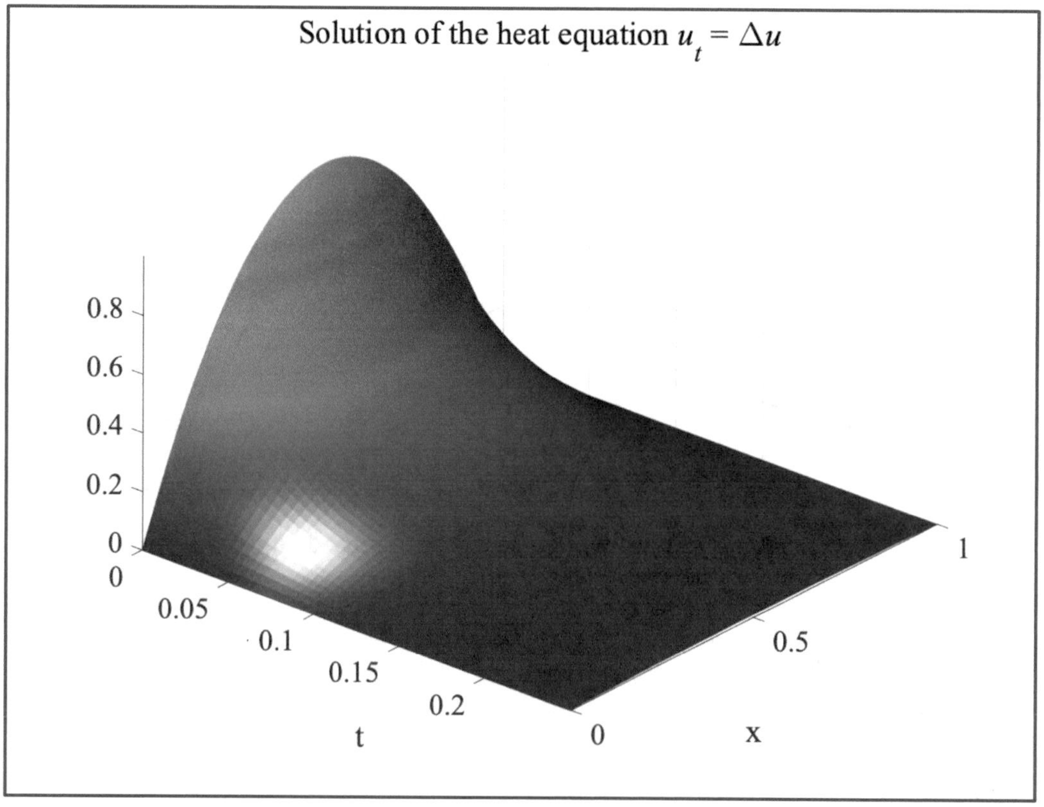

Figure 3.7. Approximate solution of (3.5.1) $u_7(x, t)$.

An approximate solution of the heat equation using the first 7 non-zero terms of the series is illustrated in Figure 3.7 over the time interval $t \in [0, ¼]$ and the spatial interval $x \in [0, L]$ with $L = 1$ and $\sigma^2 = 2$: $u_7(x,t) = \sum_{k=1}^{7} \frac{2^5 L^2}{([2k-1]\cdot\pi)^3} \exp\left(-\left(\frac{[2k-1]\cdot\pi\cdot\sigma}{L}\right)^2 t\right) \sin\left(\frac{[2k-1]\cdot\pi}{L} x\right)$. Notice how rapidly in time the heat intensity descends toward zero.

Example 3.5.2. Solve the heat equation for the rectangular plate of length L_1, width L_2, and conductivity $\kappa = \sigma^2$. Assume that the rectangular frame of the plate is held at 0° temperature and that the initial temperature distribution across the interior of the rectangle is known. This is illustrated in Figure 3.8. Then the task is to solve

$$\left.\begin{array}{ll} \frac{\partial u}{\partial t}(\boldsymbol{x},t) = \sigma^2\left(\frac{\partial^2 u}{\partial x^2}(\boldsymbol{x},t) + \frac{\partial^2 u}{\partial y^2}(\boldsymbol{x},t)\right) & \\ \boldsymbol{x} = (x,y) \in \Omega = [0,L_1]\times[0,L_2] \text{ and } t > 0 & \\ u(0,y,t) = 0 = u(L_1,y,t) & \text{for } y \in [0,L_2] \text{ and } t > 0 \\ u(x,0,t) = 0 = u(x,L_2,t) & \text{for } x \in [0,L_1] \text{ and } t > 0 \\ u(x,y,0) = f(x,y) & \text{(Cauchy data)} \end{array}\right\}. \tag{3.5.2}$$

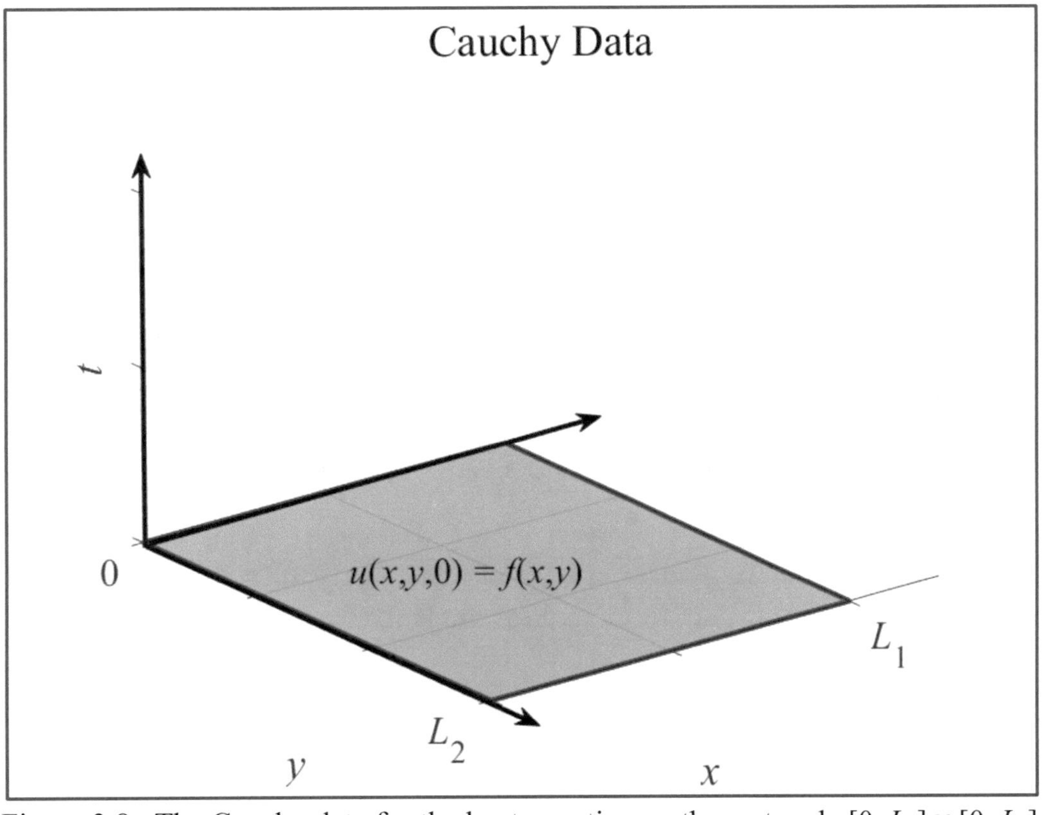

Figure 3.8. The Cauchy data for the heat equation on the rectangle $[0, L_1] \times [0, L_2]$.

Solution: Once again the classical approach, separation of variables, is utilized. For $u(x, y, t) = X(x)\cdot Y(y)\cdot T(t)$, applying the heat operator $\frac{\partial}{\partial t} - \sigma^2\Delta$ to u implies that $\frac{T'(t)}{\sigma^2 T(t)} = \frac{X''(x)}{X(x)} + \frac{Y''(y)}{Y(y)} = \lambda$ (constant). The right-hand portion of the equation above produces $\frac{X''(x)}{X(x)} = \lambda - \frac{Y''(y)}{Y(y)}$ so that these ratios must be constant. Therefore, the equations and boundary conditions give way to the series of *ODE*s below.

$$(i)\ \begin{cases} X''(x) - \mu X(x) = 0 \\ X(0) = 0 = X(L_1) \end{cases} \qquad (ii)\ \begin{cases} Y''(x) - (\lambda - \mu)Y(y) = 0 \\ Y(0) = 0 = Y(L_2) \end{cases} \qquad (iii)\ T'(t) = \lambda\sigma^2\, T(t)$$

Using the methodology established in Examples 2.2 and 3.51, equation (i) has the eigenvalues $\mu_n = -\left(\frac{n\pi}{L_1}\right)^2$ and corresponding family of solutions $X_n(x) = c_1 \sin\left(\frac{n\pi}{L_1}x\right)$ for $n \in \mathbb{Z}^+$. Similarly in equation (ii), set $\Lambda = (\lambda - \mu)$. Then ($ii$) has eigenvalues $\Lambda_m = -\left(\frac{m\pi}{L_2}\right)^2$ and the associated solution family, $Y_m(y) = c_2 \sin\left(\frac{m\pi}{L_2}y\right)$ with $m \in \mathbb{Z}^+$. This means that $\lambda_{n,m} = \Lambda_m + \mu_n = -\left[\left(\frac{m\pi}{L_2}\right)^2 + \left(\frac{n\pi}{L_1}\right)^2\right]$, and by ($iii$) $T_{n,m}(t) = c_3 \exp\left(\lambda_{n,m}\sigma^2 t\right)$. Thus, the primary solutions of (3.5.2) are

$$u_{n,m}(x,y,t) = C_{n,m} \exp\left(-\left[\left(\tfrac{n\pi}{L_1}\right)^2 + \left(\tfrac{m\pi}{L_2}\right)^2\right]\sigma^2 t\right)\sin\left(\tfrac{n\pi}{L_1}x\right)\sin\left(\tfrac{m\pi}{L_2}y\right).$$

By the superposition principle,

$$\begin{aligned} u(x,y,t) &= \sum_{n=1}^{\infty}\sum_{m=1}^{\infty} u_{n,m}(x,y,t) \\ &= \sum_{n=1}^{\infty}\sum_{m=1}^{\infty} C_{n,m} \exp\left(-\left[\left(\tfrac{n\pi}{L_1}\right)^2 + \left(\tfrac{m\pi}{L_2}\right)^2\right]\sigma^2 t\right)\sin\left(\tfrac{n\pi}{L_1}x\right)\sin\left(\tfrac{m\pi}{L_2}y\right) \end{aligned}$$

solves the heat equation (3.5.2). Further, the Cauchy data $u(x, y, 0) = f(x, y)$ combined with the change of variables formulae (3.2.2ℓ) result in the formula for $C_{n,m}$.

$$C_{n,m} = \tfrac{4}{L_1 L_2}\int_0^{L_2}\int_0^{L_1} f(x,y)\sin\left(\tfrac{n\pi}{L_1}x\right)\sin\left(\tfrac{m\pi}{L_2}y\right)dx\,dy$$

For fixed y, the solution surface will resemble Figure 3.7. ■

Example 3.5.3. Solve the heat equation on the parallelepiped of dimension $L_1 \times L_2 \times L_3$ and conductivity $\kappa = \sigma^2$. Assume that the cubic frame of the plate is held at 0° temperature and that the initial temperature distribution across the interior solid is known. Then the heat distribution is modeled by

$$\left.\begin{array}{ll}
\frac{\partial u}{\partial t}(\boldsymbol{x},t) = \sigma^2\left(\frac{\partial^2 u}{\partial x^2}(\boldsymbol{x},t) + \frac{\partial^2 u}{\partial y^2}(\boldsymbol{x},t) + \frac{\partial^2 u}{\partial z^2}(\boldsymbol{x},t)\right) & \\
\boldsymbol{x} = (x,y,z) \in \Omega = [0,L_1]\times[0,L_2]\times[0,L_3] \text{ and } t>0 & \\
u(0,y,z,t) = 0 = u(L_1,y,z,t) & \text{for } (y,z)\in[0,L_2]\times[0,L_3] \text{ and } t>0 \\
u(x,0,z,t) = 0 = u(x,L_2,z,t) & \text{for } (x,z)\in[0,L_1]\times[0,L_3] \text{ and } t>0 \\
u(x,y,0,t) = 0 = u(x,y,L_3,t) & \text{for } (x,y)\in[0,L_1]\times[0,L_2] \text{ and } t>0 \\
u(x,y,z,0) = f(x,y,z) & \text{(Cauchy data)}
\end{array}\right\}. \tag{3.5.3}$$

Solution: As is now evident from Examples 3.5.1 and 3.5.2, the classical method of separation of variables $u(x, y, z, t) = X(x)\cdot Y(y)\cdot Z(z)\cdot T(t)$ produces the primary solutions

$$u_{n,m,k}(x,y,z,t) = C_{n,m,k}\exp\left(-\left[(\tfrac{n\pi}{L_1})^2 + (\tfrac{m\pi}{L_2})^2 + (\tfrac{k\pi}{L_3})^2\right]\sigma^2 t\right)\sin\left(\tfrac{n\pi}{L_1}x\right)\sin\left(\tfrac{m\pi}{L_2}y\right)\sin\left(\tfrac{k\pi}{L_3}z\right).$$

By the superposition principle, the solution of (3.5.3) is

$$\begin{aligned}
u(x,y,t) &= \sum_{n=1}^{\infty}\sum_{m=1}^{\infty}\sum_{k=1}^{\infty} u_{n,m,k}(x,y,z,t) \\
&= \sum_{n=1}^{\infty}\sum_{m=1}^{\infty}\sum_{k=1}^{\infty} C_{n,m,k}\exp\left(-\left[(\tfrac{n\pi}{L_1})^2 + (\tfrac{m\pi}{L_2})^2 + (\tfrac{k\pi}{L_3})^2\right]\sigma^2 t\right)\sin\left(\tfrac{n\pi}{L_1}x\right)\sin\left(\tfrac{m\pi}{L_2}y\right)\sin\left(\tfrac{k\pi}{L_3}z\right).
\end{aligned}$$

The Fourier coefficients $C_{n,m,k}$ are the natural extension of (3.2.2ℓ).

$$C_{n,m,k} = \tfrac{8}{L_1L_2L_3}\int_0^{L_3}\int_0^{L_2}\int_0^{L_1} f(x,y,z)\sin\left(\tfrac{n\pi}{L_1}x\right)\sin\left(\tfrac{m\pi}{L_2}y\right)\sin\left(\tfrac{k\pi}{L_3}z\right)dx\,dy\,dz$$ ■

Example 3.5.4. Solve the heat equation on the (horizontal) cylindrical rod $C_{\rho,L}$ of length L and radius ρ and conductivity $\kappa = \sigma^2$. The boundary assumptions are similar to Examples 3.5.1–3.5.3. That is, the surface of the cylinder is held at 0° temperature and the initial temperature distribution at time $t = 0$ is known. In this case, it is sage to model the heat equation in the cylindrical coordinates $x = r\cos(\theta)$, $y = y$, and $z = r\sin(\theta)$. The heat operator $\frac{\partial}{\partial t} - \sigma^2\Delta$ with respect to these cylindrical coordinates yields the equation

$$\left.\begin{array}{l}\dfrac{\partial u}{\partial t}(\boldsymbol{x},t)=\sigma^2\left(\dfrac{\partial^2 u}{\partial r^2}(\boldsymbol{x},t)+\dfrac{1}{r}\dfrac{\partial u}{\partial r}(\boldsymbol{x},t)+\dfrac{1}{r^2}\dfrac{\partial^2 u}{\partial \theta^2}(\boldsymbol{x},t)+\dfrac{\partial^2 u}{\partial y^2}(\boldsymbol{x},t)\right)\\ \boldsymbol{x}=(r,y,\theta)\in C_{\rho,L} \text{ and } t>0\end{array}\right\}. \tag{3.5.4}$$

Make a periodicity assumption on the angle θ so that $u(r, y, \theta, t) = u(r, y, \theta + 2\pi, t)$ for all r, y, θ, and t. The 0° temperature on the boundary requires that $u(r, y, 0, t) = 0$, $u(r, 0, \theta, t) = 0 = u(r, L, \theta, t)$, and $u(\rho, y, \theta, t) = 0$ for all r, y, θ, and t. The Cauchy initial condition is $u(r, y, \theta, 0) = \phi(r, y, \theta)$.

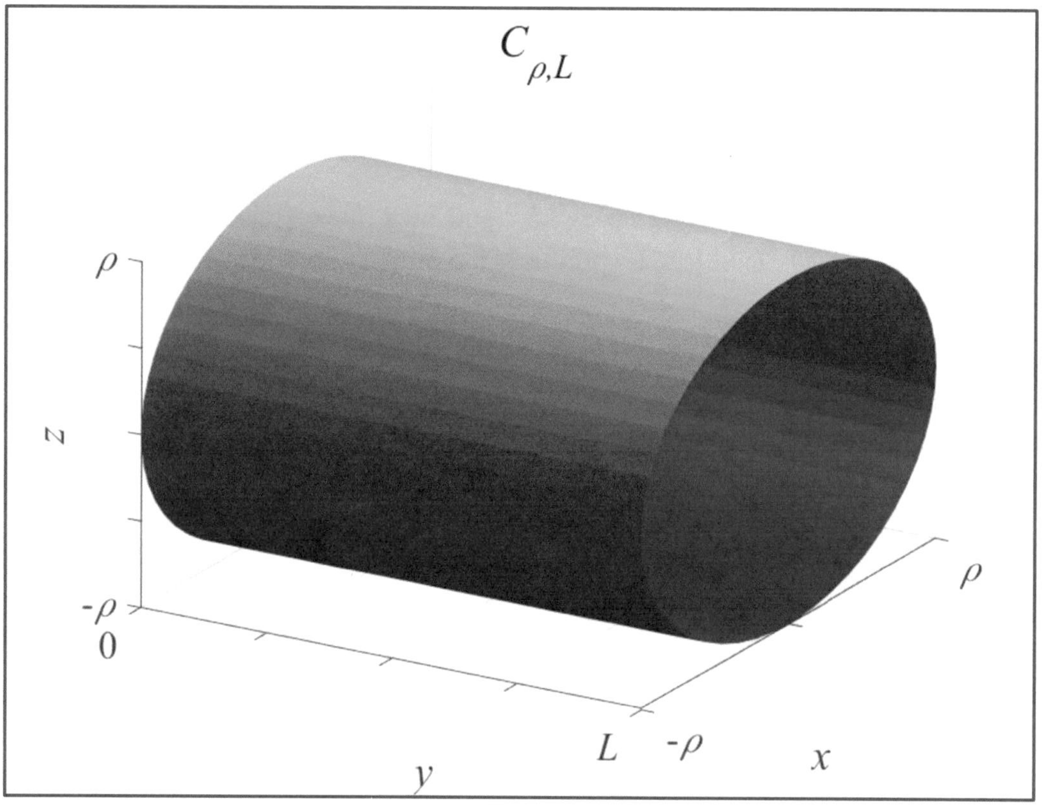

Figure 3.9. $C_{\rho,L}$, the cylindrical rod of length L and radius ρ.

Solution: In the cylindrical coordinates, the classical separation of variables method sets $u(r, y, \theta, t) = R(r)\cdot Y(y)\cdot\Theta(\theta)\cdot T(t)$. As the cylinder is the cross product of a circle and an interval, then it is reasonable to assume Θ is 2π–periodic: $\Theta(\theta) = \Theta(\theta + 2\pi)$. In particular, this means $\Theta(0) = \Theta(2\pi)$. Applying the heat operator $\dfrac{\partial}{\partial t} - \sigma^2\Delta$ to u and setting the result to 0 results in $0 = \left(\dfrac{\partial}{\partial t} - \sigma^2\Delta\right)u =$

$$T'(t)\cdot R(r)\cdot Y(y)\cdot\Theta(\theta) - \sigma^2\Big(T(t)\cdot R''(r)\cdot Y(y)\cdot\Theta(\theta) + \frac{1}{r}\cdot T(t)\cdot R'(r)\cdot Y(y)\cdot\Theta(\theta) +$$

$$\frac{1}{r^2}\cdot T(t)\cdot R(r)\cdot Y(y)\cdot\Theta''(\theta) + T(t)\cdot R(r)\cdot Y''(y)\cdot\Theta(\theta)\Big).$$

After the usual division by $R(r)\cdot Y(y)\cdot\Theta(\theta)\cdot T(t)$ and some algebra, the equation above reduces to (constant) $\lambda = \dfrac{T'(t)}{\sigma^2 T(t)} = \dfrac{R''(r)}{R(r)} + \dfrac{R'(r)}{r\cdot R(r)} + \dfrac{\Theta''(\theta)}{r^2\cdot\Theta(\theta)} + \dfrac{Y''(y)}{Y(y)}$. Thus, $\dfrac{R''(r)}{R(r)} + \dfrac{R'(r)}{r\cdot R(r)} + \dfrac{\Theta''(\theta)}{r^2\cdot\Theta(\theta)} =$ $\lambda - \dfrac{Y''(y)}{Y(y)} = \mu$ (constant) which, in turn, means that $\dfrac{r^2\cdot R''(r) + r\cdot R'(r)}{R(r)} - r^2\mu = -\dfrac{\Theta''(\theta)}{\Theta(\theta)} = \gamma$ (constant). Recalling that the surface of the cylinder is held at 0° temperature results in the system of *ODE*s below.

$$(i)\ \begin{cases} Y''(y) - (\lambda-\mu)Y(y) = 0 \\ Y(0) = 0 = Y(L) \end{cases} \qquad (ii)\ \begin{cases} \Theta''(\theta) + \gamma\cdot\Theta(\theta) = 0 \\ \Theta(0) = \Theta(2\pi) \end{cases}$$

$$(iii)\ \begin{cases} r^2\cdot R''(r) + r\cdot R'(r) - (\mu r^2 + \lambda)\cdot R(r) = 0 \\ R(\rho) = 0 \end{cases} \qquad (iv)\ T'(t) = \lambda\sigma^2 T(t)$$

Set $\nu = \lambda - \mu$. Then, as is demonstrated in the solution of (*i*) of Example 3.5.1, the eigenvalues of (*i*) above are $\nu_m = -(m\cdot\pi/L)^2$ with eigenfunction solutions $Y_m(y) = C_2 \sin(y\cdot m\cdot\pi/L)$ for $m \in \mathbb{Z}^+$. If $\gamma = \beta^2$, then (*ii*) above is solved via $\Theta(\theta) = d_1\cos(\theta\cdot\beta) + d_2\sin(\theta\cdot\beta)$. The initial conditions $\Theta(0) = \Theta(2\pi)$ imply $d_1 = \Theta(0) = d_1\cos(2\pi\cdot\beta) + d_2\sin(2\pi\cdot\beta) \Rightarrow \beta = n \in \mathbb{Z}$. As the cosine is an even function and sine is odd, then n is restricted to the non-zero integers $n \in \mathbb{Z}_o^+$. Therefore, $\gamma_n = \beta^2 = n^2$ and $\Theta_n(\theta) = d_1\cos(n\theta) + d_2\sin(n\theta)$ are the eigenfunctions of (*ii*). To solve (*iii*), set $\xi = \sqrt{-\mu}\, r$ so that R satisfies the *ODE*

$$\xi^2\frac{d^2R}{d\xi^2} + \xi\frac{dR}{d\xi} + (\xi^2 - n^2)R = 0. \tag{3.5.5}$$

Equation (3.5.5) is known as *Bessel's equation of order n* and is solved by the eponymous functions of the same name. To see how the *Bessel functions* are constructed, assume that $n = 0$ and that R is an analytic function (i.e., that R has a convergent Taylor series) on $[0, \rho]$. Then $R(\xi) = \sum_{k=0}^{\infty} b_k\xi^k$ for $\xi \in [0, \sqrt{-\mu}\,\rho]$. Substituting this power series for R into (3.5.5) and recalling that $n = 0$ results in $0 = \sum_{k=2}^{\infty}\left([k(k-1)+k]b_k + b_{k-1}\right)\xi^k + b_1\xi$. This means that $b_1 = 0$ and $b_k = -b_{k-2}/k^2$ for $k \geq 2$. Set $b_0 = 1$ and obtain the relations $b_2 = -\dfrac{1}{2^2}$, $b_4 = \dfrac{1}{2^2 4^2}$, $b_6 = -\dfrac{1}{2^2 4^2 6^2}$, ..., $b_{2k} = \dfrac{(-1)^k}{2^2 4^2 6^2 \cdots (2k)^2} = \dfrac{(-1)^k}{2^{2k}(k!)^2}$ and $0 = b_1 = b_3 = \ldots = b_{2k+1}$. Therefore, $\boxed{J_0(\xi) = \sum_{k=0}^{\infty}\dfrac{(-1)^k}{2^{2k}(k!)^2}\xi^{2k}}$ is the *Bessel function of order 0*.

Note: $J_n(\xi)$ is the traditional notation for the Bessel functions. A considerable effort has been expended in defining, explaining, and quantifying the Bessel functions as per Abramowitz and Stegun [2]; Olver, Lozier, Boisvert, and Clark [57]; and especially Watson [81]. Strang [74] has developed an intuitive and clear explanation of the relationship between $J_n(r)$ and $\cos(\,(k-½)\pi x\,)$ and the square vs. circular drum.

The *Bessel function of order n* is defined as

$$\boxed{J_n(\xi)=\sum_{k=0}^{\infty}\frac{(-1)^k}{2^{(2k+n)}(k!)\Gamma(n+k+1)}\xi^{(2k+n)}}\,. \tag{3.5.6}$$

A few remarks about the Bessel functions are in order. First, when n and k are non-negative integers, $\Gamma(n+k+1)=(n+k)!$. Second, the use of the gamma function $\Gamma(n+k+1)$ permits *non-integer* values for n. Graphs of select Bessel functions are provided in Figures 3.10–3.11.

Therefore, $e_o \cdot J_n(\xi)=e_o \cdot J_n(\sqrt{-\mu}\,r)$ solves the *ODE* (*iii*) above for arbitrary constant E_o. To satisfy the initial condition $R(\rho)=0$, it must be the case that $J_n(\sqrt{-\mu}\,\rho)=0$. As can be seen in Figures 3.10–3.11, the Bessel functions, while not periodic, have a countable number of zeros. It turns out that these zeros are countably infinite. Let $\zeta_{n,1}<\zeta_{n,2}<\zeta_{n,3}<\ldots<\zeta_{n,k}<\ldots$ be the zeros of the Bessel function J_n of order n in ascending order. Then $J_n(\sqrt{-\mu}\,\rho)=0$ is satisfied provided $\sqrt{-\mu}\,\rho=\zeta_{n,k}$ or $\mu_{n,k}=-\left(\frac{\zeta_{n,k}}{\rho}\right)^2$ for $k\in\mathbb{Z}^+$ and so $R_{n,k}(r)=E_o\cdot J_n\left(\frac{\zeta_{n,k}}{\rho}r\right)$ is the solution of (*iii*).

As $\lambda=\nu+\mu$, then $\lambda_{m,n,k}=\nu_m+\mu_{n,k}=-\left[\left(\frac{m\pi}{L}\right)^2+\left(\frac{\zeta_{n,k}}{\rho}\right)^2\right]<0$. Hence, $T_{m,n,k}(t)=G_o\cdot\exp(\lambda_{m,n,k}\cdot\sigma^2 t)$ satisfies (*iv*) for arbitrary constant G_o. Consequently, the primary solutions of (3.5.4) are

$$u_{m,n,k}(r,y,\theta,t)=C_{m,n,k}\cdot J_n\left(\tfrac{\zeta_{n,k}}{\rho}r\right)\cdot\sin\left(\tfrac{m\pi}{L}y\right)\cdot(\cos(n\theta)+\sin(n\theta))\cdot e^{\lambda_{m,n,k}\cdot\sigma^2\cdot t}\,.$$

This means that

$$u(r,y,\theta,t)=\sum_{m=1}^{\infty}\sum_{n=0}^{\infty}\sum_{k=1}^{\infty}C_{m,n,k}\cdot u_{m,n,k}(r,y,\theta,t)$$

and the associated Fourier coefficients

$$C_{m,n,k}=\frac{1}{\gamma_{n,k}}\int_0^{\rho}\int_0^{L}\int_0^{2\pi}\phi(r,y,\theta)\cdot J_n\left(\tfrac{\zeta_{n,k}}{\rho}r\right)\cdot\sin\left(\tfrac{m\pi}{L}y\right)\cdot\left[\cos(n\theta)+\sin(n\theta)\right]d\theta\,dy\,r dr$$

where $\gamma_{n,k} = \int_0^{\rho}\left(R_{n,k}(r)\right)^2 r\,dr \cdot \int_0^{2\pi}(\Theta_n(\theta))^2\,d\theta \cdot \int_0^{L}(Y_m(y))^2\,dy$ form the solution of (3.5.4). A direct calculation shows that $\int_0^{2\pi}(\Theta_n(\theta))^2\,d\theta = \int_0^{2\pi}(\cos(n\theta)+\sin(n\theta))^2\,d\theta = 2\pi$ and $\int_0^{L}(Y_m(y))^2\,dy = \int_0^{L}\sin^2\left(\frac{m\pi}{L}y\right)dy = L/2$ for all n and m. The first integral $\int_0^{\rho}\left(R_{n,k}(r)\right)^2 r\,dr = \int_0^{\rho}\left(J_n(\frac{\zeta_{n,k}}{\rho}r)\right)^2 r\,dr$ is difficult to derive but can readily be determined via a computer algebra system (such as the Symbolic Math Toolbox) and/or a table of integrals such as Gradshteyn and Ryzhik [29, formula 6.521]. It is $\int_0^{\rho}\left(J_n(\frac{\zeta_{n,k}}{\rho}r)\right)^2 r\,dr = ½\,\rho^2\cdot\left(J_{n+1}(\zeta_{n,k})\right)^2$. This is the case as $\int_0^{\rho}\left(J_n(\frac{\zeta_{n,k}}{\rho}r)\right)^2 r\,dr = \frac{\rho^2}{2}\left(J_n(\zeta_{n,k})^2 + J_{n+1}(\zeta_{n,k})^2\right) - \frac{n\rho^2}{\zeta_{n,k}}J_n(\zeta_{n,k})\cdot J_{n+1}(\zeta_{n,k}) = ½\,\rho^2\cdot\left(J_{n+1}(\zeta_{n,k})\right)^2$ and, by construction, $J_n(\zeta_{n,k}) = 0$. Therefore, $\gamma_{n,k} = ½\cdot\rho^2\cdot\left(J_{n+1}(\zeta_{n,k})\right)^2\cdot½\cdot L\cdot 2\pi = ½\,L\,\pi\,\rho^2\,\left(J_{n+1}(\zeta_{n,k})\right)^2$ and the formulae above comprise the solution of the heat equation (3.5.4) on the cylinder of length L and radius ρ. ■

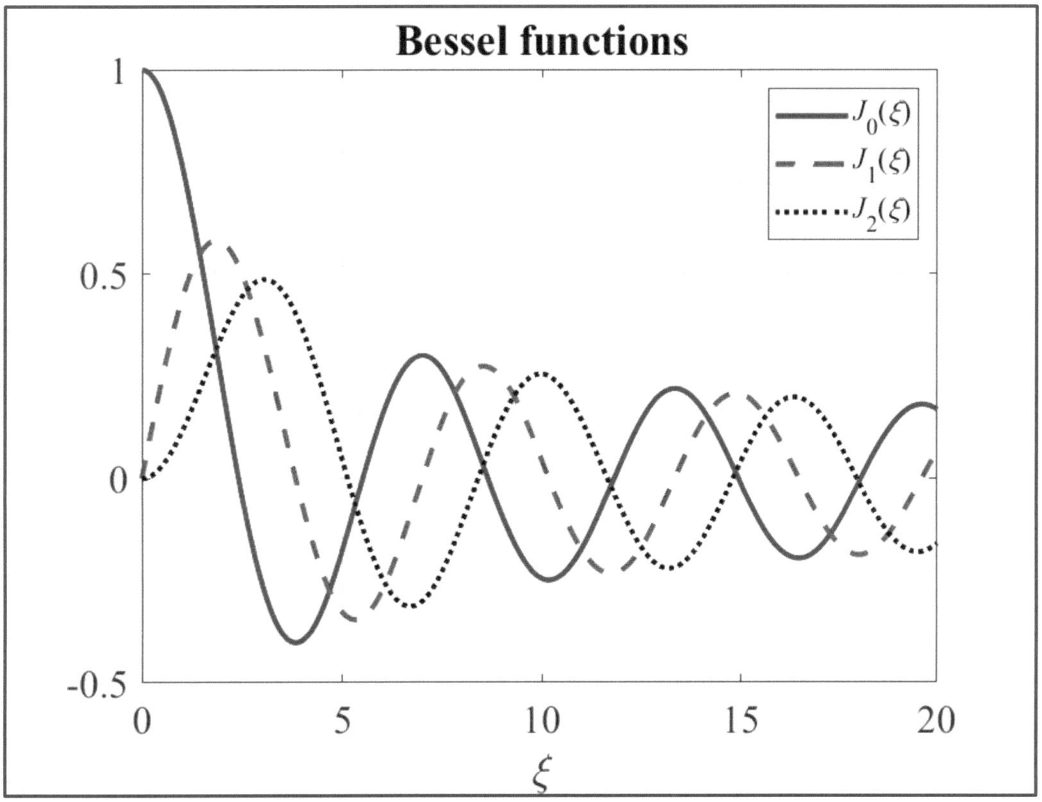

Figure 3.10. Bessel functions for $n = 0, 1, 2$.

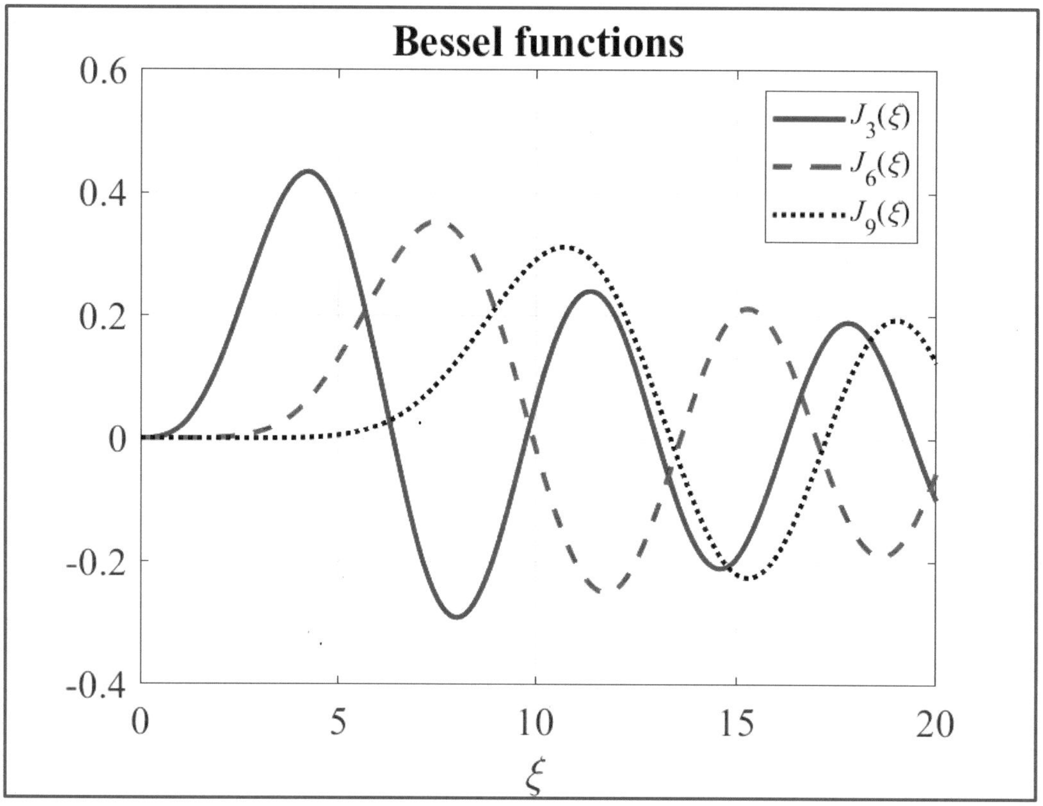

Figure 3.11. Bessel functions for $n = 3, 6, 9$.

Remark 3.1: Since the example above shows how to use polar coordinates to solve the heat equation on the cylinder, it is natural to ask whether spherical coordinates can be used to solve the heat equation on the sphere. The short answer to this question is "not easily." An excursion back to Laplace's equation on the sphere is in order. That is, consider $\Delta u = 0$ for $\boldsymbol{x} = (x, y, z) \in \mathbb{S}(\boldsymbol{0},r_o)= \{\boldsymbol{x}|\ x^2 + y^2 + z^2 = r_o^2\}$. Using spherical coordinates $x = r\sin(\theta)\cos(\omega)$, $y = r\sin(\theta)\sin(\omega)$, $z = r\cos(\theta)$, it is a straightforward exercise in multivariable calculus to show that $\Delta u = \dfrac{\partial^2 u}{\partial x^2} + \dfrac{\partial^2 u}{\partial y^2} + \dfrac{\partial^2 u}{\partial z^2}$ $= \dfrac{\partial^2 u}{\partial r^2} + \dfrac{2}{r}\dfrac{\partial u}{\partial r} + \dfrac{1}{r^2}\left(\dfrac{\partial^2 u}{\partial \theta^2} + \cot(\theta)\dfrac{\partial u}{\partial \theta} + \csc^2(\theta)\dfrac{\partial^2 u}{\partial \omega^2}\right)$. See Exercise 3.8. The domain of definition of the sphere is $0 \le r \le r_o$, $0 \le \theta \le \pi$, and $0 \le \omega \le 2\pi$ so that the periodicity conditions on $u = u(r, \theta, \omega)$ are $u(r_o, \theta, \omega) = u(r_o, \theta + \pi, \omega)$ and $u(r_o, \theta, \omega) = u(r_o, \theta, \omega + 2\pi)$. Now the associated separation of variables transformation $u(r, \theta, \omega) = R(r)\cdot\Theta(\theta)\cdot\Omega(\omega)$ means that $0 = \Delta u =$ $R''\cdot\Theta\cdot\Omega + \dfrac{2}{r}R'\cdot\Theta\cdot\Omega + \dfrac{1}{r^2}\left(R\cdot\Theta''\cdot\Omega + \cot(\theta)R\cdot\Theta'\cdot\Omega\csc^2(\theta)R\cdot\Theta\cdot\Omega''\right)$. Dividing through by the non-zero product $R\cdot\Theta\cdot\Omega$ results in $r^2\dfrac{R''}{R} + 2r\dfrac{R'}{R} = -\left(\dfrac{\Theta''}{\Theta} + \cot(\theta)\dfrac{\Theta'}{\Theta} + \csc^2(\theta)\dfrac{\Omega''}{\Omega}\right) \equiv \lambda$ (constant).

This yields the *ODE* $r^2R'' + 2rR' - \lambda R = 0$ along with

$\sin^2(\theta)\dfrac{\Theta''}{\Theta}+\sin(\theta)\cos(\theta)\dfrac{\Theta'}{\Theta}+\lambda\sin^2(\theta)=-\dfrac{\Omega''}{\Omega}\equiv\mu\,(\text{constant})$. The last equation $\Omega''+\mu\,\Omega=0$ with periodicity condition $\Omega(\omega) = \Omega(\omega + 2\pi)$ has solution $\boxed{\Omega_n(\omega)=a_1\cdot\cos(n\omega)+a_2\cdot\sin(n\omega)}$ provided $\mu_n = n^2$, $n \in \mathbb{Z}_o^+$ are the associated eigenvalues. Now the *ODE* in the radial function can be solved via the substitution $R(r) = r^m$. In this case, $0 = m(m-1)\, r^m + 2m\, r^m - \lambda\, r^m$ or $0 = m(m-1)\, r^m + 2m\, r^m - \lambda$ which has roots $m = m_1 = \frac{1}{2}\left(-1+\sqrt{1+4\lambda}\right)$ and $m_2 = \frac{1}{2}\left(-1-\sqrt{1+4\lambda}\right) = -(1+ m)$. Therefore, $\boxed{R_m(r) = b_1\, r^m + b_2\, r^{-(1+m)}}$ with eigenvalues $\lambda_m = m(m+1)$, $m \in \mathbb{Z}_o^+$. This means that the *ODE* for $\Theta(\theta)$ becomes $\sin^2(\theta)\,\Theta''(\theta) + \sin(\theta)\cos(\theta)\,\Theta'(\theta) + (\lambda_m \sin^2(\theta) - \mu_n)\,\Theta(\theta) = 0$. Now let $\psi = \cos(\theta)$ so that $\theta = \arccos(\psi)$, $\sin^2(\theta) = 1 - \psi^2$, $\dfrac{d\theta}{d\psi}=-\dfrac{1}{\sqrt{1-\psi^2}}$. By letting $\Psi(\psi) = \Theta(\arccos(\theta))$ and performing a substantial amount of calculus, the *ODE* in Θ becomes $\boxed{(1-\psi^2)\cdot\Psi''(\psi)-2\psi\cdot\Psi'(\psi)+\left(\lambda_m-\dfrac{\mu_n}{1-\psi^2}\right)\Psi(\psi)=0}$. This is called the *associated Legendre equation* which has solution $P_m^n(\psi)=P_m^n(\cos(\theta))$ which is the *associated Legendre function of the first kind*. It is noted that the Legendre polynomials are orthogonal and can be expressed via the hypergeometric function ${}_2F_1$

$$P_m^n(z)=\frac{(-1)^n}{n!}\frac{\Gamma(m+n+1)}{\Gamma(m-n+1)}\left(\frac{z+1}{z-1}\right)^{n/2}{}_2F_1\left(m+1,-m,n+1,\tfrac{1}{2}(1-z)\right).$$

See Olver at al. [57] for details. Consequently, $\boxed{\Theta_{m,n}(\theta)=c\;P_m^n(\cos(\theta))}$ so that

$$\boxed{u_{m,n}(r,\theta,\omega)=r^m P_m^n(\cos(\theta))\left[a_{m,n}\cos(n\omega)+b_{m,n}\sin(n\omega)\right]}$$

are the fundamental solutions of Laplace's equation on the sphere $\mathbb{S}_{r_o}$. If the Cauchy data $u(r_o, \theta, \omega) = \phi(\theta, \omega)$ is prescribed, then the following formulae hold.

$$\boxed{\begin{aligned}
u(r,\theta,\omega)&=\sum_{m=0}^{\infty}\sum_{n=0}^{m}u_{m,n}(r,\theta,\omega)\\
\phi(\theta,\omega)&=\sum_{m=0}^{\infty}\sum_{n=0}^{m}r_o^m P_m^n(\cos(\theta))\left[a_{m,n}\cos(n\omega)+b_{m,n}\sin(n\omega)\right]\\
a_{m,n}&=\frac{(2m+1)\cdot(m-n)!}{2\pi r_o^m\cdot(m+n)!}\int_0^{2\pi}\int_0^{\pi}\phi(\theta,\omega)P_m^n(\cos(\theta))\cos(n\omega)\sin(\theta)\,d\theta\,d\omega\\
b_{m,n}&=\frac{(2m+1)\cdot(m-n)!}{2\pi r_o^m\cdot(m+n)!}\int_0^{2\pi}\int_0^{\pi}\phi(\theta,\omega)P_m^n(\cos(\theta))\sin(n\omega)\sin(\theta)\,d\theta\,d\omega
\end{aligned}}$$

The formulae above constitute the solution of Laplace's equation in spherical coordinates. Can this methodology then be used to solve the heat equation on the sphere of radius r_o? Regrettably, the answer is *no*. Why? First, let Δ_S be the Laplacian in spherical coordinates. For the heat equation, the separation of variables transformation is $u(r, \theta, \omega, t) = T(t)\cdot U(r, \theta, \omega)$ and the application of the heat operator $\frac{\partial}{\partial t} - \sigma^2\Delta_S$ produces the equation $-\frac{T'(t)}{\sigma^2 T(t)} = \frac{\Delta_S U(r,\theta,\omega)}{U(r,\theta,\omega)} \equiv \lambda$ (constant). This produces the first order *ODE* in the temporal function $T(t) + \sigma^2\,\lambda\, T(t) = 0$ which has solution $T(t) = A\, e^{-\sigma^2 \lambda t}$. Since heat intensity decreases with time, by physical necessity $\lambda > 0$. So far, so good. But the resulting *ODE* for the radial function changes from the Laplace equation case of $r^2R'' + 2rR' - \lambda R = 0$ to $r^2R'' + 2rR' - \lambda r^2 R = 0$. This means that the substitution $R(r) = r^m$ results in the algebraic equation $m(m+1) - \lambda\cdot r^2 = 0$ or $\lambda = m(m+1)/r^2$ which is *not* a constant. This is a contradiction to the separation of variables formation of λ. If λ is selected to be a positive number $\lambda = n^2$, then the *ODE* yields the solution $R(r) = \frac{\sqrt{2}}{r\sqrt{i\pi n}}\left[C_1 \sinh(nr) - C_2 \cosh(nr)\right]$, a complex valued function. This also contradicts the physics of the problem. Therefore, it must be the case that the separation of variables method cannot be applied as an analytic method to solve the heat equation on the sphere. ■

This chapter has developed solutions to Laplace's, Poisson's, and the heat equation for various finite geometries; that is, $\Omega \subset \mathbb{R}^n$ but $\Omega \neq \mathbb{R}^n$. Bessel functions and the idea of superposition of solutions are described herein. The limitations of the separation of variables transform on more challenging geometries is demonstrated as well. To handle the case of infinite domains, different techniques are required: Specifically, the *Fourier Transform.* As a consequence, the solution of the *wave equation* will be postponed until the next chapter so that properties and applications of the Fourier transform can be detailed. Then, as it is one of the most important linear *PDE*s in mathematical physics, the solution of the three–dimensional wave equation is reserved for Chapter 5.

Exercises

3.1 Solve the *Robin problem* on the cube of length π subject to zero–value mixed conditions everywhere but the top and bottom faces of the cube $C^3(\pi)$. That is, solve

$$\left.\begin{array}{ll}
\Delta u(\boldsymbol{x}) = 0 & \text{for } \boldsymbol{x} \in C^3(\pi) = \left\{\boldsymbol{x} \in \mathbb{R}^3 : 0 \le x, y, z \le \pi\right\} \\
\frac{\partial u}{\partial \boldsymbol{n}}(0, y, z) = 0 = u(\pi, y, z) & \text{for } 0 \le y, z \le \pi \\
\frac{\partial u}{\partial \boldsymbol{n}}(x, 0, z) = 0 = u(x, \pi, z) & \text{for } 0 \le x, z \le \pi \\
\multicolumn{2}{l}{\frac{\partial u}{\partial \boldsymbol{n}}(x, y, 0) = 0 \text{ for } 0 \le x, y \le \pi \text{ and } u(x, y, \pi) = f(x, y) \neq 0 \text{ for } 0 \le x, y \le \pi}
\end{array}\right\}. \tag{E.3.1}$$

3.2 Show that $\{1, \cos(x), \sin(x), \cos(2x), \sin(2x), \cos(3x), \sin(3x), \ldots\}$ is orthonormal with respect to the inner product $<f, g> \equiv \frac{1}{\pi}\int_{-\pi}^{\pi} f(x)\cdot g(x)\,dx$. That is, show $\frac{1}{\pi}\int_{-\pi}^{\pi}\cos^2(nx)\,dx = 1$, $\frac{1}{\pi}\int_{-\pi}^{\pi}\sin^2(nx)\,dx = 1$, and $\frac{1}{\pi}\int_{-\pi}^{\pi}\cos(mx)\cdot\sin(nx)\,dx = 0$ for all $m, n \in \mathbb{Z}^+$.

3.3 Calculate the Fourier series of the 2π–periodic extensions of the functions $f(x) = x$ for $-\pi \le x \le \pi$ and $g(x) = \begin{cases} -\pi & \text{for } -\pi \le x < 0 \\ x & \text{for } 0 < x < \pi \end{cases}$. Construct plots comparable to Figures 3.2 and 3.3 to illustrate the results.

3.4 Show that, if G satisfies (3.4.2), w satisfies (3.4.6), and $v(\mathbf{x}, \mathbf{x}_o) = w(\mathbf{x}) + G(\mathbf{x}, \mathbf{x}_o)$, then (3.4.8) solves (3.4.7). *Hint*: Follow the proof of Theorem 3.1.

3.5 Show that, if G satisfies (3.4.2), w satisfies (3.4.9), and $v(\mathbf{x}, \mathbf{x}_o) = w(\mathbf{x}) + G(\mathbf{x}, \mathbf{x}_o)$, then (3.4.8) solves (3.4.10). *Hint*: Follow the proof of Theorem 3.1.

3.6 Show that the constants γ_1 and γ_2 from Example 3.5.1 are equal to $(c_1 + c_2)$ and $i(c_1 - c_2)$, respectively. *Hint*: Recall *Euler's formula* $e^{ikx} = \cos(kx) + i\cdot\sin(kx)$.

3.7 Use formula 6.521 of Gradshteyn and Ryzhik [29], $\int_0^1 J_n(\alpha x)\cdot J_n(\beta x)\cdot x\,dx = \frac{1}{2}\left(J_{n+1}(\alpha x)\right)^2$ when $\alpha = \beta$ and $J_n(\alpha) = 0 = J_n(\beta)$ to verify $\int_0^{\rho}\left(J_n\left(\frac{\zeta_{n,k}}{\rho}r\right)\right)^2 r\,dr = ½\,\rho^2\cdot\left(J_{n+1}(\zeta_{n,k})\right)^2$.

3.8 Show that in spherical coordinates $x = r\sin(\theta)\cos(\omega)$, $y = r\sin(\theta)\sin(\omega)$, $z = r\cos(\theta)$, the Laplacian $\Delta u = \dfrac{\partial^2 u}{\partial x^2} + \dfrac{\partial^2 u}{\partial y^2} + \dfrac{\partial^2 u}{\partial z^2} = \dfrac{\partial^2 u}{\partial r^2} + \dfrac{2}{r}\dfrac{\partial u}{\partial r} + \dfrac{1}{r^2}\left(\dfrac{\partial^2 u}{\partial \theta^2} + \cot(\theta)\dfrac{\partial u}{\partial \theta} + \csc^2(\theta)\dfrac{\partial^2 u}{\partial \omega^2}\right)$.

Chapter 4

The Fourier Transform, Heat Conduction, and the Wave Equation

The numerous examples of Chapters 2 and 3 demonstrate how the classical method of separation of variables is used to generate solutions of Laplace's and the heat equation over specified finite domains. Utilizing the modern methods inherent in Green's functions and Green's Theorem, solutions of Poisson's equation over an arbitrary finite domain are established. Can such modern methods be applied to the heat equation? Unlike Laplace's or Poisson's equations which are static in time, the heat equation evolves in time. Indeed, the heat equation is the prototype for what are commonly called *evolution equations*. Nonlinear evolution equations are examined in Chapter 6 and a special nonlinear system is detailed in Chapter 7. For now, the focus will be on the linear heat and wave equations. Instead of Green's Theorem, one of the most powerful ideas in modern mathematics is applied: *The Fourier transform*. The program of study for this chapter then is to define the Fourier transform, develop its properties, apply it to the heat and wave equations, and derive analytic solutions.

§4.1. Fourier Transform

Let $f \in L_1\left(\mathbb{R}^n\right) = \left\{ f : \int\limits_{\mathbb{R}^n} |f(\boldsymbol{x})|\, d\boldsymbol{x} < \infty \right\}$ be an integrable function that vanishes as the norm of $\boldsymbol{x}$ approaches infinity. That is, $\lim\limits_{\|\boldsymbol{x}\|\to\infty} f(\boldsymbol{x}) = 0$. The *Fourier transform* is the integral operator which maps a function $f \in L_1\left(\mathbb{R}^n\right)$ of the n–dimensional spatial variable $\boldsymbol{x} = [x_1, x_2, \ldots, x_n]$ into a function F of n–dimensional frequencies $\boldsymbol{\omega} = [\omega_1, \omega_2, \ldots, \omega_n]$ via the formula (4.1.1).

$$\boxed{F(\boldsymbol{\omega}) \equiv \mathfrak{F}\left[f(\boldsymbol{x})\right](\boldsymbol{\omega}) = \frac{1}{\left(\sqrt{2\pi}\right)^n} \int\limits_{\mathbb{R}^n} f(\boldsymbol{x}) e^{i(\boldsymbol{x}\bullet\boldsymbol{\omega})}\, d\boldsymbol{x}} \qquad (4.1.1)$$

The symbols $\mathfrak{F}[f(\boldsymbol{x})](\boldsymbol{\omega})$ read as "the Fourier transform of the function $f(\boldsymbol{x})$ evaluated at $\boldsymbol{\omega}$." Notice that $\boldsymbol{x}\bullet\boldsymbol{\omega} = \sum_{j=1}^{n} x_j \cdot \omega_j$ is the inner product of the two vectors $\boldsymbol{x}$ and $\boldsymbol{\omega}$. The Fourier transform is an invertible operator that maps frequency–valued functions into spatial–valued functions via the *inverse Fourier transform* (4.1.2).

$$g(\boldsymbol{x}) = \mathfrak{F}^{-1}\left[G(\boldsymbol{\omega})\right](\boldsymbol{x}) = \frac{1}{\left(\sqrt{2\pi}\right)^n} \int_{\mathbb{R}^n} G(\boldsymbol{\omega})\, e^{-i(\boldsymbol{x}\bullet\boldsymbol{\omega})}\, d\boldsymbol{\omega} \tag{4.1.2}$$

Remark 4.1: It should be noted that there are *many* forms of the Fourier transform/inverse Fourier transform. Some texts remove the integral normalization $\frac{1}{\left(\sqrt{2\pi}\right)^n}$ and make the exponential kernel $e^{\pm 2\pi i(\boldsymbol{x}\bullet\boldsymbol{\omega})}$ in the respective transform/inverse transform pairing. Other authors switch the signs of the exponential in the transform/inverse transform pairs. Equations (4.1.1)–(4.1.2) conform to the National Institutes of Standards and Technologies' [57] definition.

By combining the notation and definitions inherent in (4.1.1)–(4.1.2), the following inversion formulas are obtained.

$$f(\boldsymbol{x}) = \mathfrak{F}^{-1}\left[F(\boldsymbol{\omega})\right](\boldsymbol{x}) = \frac{1}{\left(\sqrt{2\pi}\right)^n} \int_{\mathbb{R}^n} \mathfrak{F}[f(\boldsymbol{x})](\boldsymbol{\omega})\, e^{-i(\boldsymbol{x}\bullet\boldsymbol{\omega})}\, d\boldsymbol{\omega} = \frac{1}{\left(\sqrt{2\pi}\right)^n} \int_{\mathbb{R}^n} F(\boldsymbol{\omega})\, e^{-i(\boldsymbol{x}\bullet\boldsymbol{\omega})}\, d\boldsymbol{\omega} \tag{4.1.3}$$

$$G(\boldsymbol{\omega}) = \mathfrak{F}\left[g(\boldsymbol{x})\right](\boldsymbol{\omega}) = \frac{1}{\left(\sqrt{2\pi}\right)^n} \int_{\mathbb{R}^n} \mathfrak{F}^{-1}\left[G(\boldsymbol{\omega})\right](\boldsymbol{x})\, e^{i(\boldsymbol{x}\bullet\boldsymbol{\omega})}\, d\boldsymbol{x} = \frac{1}{\left(\sqrt{2\pi}\right)^n} \int_{\mathbb{R}^n} g(\boldsymbol{x})\, e^{i(\boldsymbol{x}\bullet\boldsymbol{\omega})}\, d\boldsymbol{x} \tag{4.1.4}$$

Just as Fourier series are used to find solutions of linear *PDE*s over specified *finite domains*, the Fourier transform is used to construct solutions over infinite domains. In some sense then, the Fourier transform can be thought of as the infinite dimensional extension of Fourier series. Before putting the transform to work on *PDE*s, however, several critical properties are developed.

Fourier Calculus

The Fourier transform obeys the following distinctive and valuable properties.

§4.1.1. Linearity

For any constants c_1 and c_2 and functions $f_1, f_2 \in L_1(\mathbb{R}^n)$,

$$\mathfrak{F}[c_1 \cdot f_1(\boldsymbol{x}) + c_2 \cdot f_2(\boldsymbol{x})](\boldsymbol{\omega}) = c_1 \cdot \mathfrak{F}[f_1(\boldsymbol{x})](\boldsymbol{\omega}) + c_2 \cdot \mathfrak{F}[f_2(\boldsymbol{x})](\boldsymbol{\omega}).$$

§4.1.2. Identity

Recall the *Dirac delta function* $\delta(\boldsymbol{x})$ from §2.2. This function acts as the identity element with respect to the Fourier transform.

$$\mathfrak{F}[(\sqrt{2\pi})^n\,\delta(\boldsymbol{x})](\boldsymbol{\omega}) = 1 \text{ and } \mathfrak{F}^{-1}\left[\tfrac{1}{(\sqrt{2\pi})^n}\right](\boldsymbol{x}) = \delta(\boldsymbol{x})$$

Proof: Proceeding directly,

$$\mathfrak{F}[(\sqrt{2\pi})^n\,\delta(\boldsymbol{x})](\boldsymbol{\omega}) = \tfrac{1}{(\sqrt{2\pi})^n}\int_{\mathbb{R}^n}(\sqrt{2\pi})^n\,\delta(\boldsymbol{x})e^{i(\boldsymbol{x}\cdot\boldsymbol{\omega})}\,d\boldsymbol{x} = \frac{(\sqrt{2\pi})^n}{(\sqrt{2\pi})^n}e^{i(\boldsymbol{0}\cdot\boldsymbol{\omega})} = 1\cdot e^0 = 1.$$

By (4.1.3), $\mathfrak{F}^{-1}[1](\boldsymbol{x}) = \mathfrak{F}^{-1}\left[\mathfrak{F}\left[(\sqrt{2\pi})^n\,\delta(\boldsymbol{x})\right](\boldsymbol{\omega})\right](\boldsymbol{x}) = (\sqrt{2\pi})^n\,\delta(\boldsymbol{x})$. ■

§4.1.3. Scaling

For any scalar $a \in \mathbb{R}$,

$$\mathfrak{F}[f(a\boldsymbol{x})](\boldsymbol{\omega}) = \tfrac{1}{(\sqrt{2\pi})^n}\int_{\mathbb{R}^n} f(a\cdot\boldsymbol{x})e^{i(\boldsymbol{x}\cdot\boldsymbol{\omega})}\,d\boldsymbol{x} \overset{a\cdot\boldsymbol{x}\mapsto\boldsymbol{z}}{=} \tfrac{1}{(\sqrt{2\pi})^n}\int_{\mathbb{R}^n}\tfrac{1}{|a|}f(\boldsymbol{z})e^{i(\boldsymbol{z}\cdot\boldsymbol{\omega}/a)}\,d\boldsymbol{z} = \tfrac{1}{|a|}\,\mathfrak{F}[f(\boldsymbol{x})](\boldsymbol{\omega}).$$

The reason that the Fourier transform is scaled by $|a|$ is due to the substitution $\boldsymbol{z} = a\cdot\boldsymbol{x}$. When $a < 0$, then the limits of integration are $+\infty$ (lower limit) to $-\infty$ (upper limit). Thus, to flip the limits from $-\infty$ to $+\infty$, multiply by -1 so that $-1\cdot a > 0$. When $a > 0$, the limits range from $-\infty$ to $+\infty$ and the order of integration does not need to be reversed.

§4.1.4. Shifting/Translation

For any $\boldsymbol{a} \in \mathbb{R}^n$,

$$\mathfrak{F}[e^{i(\boldsymbol{a}\cdot\boldsymbol{x})}f(\boldsymbol{x})](\boldsymbol{\omega}) = \tfrac{1}{(\sqrt{2\pi})^n}\int_{\mathbb{R}^n} e^{i(\boldsymbol{a}\cdot\boldsymbol{x})}f(\boldsymbol{x})e^{i(\boldsymbol{x}\cdot\boldsymbol{\omega})}\,d\boldsymbol{x} = \tfrac{1}{(\sqrt{2\pi})^n}\int_{\mathbb{R}^n} f(\boldsymbol{x})e^{i(\boldsymbol{x}\cdot[\boldsymbol{\omega}+\boldsymbol{a}])}\,d\boldsymbol{x} = \mathfrak{F}[f(\boldsymbol{x})](\boldsymbol{\omega}+\boldsymbol{a})$$

and

$$\mathfrak{F}[f(\boldsymbol{x}+\boldsymbol{a})](\boldsymbol{\omega}) = \tfrac{1}{(\sqrt{2\pi})^n}\int_{\mathbb{R}^n} f(\boldsymbol{x}-\boldsymbol{a})e^{i(\boldsymbol{x}\cdot\boldsymbol{\omega})}\,d\boldsymbol{x} \overset{\boldsymbol{x}-\boldsymbol{a}\mapsto\boldsymbol{z}}{=} \tfrac{1}{(\sqrt{2\pi})^n}\int_{\mathbb{R}^n} f(\boldsymbol{z})e^{i([\boldsymbol{z}+\boldsymbol{a}]\cdot\boldsymbol{\omega})}\,d\boldsymbol{z}$$

$$= e^{i(\boldsymbol{a}\cdot\boldsymbol{\omega})}\tfrac{1}{(\sqrt{2\pi})^n}\int_{\mathbb{R}^n} f(\boldsymbol{x})e^{i(\boldsymbol{x}\cdot\boldsymbol{\omega})}\,d\boldsymbol{x} = e^{i(\boldsymbol{a}\cdot\boldsymbol{\omega})}\mathfrak{F}[f(\boldsymbol{x})](\boldsymbol{\omega}).$$

§4.1.5. Differentiation (with respect to the spatial variable)

$$\mathfrak{F}\left[\frac{\partial f}{\partial x_j}(\boldsymbol{x})\right](\boldsymbol{\omega}) = (-i\cdot\omega_j)\mathfrak{F}[f(\boldsymbol{x})](\boldsymbol{\omega}) \text{ and more generally,}$$

$$\mathfrak{F}\left[\frac{\partial^k f}{\partial x_j^k}(\boldsymbol{x})\right](\boldsymbol{\omega}) = (-i\cdot\omega_j)^k \mathfrak{F}[f(\boldsymbol{x})](\boldsymbol{\omega}).$$

Proof: Observe that

$$\mathfrak{F}\left[\frac{\partial f}{\partial x_j}(\boldsymbol{x})\right](\boldsymbol{\omega}) = \frac{1}{(\sqrt{2\pi})^n}\int_{\mathbb{R}^n}\frac{\partial f}{\partial x_j}(\boldsymbol{x})e^{i(\boldsymbol{x}\bullet\boldsymbol{\omega})}\,d\boldsymbol{x}$$

$$= \frac{1}{(\sqrt{2\pi})^n}\int_{\mathbb{R}} e^{i(x_1\cdot\omega_1)}\,dx_1\int_{\mathbb{R}} e^{i(x_2\cdot\omega_2)}\,dx_2\cdots\int_{\mathbb{R}}\frac{\partial f}{\partial x_j}(\boldsymbol{x})e^{i(x_j\cdot\omega_j)}\,dx_j\cdots\int_{\mathbb{R}} e^{i(x_n\cdot\omega_n)}\,dx_n.$$

This arrangement of the order of integration is a result of Fubini's Theorem (see, e.g., Bartle [7]). Now apply integration by parts on the integral containing the partial derivative with $u = e^{i(x_j\cdot\omega_j)}$ and $dv = \partial f(\boldsymbol{x})/\partial x_j$. Then,

$$\int_{\mathbb{R}}\frac{\partial f}{\partial x_j}(\boldsymbol{x})e^{i(x_j\cdot\omega_j)}\,dx_j = e^{i(x_j\cdot\omega_j)}f(\boldsymbol{x})\Big|_{x_j=-\infty}^{+\infty} - (i\omega_j)\int_{\mathbb{R}} f(\boldsymbol{x})e^{i(x_j\cdot\omega_j)}\,dx_j = 0-(i\omega_j)\int_{\mathbb{R}} f(\boldsymbol{x})e^{i(x_j\cdot\omega_j)}\,dx_j.$$

This is the case since $|e^{i(x_j\cdot\omega_j)}| = 1$ and, by supposition, $\lim_{\|\boldsymbol{x}\|\to\infty} f(\boldsymbol{x}) = 0$. Inserting this result into the original equation produces

$$\mathfrak{F}\left[\frac{\partial f}{\partial x_j}(\boldsymbol{x})\right](\boldsymbol{\omega}) = (-i\omega_j)\frac{1}{(\sqrt{2\pi})^n}\int_{\mathbb{R}} e^{i(x_1\cdot\omega_1)}\,dx_1\int_{\mathbb{R}} e^{i(x_2\cdot\omega_2)}\,dx_2\cdots\int_{\mathbb{R}}\frac{\partial f}{\partial x_j}(\boldsymbol{x})e^{i(x_j\cdot\omega_j)}\,dx_j\cdots\int_{\mathbb{R}} e^{i(x_n\cdot\omega_n)}\,dx_n$$

$$= (-i\omega_j)\mathfrak{F}\left[f(\boldsymbol{x})\right](\boldsymbol{\omega}).$$

The general case of $\mathfrak{F}\left[\frac{\partial^k f}{\partial x_j^k}(\boldsymbol{x})\right](\boldsymbol{\omega})$ is now demonstrated by repeated applications of integration by parts on the integral containing the partial derivative $\int_{\mathbb{R}}\frac{\partial^k f}{\partial x_j^k}(\boldsymbol{x})e^{i(x_j\cdot\omega_j)}\,dx_j$. ■

Before listing the next and in many ways most valuable property of the Fourier transform, the *convolution* of two functions in $L_2(\mathbb{R}^n) = \left\{f : \int_{\mathbb{R}^n}|f(\boldsymbol{x})|^2\,d\boldsymbol{x} < \infty\right\}$ is defined.

Definition 4.1: For f, $g \in L_2(\mathbb{R}^n)$, the *convolution* of f with g at $\boldsymbol{x} \in \mathbb{R}^n$ is

$$\boxed{(f\odot g)(\boldsymbol{x}) = \frac{1}{(\sqrt{2\pi})^n}\int_{\mathbb{R}^n} f(\xi)\cdot g(\boldsymbol{x}-\xi)\,d\xi}.$$

The factor of $\frac{1}{(\sqrt{2\pi})^n}$ is required for normalization and consistency with the definition of the n–dimensional Fourier transform. Just as the exponential maps multiplication into addition via $e^x \cdot e^y = e^{x+y}$, the Fourier transform converts multiplication into convolution.

§4.1.6. Convolution

$$\mathfrak{F}[f(\boldsymbol{x})](\boldsymbol{\omega})\cdot \mathfrak{F}[g(\boldsymbol{x})](\boldsymbol{\omega}) = \mathfrak{F}\left[(f \odot g)(\boldsymbol{x})\right](\boldsymbol{\omega}).$$

Proof: Proceeding directly, it is seen that

$$\begin{aligned}\mathfrak{F}\left[(f \odot g)(\boldsymbol{x})\right](\boldsymbol{\omega}) &= \frac{1}{(\sqrt{2\pi})^n}\int\limits_{\mathbb{R}^n}\left[\frac{1}{(\sqrt{2\pi})^n}\int\limits_{\mathbb{R}^n} f(\xi)g(\boldsymbol{x}-\xi)\,d\xi\right]e^{i(\boldsymbol{x}\cdot\boldsymbol{\omega})}\,d\boldsymbol{\omega}\\ &= \frac{1}{(\sqrt{2\pi})^n}\frac{1}{(\sqrt{2\pi})^n}\int\limits_{\mathbb{R}^n}\int\limits_{\mathbb{R}^n} f(\xi)g(\boldsymbol{x}-\xi)\,d\xi\, e^{i(\boldsymbol{x}\cdot\boldsymbol{\omega})}\,d\boldsymbol{\omega}.\end{aligned}$$

Let $\boldsymbol{u} = \boldsymbol{x} - \boldsymbol{\xi}$ so that,

$$\begin{aligned}\mathfrak{F}\left[(f \odot g)(\boldsymbol{x})\right](\boldsymbol{\omega}) &= \frac{1}{(\sqrt{2\pi})^n}\frac{1}{(\sqrt{2\pi})^n}\int\limits_{\mathbb{R}^n}\int\limits_{\mathbb{R}^n} f(\xi)g(\boldsymbol{u})e^{i(\xi\cdot\boldsymbol{\omega})}d\xi\, e^{i(\boldsymbol{u}\cdot\boldsymbol{\omega})}\,d\boldsymbol{u}\\ &= \frac{1}{(\sqrt{2\pi})^n}\int\limits_{\mathbb{R}^n} f(\xi)e^{i(\xi\cdot\boldsymbol{\omega})}\,d\xi\;\frac{1}{(\sqrt{2\pi})^n}\int\limits_{\mathbb{R}^n} g(\boldsymbol{u})e^{i(\boldsymbol{u}\cdot\boldsymbol{\omega})}\,d\boldsymbol{u}\ \text{(by Fubini's Theorem)} \quad \blacksquare\\ &= \mathfrak{F}\left[f(\boldsymbol{x})\right](\boldsymbol{\omega})\cdot\mathfrak{F}\left[g(\boldsymbol{x})\right](\boldsymbol{\omega}).\end{aligned}$$

The final item in the list of Fourier transform properties is the invariance of the transform of functions of spatial and temporal variables under differentiation with respect to time. That is, if f is a function of $\boldsymbol{x} \in \mathbb{R}^n$ and t, then as operators the derivative $\partial/\partial t$ and $\mathfrak{F}$ commute. This is a crucial property of the Fourier transform as it can turn a *linear PDE* into a linear *ODE*.

§4.1.7. Invariance with respect to temporal differentiation

Let f be a function of $\boldsymbol{x} \in \mathbb{R}^n$ and t so that $f(\boldsymbol{x}, t)$, $\frac{\partial f}{\partial t}(\boldsymbol{x},t) \in L_1\left(\mathbb{R}^n\right)\cap C^1\left([0,+\infty)\right)$. That is, as a function of $\boldsymbol{x}$, $f(\boldsymbol{x}, t)$ and $\frac{\partial f}{\partial t}(\boldsymbol{x},t)$ are in $L_1\left(\mathbb{R}^n\right)$ while as a function of t, $f(\boldsymbol{x}, t)$ and $\frac{\partial f}{\partial t}(\boldsymbol{x},t)$ are continuously differentiable over the non–negative real line, i.e., in $C^1\left([0,+\infty)\right)$. Under these conditions, $\frac{\partial}{\partial t}\mathfrak{F}\left[f(\boldsymbol{x},t)\right](\boldsymbol{\omega}) = \mathfrak{F}\left[\frac{\partial f}{\partial t}(\boldsymbol{x},t)\right](\boldsymbol{\omega})$.

Proof: Let $\Omega = [a_1, b_1] \times [a_2, b_2] \times \ldots \times [a_n, b_n]$ be an n–dimensional hypercube. The notation $\lim_{|\Omega|\to\infty}$ shall mean that the component limits a_j and b_j tend to $-\infty$ and $+\infty$. That is,

$\lim_{|\Omega|\to\infty} = \lim_{a_1\to-\infty}\lim_{a_2\to-\infty}\cdots\lim_{a_n\to-\infty}\lim_{b_1\to+\infty}\lim_{b_2\to+\infty}\cdots\lim_{b_n\to+\infty}$. Heuristically, as $|\Omega| \to \infty$ then $\Omega \to \mathbb{R}^n$. As Ω is a finite domain and f and $\frac{\partial f}{\partial t}$ are continuous functions of t, then by Leibniz's formula,

$\frac{\partial}{\partial t}\frac{1}{(\sqrt{2\pi})^n}\int_\Omega f(\boldsymbol{x},t)e^{i(\boldsymbol{x}\bullet\boldsymbol{\omega})}\,d\boldsymbol{x} = \frac{1}{(\sqrt{2\pi})^n}\int_\Omega \frac{\partial f}{\partial t}(\boldsymbol{x},t)e^{i(\boldsymbol{x}\bullet\boldsymbol{\omega})}\,d\boldsymbol{x}$. Therefore,

$$\frac{\partial}{\partial t}\frac{1}{(\sqrt{2\pi})^n}\int_{\mathbb{R}^n} f(\boldsymbol{x},t)e^{i(\boldsymbol{x}\bullet\boldsymbol{\omega})}\,d\boldsymbol{x} = \frac{\partial}{\partial t}\frac{1}{(\sqrt{2\pi})^n}\lim_{|\Omega|\to\infty}\int_\Omega f(\boldsymbol{x},t)e^{i(\boldsymbol{x}\bullet\boldsymbol{\omega})}\,d\boldsymbol{x}$$

$$= \lim_{|\Omega|\to\infty}\frac{\partial}{\partial t}\frac{1}{(\sqrt{2\pi})^n}\int_\Omega f(\boldsymbol{x},t)e^{i(\boldsymbol{x}\bullet\boldsymbol{\omega})}\,d\boldsymbol{x} = \lim_{|\Omega|\to\infty}\frac{1}{(\sqrt{2\pi})^n}\int_\Omega \frac{\partial f}{\partial t}(\boldsymbol{x},t)e^{i(\boldsymbol{x}\bullet\boldsymbol{\omega})}\,d\boldsymbol{x}$$

$$= \int_{\mathbb{R}^n}\frac{\partial f}{\partial t}(\boldsymbol{x},t)e^{i(\boldsymbol{x}\bullet\boldsymbol{\omega})}\,d\boldsymbol{x}.$$ ■

Next, the properties of the Fourier transform are used to produce some useful formulae.

§4.2. Examples of the Fourier Transform

As this title of this section indicates, the Fourier transform of a select set of functions will be calculated. Naturally, as the astute reader will surmise, these examples will be of considerable importance when discussing the heat and wave equations over infinite domains.

Example 4.2.1. Compute the Fourier transform of $e^{-(a\cdot x)^2}$.

Process: Observe that $(ax)^2 - i\cdot\omega\,x = \left(ax - i\frac{\omega}{2a}\right)^2 + \left(\frac{\omega}{2a}\right)^2$. The substitution $z = ax - i\frac{\omega}{2a}$ yields the expression

$$\mathfrak{F}\left[e^{-(ax)^2}\right](\omega) = \frac{1}{\sqrt{2\pi}}\int_{\mathbb{R}} e^{-(ax)^2}e^{i(x\cdot\omega)}\,dx = \frac{1}{\sqrt{2\pi}}\int_{\mathbb{R}} e^{-[(ax)^2-ix\omega]}\,dx$$

$$= \frac{1}{\sqrt{2\pi}}\int_{\mathbb{R}} e^{-(ax-i\omega/2a)^2}e^{-(\omega/2a)^2}\,dx = e^{-(\omega/2a)^2}\frac{1}{\sqrt{2\pi}}\int_{\mathbb{R}}\frac{1}{|a|}e^{-z^2}\,dz = \frac{e^{-(\omega/2a)^2}}{|a|\sqrt{2}}.$$

Notice that the denominator of this computation contains $|a|$. This is because the substitution $z = ax - i\frac{\omega}{2a}$ compels z to range from $-\infty$ to $+\infty$ as x ranges from $-\infty$ to $+\infty$ when $a > 0$ during the integration. If $a < 0$, then z ranges from $+\infty$ to $-\infty$ as x ranges from $-\infty$ to $+\infty$. Therefore, the sign of the integral changes from positive to negative as a changes from positive to negative. The term

$\frac{1}{|a|}$ covers both cases. If the x^2 term in $e^{-(a\cdot x)^2}$ is replaced by $\|\boldsymbol{x}\|^2 = \sum_{j=1}^{n} x_j^2$ and $\boldsymbol{\omega} = [\omega_1, \omega_2, \ldots, \omega_n]$, then the above calculation can be extended to show that the multi–dimensional Fourier transform of the Gaussian function $e^{-(a\cdot\|\boldsymbol{x}\|)^2}$ has a comparable form. That is, $\mathfrak{F}\left[e^{-(a\cdot\|\boldsymbol{x}\|)^2}\right](\boldsymbol{\omega}) = \frac{e^{-(\|\boldsymbol{\omega}\|/2a)^2}}{\left(|a|\sqrt{2}\right)^n}$. ■

Example 4.2.2. Compute the Fourier transform of $\cos(\omega_o x)$.

Process: First note that $f(x) = \cos(\omega_o x) \notin L_1(\mathbb{R})$ so that the Fourier transform of f is not even well defined. Is there then any hope of computing $\mathfrak{F}[\cos(\omega_o x)](\omega)$? Yes, but the calculation must be done "in distribution." This means $f(x) = \cos(\omega_o x)$ must be treated as a *generalized function.* Define the truncated cosine function $f_a(x) = \cos(\omega_o x)\cdot\mathfrak{I}_{[-a,a]}(x)$ where $\mathfrak{I}$ is the indicator function of §2.2 and $a > 0$. Then

$$\begin{aligned}\mathfrak{F}\left[f_a(x)\right](\omega) &= \tfrac{1}{\sqrt{2\pi}}\int_{-a}^{a}\cos(\omega_o x)\,e^{i(x\cdot\omega)}\,dx = \tfrac{1}{\sqrt{2\pi}}\int_{-a}^{a}\left[\cos(\omega_o x)\cos(\omega x) + i\cos(\omega_o x)\sin(\omega x)\right]dx \\ &= \tfrac{1}{\sqrt{2\pi}}\cdot\tfrac{1}{2}\int_{-a}^{a}\left(\cos([\omega_o+\omega]x) + \cos([\omega_o-\omega]x)\right)\,dx \\ &+ \tfrac{1}{\sqrt{2\pi}}\cdot\tfrac{i}{2}\int_{-a}^{a}\left(\sin([\omega_o+\omega]x) + \sin([\omega_o-\omega]x)\right)\,dx.\end{aligned}$$

Now as *sine* is an odd function, then its integral over the symmetric interval $[-a, a]$ is 0. Thus,

$$\begin{aligned}\mathfrak{F}\left[f_a(x)\right](\omega) &= \tfrac{1}{\sqrt{2\pi}}\cdot\tfrac{1}{2}\int_{-a}^{a}\left(\cos([\omega_o+\omega]x) + \cos([\omega_o-\omega]x)\right)\,dx \\ &= \tfrac{1}{\sqrt{2\pi}}\left[\frac{\sin([\omega_o+\omega]a)}{\omega_o+\omega} + \frac{\sin([\omega_o-\omega]a)}{\omega_o-\omega}\right].\end{aligned}$$

The full *cosine*, however, is the limit as $a \to \infty$ of the truncated function $f_a(x)$. Since the Fourier transform is a continuous operator, then

$$\begin{aligned}\mathfrak{F}\left[\cos(\omega_o x)\right](\omega) &= \mathfrak{F}\left[\lim_{a\to\infty} f_a(x)\right](\omega) = \lim_{a\to\infty}\mathfrak{F}\left[f_a(x)\right](\omega) \\ &= \lim_{a\to\infty}\tfrac{1}{\sqrt{2\pi}}\left[\frac{\sin([\omega+\omega_o]a)}{\omega+\omega_o} + \frac{\sin([\omega-\omega_o]a)}{\omega-\omega_o}\right]\end{aligned}$$

provided this limit exists. It has been shown (see, e.g., Andrews and Shivamoggi [4]) that

$$\phi(x) = \tfrac{1}{\pi}\int_{-\infty}^{\infty} \phi(\xi)\left[\lim_{a\to\infty}\frac{\sin([\xi - x]a)}{\xi - x}\right]d\xi .$$

But this is precisely the property of the Dirac delta function so that $\tfrac{1}{\pi}\lim_{a\to\infty}\frac{\sin([\xi - x]a)}{\xi - x} \equiv \delta(\xi - x)$ or $\pi\cdot\delta(\xi - x) = \lim_{a\to\infty}\frac{\sin([\xi - x]a)}{\xi - x}$. Therefore,

$$\begin{aligned}\mathfrak{F}\left[\cos(\omega_o x)\right](\omega) &= \lim_{a\to\infty}\tfrac{1}{\sqrt{2\pi}}\left[\frac{\sin([\omega+\omega_o]a)}{\omega+\omega_o} + \frac{\sin([\omega-\omega_o]a)}{\omega-\omega_o}\right] \\ &= \tfrac{1}{\sqrt{2\pi}}\left[\pi\cdot\delta(\omega+\omega_o) + \pi\cdot\delta(\omega-\omega_o)\right] = \sqrt{\tfrac{\pi}{2}}\left[\delta(\omega+\omega_o)+\delta(\omega-\omega_o)\right].\end{aligned}$$

Following the same procedure produces $\mathfrak{F}^{-1}\left[\cos(x_o\omega)\right](x) = \sqrt{\tfrac{\pi}{2}}\left[\delta(x+x_o)+\delta(x-x_o)\right]$.

More generally, observe that $\mathfrak{F}[\delta(\mathbf{x} - \mathbf{x}_o)](\boldsymbol{\omega}) = \frac{1}{(\sqrt{2\pi})^n}\int_{\mathbb{R}^n} \delta(\mathbf{x}-\mathbf{x}_o)e^{i(\mathbf{x}\bullet\boldsymbol{\omega})}\,d\mathbf{x} = \frac{1}{(\sqrt{2\pi})^n}e^{i(\mathbf{x}_o\bullet\boldsymbol{\omega})}$ and $\mathfrak{F}[\delta(\mathbf{x}+\mathbf{x}_o)](\boldsymbol{\omega}) = \frac{1}{(\sqrt{2\pi})^n}e^{-i(\mathbf{x}_o\bullet\boldsymbol{\omega})}$. Hence, by the linearity property 4.1.1, $\mathfrak{F}[\delta(\mathbf{x}-\mathbf{x}_o) + \delta(\mathbf{x}+\mathbf{x}_o)](\boldsymbol{\omega})$ $= \mathfrak{F}[\delta(\mathbf{x}-\mathbf{x}_o)](\boldsymbol{\omega}) + \mathfrak{F}[\delta(\mathbf{x}+\mathbf{x}_o)](\boldsymbol{\omega}) = \frac{2}{(\sqrt{2\pi})^n}\tfrac{1}{2}\left(e^{i(\mathbf{x}_o\bullet\boldsymbol{\omega})} + e^{-i(\mathbf{x}_o\bullet\boldsymbol{\omega})}\right) = \frac{2}{(\sqrt{2\pi})^n}\cos(\mathbf{x}_o\bullet\boldsymbol{\omega}) \Leftrightarrow$ $\mathfrak{F}^{-1}[\cos(\mathbf{x}_o\bullet\boldsymbol{\omega})](\mathbf{x}) = \frac{(\sqrt{2\pi})^n}{2}(\delta(\mathbf{x}-\mathbf{x}_o) + \delta(\mathbf{x}+\mathbf{x}_o))$. When $n = 1$, this is precisely the result above. By following this process, it is a straightforward matter to arrive at the formulae $\mathfrak{F}[\delta(\mathbf{x}+\mathbf{x}_o) - \delta(\mathbf{x}-\mathbf{x}_o)](\boldsymbol{\omega}) = \frac{2}{(\sqrt{2\pi})^n}\sin(\mathbf{x}_o\bullet\boldsymbol{\omega}) \Leftrightarrow \mathfrak{F}^{-1}[\sin(\mathbf{x}_o\bullet\boldsymbol{\omega})](\mathbf{x}) = \frac{(\sqrt{2\pi})^n}{2}(\delta(\mathbf{x}+\mathbf{x}_o) - \delta(\mathbf{x}-\mathbf{x}_o))$. ■

Example 4.2.3. Compute the Fourier transform of the indicator function $\mathfrak{I}_{[-a,\,a]}(x) = 1$ for $x \in [-a, a]$ and $\mathfrak{I}_{[-a,\,a]}(x) = 0$ for $x \notin [-a, a]$.

Process: A direct computation shows $\mathfrak{F}[\mathfrak{I}_{[-a,\,a]}(x)](\omega) = \frac{1}{\sqrt{2\pi}}\int_{\mathbb{R}}\mathfrak{I}_{[-a,a]}(x)e^{i(x\cdot\omega)}\,dx = \frac{1}{\sqrt{2\pi}}\int_{-a}^{a}e^{i(x\cdot\omega)}\,dx$ $= \frac{1}{\sqrt{2\pi}}\int_{-a}^{a}\left[\cos(x\omega)+i\sin(x\omega)\right]dx = \frac{1}{\sqrt{2\pi}}\frac{1}{\omega}\left[\sin(x\omega)-i\cos(x\omega)\right]\Big|_{-a}^{a} = \sqrt{\tfrac{2}{\pi}}\frac{\sin(a\omega)}{\omega}$ or $\mathfrak{F}[\tfrac{1}{a}\mathfrak{I}_{[-a,\,a]}(x)](\omega) = \sqrt{\tfrac{2}{\pi}}\frac{\sin(a\omega)}{a\omega} \Leftrightarrow \sqrt{\tfrac{2}{\pi}}\mathfrak{F}^{-1}\left[\frac{\sin(a\omega)}{a\omega}\right](x) = \frac{1}{a}\mathfrak{I}_{[-a,a]}(x)$. ■

Table 4.1 summarizes the results of the examples in this section. The spatial and frequency functions $\{f(\mathbf{x}), F(\boldsymbol{\omega})\}$ are known as *Fourier pairs*. More detailed lists of Fourier pairs can be found in Andrews and Shivamoggi [4], Champeney [16], Gradshteyn and Ryzhik [29], Olver, Lozier, Boisvert, and Clark [57], Körner [46], and Selby [69] among many other resources. In the next section, the power of the Fourier transform will be demonstrated as it is applied to the heat equation.

$f(\boldsymbol{x})$	$F(\boldsymbol{\omega})$
$e^{-(a\cdot x)^2}, \quad a, x \in \mathbb{R}$	$\frac{1}{\sqrt{2}\lvert a\rvert} e^{-\left(\omega/2a\right)^2}, \quad \omega \in \mathbb{R}$
$e^{-(a\cdot\lVert \boldsymbol{x}\rVert)^2}, \quad a \in \mathbb{R}, \boldsymbol{x} \in \mathbb{R}^n$	$\frac{1}{\left(\sqrt{2}\lvert a\rvert\right)^n} e^{-\left(\lVert\boldsymbol{\omega}\rVert/2a\right)^2}, \quad \boldsymbol{\omega} \in \mathbb{R}^n$
$\cos(\boldsymbol{\omega}_o \bullet \boldsymbol{x}),\ \boldsymbol{\omega}_o, \boldsymbol{x} \in \mathbb{R}^n$	$\frac{(\sqrt{2\pi})^n}{2}[\delta(\boldsymbol{\omega}+\boldsymbol{\omega}_o)+\delta(\boldsymbol{\omega}-\boldsymbol{\omega}_o)],\ \boldsymbol{\omega} \in \mathbb{R}^n$
$\sin(\boldsymbol{\omega}_o \bullet \boldsymbol{x}),\ \boldsymbol{\omega}_o, \boldsymbol{x} \in \mathbb{R}^n$	$\frac{(\sqrt{2\pi})^n}{2}[\delta(\boldsymbol{\omega}+\boldsymbol{\omega}_o)-\delta(\boldsymbol{\omega}-\boldsymbol{\omega}_o)],\ \boldsymbol{\omega} \in \mathbb{R}^n$
$\delta(\boldsymbol{x}-\boldsymbol{x}_o),\ \boldsymbol{x}_o, \boldsymbol{x} \in \mathbb{R}^n$	$\frac{1}{(\sqrt{2\pi})^n} e^{i(\boldsymbol{x}_o \bullet \boldsymbol{\omega})}, \quad \boldsymbol{\omega} \in \mathbb{R}^n$
$\frac{1}{a}\mathcal{I}_{[-a,a]}(x), \quad a > 0, x \in \mathbb{R}$	$\sqrt{\frac{2}{\pi}}\frac{\sin(a\omega)}{a\omega},\ \omega \in \mathbb{R}$

Table 4.1. Fourier pairs, $F(\boldsymbol{\omega}) \equiv \mathfrak{F}\left[f(\boldsymbol{x})\right](\boldsymbol{\omega})$

§4.3. The Fourier Transform and the Heat Equation

Consider the heat equation on an infinite domain.

$$\left.\begin{aligned} &\frac{\partial u}{\partial t}(\boldsymbol{x},t) = \sigma^2\,\Delta u(\boldsymbol{x},t), && \boldsymbol{x} \in \mathbb{R}^n, t > 0 \\ &u(\boldsymbol{x},0) = f(\boldsymbol{x}), && \boldsymbol{x} \in \mathbb{R}^n \end{aligned}\right\} \tag{4.3.1}$$

Apply the Fourier transform to (4.3.1) and use properties 4.1.5 and 4.1.7 to obtain

$$\frac{\partial}{\partial t}\mathfrak{F}\left[u(\boldsymbol{x},t)\right](\boldsymbol{\omega}) = \mathfrak{F}\left[\frac{\partial u}{\partial t}(\boldsymbol{x},t)\right](\boldsymbol{\omega}) = \sigma^2\sum_{j=1}^{n}(i\omega_j)^2\,\mathfrak{F}\left[u(\boldsymbol{x},t)\right](\boldsymbol{\omega}) \; = \; -\sigma^2\,\lVert\boldsymbol{\omega}\rVert^2\,\mathfrak{F}\left[u(\boldsymbol{x},t)\right](\boldsymbol{\omega}).$$

That is, the partial differential equation (4.3.1) has been transformed into an ordinary differential equation. If the assignments $\mathfrak{F}\left[u(\boldsymbol{x},t)\right](\boldsymbol{\omega}) \mapsto U(\boldsymbol{\omega},t)$ and $\sigma\lVert\boldsymbol{\omega}\rVert \mapsto \alpha$ are made, then the equation above is the first order *ODE* $\frac{dU}{dt} = -\alpha^2 U$ which has the solution $U(\boldsymbol{\omega},t) = U(\boldsymbol{\omega},0)e^{-\alpha^2 t}$. The initial condition $u(\boldsymbol{x}, 0) = f(\boldsymbol{x})$ implies that $U(\boldsymbol{\omega},0) = \mathfrak{F}\left[u(\boldsymbol{x},0)\right](\boldsymbol{\omega}) = \mathfrak{F}\left[f(\boldsymbol{x})\right](\boldsymbol{\omega}) \equiv F(\boldsymbol{\omega})$. Thus, $U(\boldsymbol{\omega},t) = F(\boldsymbol{\omega})e^{-\alpha^2 t} = F(\boldsymbol{\omega})e^{-(\sqrt{t}\sigma\lVert\boldsymbol{\omega}\rVert)^2}$. By Table 4.1 with $a = \frac{1}{2\sigma\sqrt{t}}$, it is seen that

$\left(\sqrt{2t}\sigma\right)^n e^{-(\sigma^2 t)\|\omega\|^2} = \mathfrak{F}\left[e^{-\|x\|^2/(\sigma^2 t)}\right](\omega)$ or $e^{-(\sigma^2 t)\|\omega\|^2} = \mathfrak{F}\left[\frac{1}{\left(\sqrt{2t}\sigma\right)^n} e^{-\|x\|^2/(\sigma^2 t)}\right](\omega)$. Set $H_\sigma(x, t) = \frac{1}{\left(\sqrt{2t}\sigma\right)^n} e^{-\|x\|^2/(\sigma^2 t)}$. Then $e^{-(\sigma^2 t)\|\omega\|^2} = \mathfrak{F}\left[H_\sigma(x,t)\right](\omega) \equiv H_\sigma(\omega,t)$ so that $u(\omega,t) = f(\omega)H_\sigma(\omega,t) = \mathfrak{F}\left[f(x)\right](\omega)\cdot\mathfrak{F}\left[H_\sigma(x,t)\right](\omega)$. Now by the convolution property 4.1.6 of the Fourier transform, this means that $\mathfrak{F}\left[u(x,t)\right](\omega) \equiv U(\omega,t) = \mathfrak{F}\left[\left(f \odot H_\sigma\right)(x,t)\right](\omega)$. Applying the inverse Fourier transform simply cancels the Fourier transform leaving

$$\boxed{u(x, t) = \left(f \odot H_\sigma\right)(x,t) = \int_{\mathbb{R}^n} f(\xi)\cdot H_\sigma(x-\xi,t)\,d\xi} \,. \tag{4.3.2}$$

Equation (4.3.2) is the general solution of the heat equation (4.3.1) and the function

$$\boxed{H_\sigma(x, t) = \frac{1}{\left(\sqrt{2t}\sigma\right)^n} e^{-\|x\|^2/(\sigma^2 t)}} \tag{4.3.3}$$

is called the *heat kernel*. The heat kernel $H_\sigma(x, t)$ satisfies the heat equation $\frac{\partial H_\sigma}{\partial t} = \sigma^2 \Delta H_\sigma$ and has many other interesting properties. For a detailed description, see John [43]. Those familiar with probability theory will recognize the heat kernel as the multivariate Gaussian (Normal) distribution. In particular, when $n = 1$, $H_\sigma(x, t)$ is the normal probability density function with mean 0 and variance $2t\sigma^2$.

Now consider the *inhomogeneous heat equation* (4.3.4) with temperature forcing function $\rho(x, t)$.

$$\left.\begin{array}{ll} \frac{\partial u}{\partial t}(x,t) = \sigma^2\, \Delta u(x,t) + \rho(x,t), & x \in \mathbb{R}^n, t > 0 \\ u(x,0) = f(x), & x \in \mathbb{R}^n \end{array}\right\} \tag{4.3.4}$$

Applying the Fourier transform to (4.3.4), using a capital letter to indicate a transformed function, and proceeding as above, produces the *ODE* $\frac{dU}{dt} = -\alpha^2 U + P$ with $\alpha = \sigma\|\omega\|$. This *ODE* has the solution $U(t) = e^{-\alpha^2 t}U(0) + \int_0^t e^{-\alpha^2(t-s)}P(s)\,ds$. Recalling that $e^{-\alpha^2 t} = e^{-(\sigma^2 t)\|\omega\|^2} = \mathfrak{F}\left[H_\sigma(x,t)\right](\omega)$, the solution above yields

$$\mathfrak{F}[u(\boldsymbol{x},t)](\boldsymbol{\omega}) = \mathfrak{F}[H_\sigma(\boldsymbol{x},t)](\boldsymbol{\omega})\mathfrak{F}[f(\boldsymbol{x})](\boldsymbol{\omega}) + \int_0^t \mathfrak{F}[H_\sigma(\boldsymbol{x},t-s)](\boldsymbol{\omega})\mathfrak{F}[f(\boldsymbol{x})](\boldsymbol{\omega})\,ds$$
$$= \mathfrak{F}[(f \odot H_\sigma)(\boldsymbol{x},t)](\boldsymbol{\omega}) + \int_0^t \mathfrak{F}[(\rho \odot H_\sigma)(\boldsymbol{x},t-s)]\,ds$$
$$= \mathfrak{F}[(f \odot H_\sigma)(\boldsymbol{x},t)](\boldsymbol{\omega}) + \mathfrak{F}\left(\int_0^t [(\rho \odot H_\sigma)(\boldsymbol{x},t-s)]\,ds\right)$$
$$= \mathfrak{F}\left((f \odot H_\sigma)(\boldsymbol{x},t) + \int_0^t [(\rho \odot H_\sigma)(\boldsymbol{x},t-s)]\,ds\right).$$

Notice by Fubini's theorem, the Fourier transform and the integral from 0 to t can be interchanged. Therefore, by applying the inverse Fourier transform to this equation produces

$$u(\boldsymbol{x}, t) = (f \odot H_\sigma)(\boldsymbol{x},t) + \int_0^t [(\rho \odot H_\sigma)(\boldsymbol{x},t-s)]\,ds$$

or

$$\boxed{u(\boldsymbol{x}, t) = \int_{\mathbb{R}^n} f(\xi)\cdot H_\sigma(\boldsymbol{x}-\xi,t)\,d\xi + \int_0^t \int_{\mathbb{R}^n} \rho(\xi,s)\cdot H_\sigma(\boldsymbol{x}-\xi,t-s)\,d\xi\,ds} \tag{4.3.5}$$

which solves the inhomogeneous heat equation (4.3.4).

Example 4.3.1. Solve the inhomogeneous heat equation (4.3.4) in $n = 1$ dimension with the Cauchy data $f(x) = e^{-x^2}$ and the forcing function $\rho(x, t) = x\cdot\cos(t)$.

Procedure: By (4.3.5), the two integrals

$$F_\sigma(x,t) = \int_{-\infty}^{\infty} f(\xi)\cdot H_\sigma(x-\xi,t)\,d\xi \text{ and } R_\sigma(x, t) = \int_0^t \int_{-\infty}^{\infty} \rho(\xi,s)\cdot H_\sigma(x-\xi,t-s)\,d\xi\,ds$$

must be calculated. With considerable effort, or using the MATLAB supplemental Symbolic Math Toolbox, it can be determined that for $\sigma > 0$

$$F_\sigma(x,t) = \frac{\sqrt{2\pi}}{2\sqrt{\sigma^2 t+1}}\cdot\exp\left(-\frac{x^2}{\sigma^2 t+1}\right) \text{ while}$$

$$R_\sigma(x,t) = \begin{cases} \frac{\sqrt{2\pi}}{2}x\sin(t) & \text{for } x \geq 0 \\ -\frac{\sqrt{2\pi}}{2}x\sin(t) & \text{for } x > 0 \end{cases} = \frac{\sqrt{2\pi}}{2}|x|\sin(t)\text{, which is independent of } \sigma.$$

Hence, $u(x, t) = F_\sigma(x, t) + R_\sigma(x, t) = \frac{\sqrt{2\pi}}{2\sqrt{\sigma^2 t+1}}\cdot\exp\left(-\frac{x^2}{\sigma^2 t+1}\right) + \frac{\sqrt{2\pi}}{2}|x|\sin(t)$. Figure 4.0 illustrates the results for $\sigma = 2$.

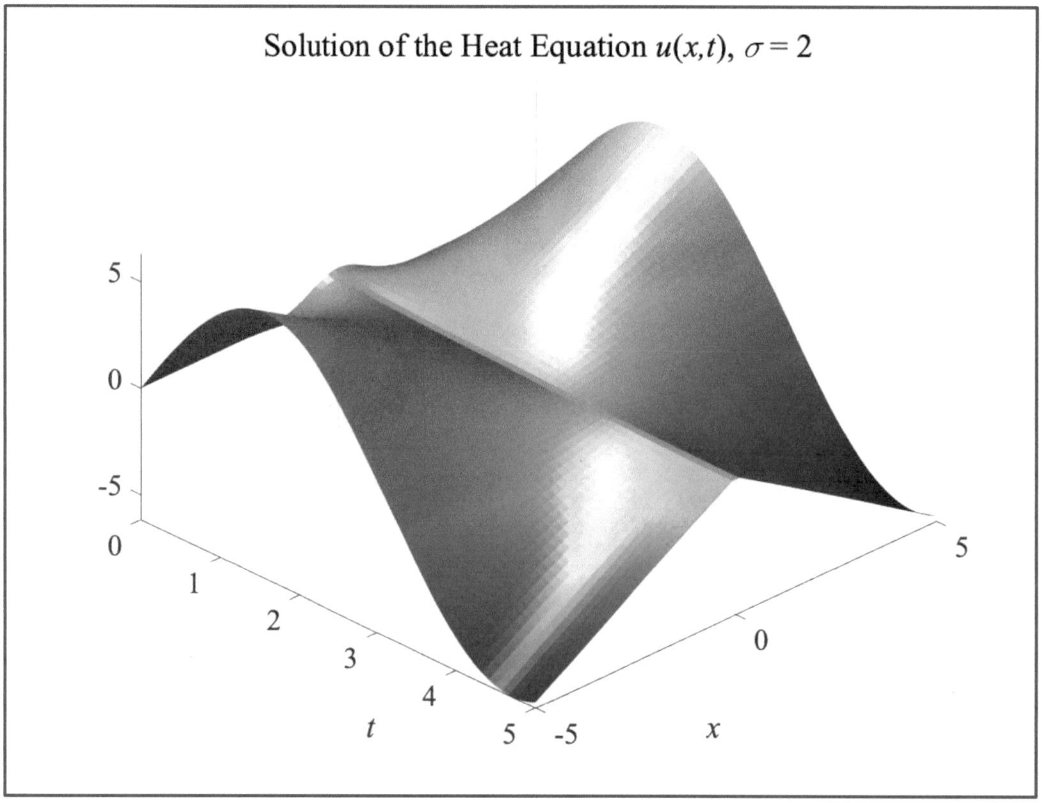

Figure 4.0. Solution of the inhomogeneous heat equation with dispersion constant $\sigma = 2$.

§4.4. The Fourier Transform and the Wave Equation

Thus far it has been demonstrated that Laplace's equation $\Delta u = 0$ can be solved via the classical method of separation of variables over finite domains and the modern method of Green's functions for arbitrary domains. Solutions of Poisson's equation $\Delta u = \rho$ are arrived at via Green's functions. The heat equation is solved over finite domains by separation of variables and by the Fourier transform over infinite domains. The same observation applies to the wave equation. Let $f \in C^2(\mathbb{R}^n)$ and $g \in C^1(\mathbb{R}^n)$ be the initial wave displacement and velocity, respectively, and $\sigma > 0$ be the wave speed. Then the *n–dimensional homogeneous wave equation* is written as (4.4.1).

$$\left.\begin{array}{ll} \dfrac{\partial^2 u}{\partial t^2}(\boldsymbol{x},t) = \sigma^2 \Delta u(\boldsymbol{x},t), & \text{for } \boldsymbol{x} \in \mathbb{R}^n, t > 0 \\ u(\boldsymbol{x},0) = f(\boldsymbol{x}), & \text{for } \boldsymbol{x} \in \mathbb{R}^n \\ \dfrac{\partial u}{\partial t}(\boldsymbol{x}) = g(\boldsymbol{x}), & \text{for } \boldsymbol{x} \in \mathbb{R}^n \end{array}\right\} \tag{4.4.1}$$

The success realized by applying the Fourier transform to the heat equation (4.3.1) indicates the approach to solving the wave equation. Applying the Fourier transform to the top equation in

(4.4.1) leads to the *ODE* $\frac{\partial^2}{\partial t^2}\mathfrak{F}[u(\boldsymbol{x},t)](\boldsymbol{\omega}) = -(\sigma\|\boldsymbol{\omega}\|)^2\,\mathfrak{F}[u(\boldsymbol{x},t)](\boldsymbol{\omega})$. Using the notation $U(t) = \mathfrak{F}[u(\boldsymbol{x},t)](\boldsymbol{\omega})$, the equation is merely the familiar second order ordinary differential equation $\frac{d^2}{dt^2}U(t) = -(\sigma\|\boldsymbol{\omega}\|)^2\,U(t)$ which has appeared in the separation of variables work on the finite domain heat and Laplace's equations. The solution of this *ODE* is $\mathfrak{F}[u(\boldsymbol{x},t)](\boldsymbol{\omega}) = U(t) = c_1\cos(\sigma\|\boldsymbol{\omega}\|t) + c_2\sin(\sigma\|\boldsymbol{\omega}\|t)$. The first initial condition implies that $\mathfrak{F}[f(\boldsymbol{x})](\boldsymbol{\omega}) = \mathfrak{F}[u(\boldsymbol{x},0)](\boldsymbol{\omega}) = c_1$. Now by property 4.1.7 of the Fourier transform,

$$\mathfrak{F}\left[\frac{\partial}{\partial t}u(\boldsymbol{x},t)\right](\boldsymbol{\omega}) = \frac{\partial}{\partial t}\mathfrak{F}[u(\boldsymbol{x},t)](\boldsymbol{\omega}) = \frac{\partial}{\partial t}\left[c_1\cos(\sigma\|\boldsymbol{\omega}\|t) + c_2\sin(\sigma\|\boldsymbol{\omega}\|t)\right]$$
$$= -\sigma\|\boldsymbol{\omega}\|\,c_1\sin(\sigma\|\boldsymbol{\omega}\|\,t) + \sigma\|\boldsymbol{\omega}\|\,c_2\cos(\sigma\|\boldsymbol{\omega}\|\,t).$$

Therefore, $\mathfrak{F}[g(\boldsymbol{x})](\boldsymbol{\omega}) = \mathfrak{F}\left[\frac{\partial}{\partial t}u(\boldsymbol{x},0)\right](\boldsymbol{\omega}) = \frac{\partial}{\partial t}\mathfrak{F}[u(\boldsymbol{x},0)](\boldsymbol{\omega}) = \sigma\|\boldsymbol{\omega}\|c_2$ or $c_2 = \frac{1}{\sigma\|\boldsymbol{\omega}\|}\mathfrak{F}[g(\boldsymbol{x})](\boldsymbol{\omega})$. Combining these calculations gives way to the equation

$$\mathfrak{F}[u(\boldsymbol{x},t)](\boldsymbol{\omega}) = \mathfrak{F}[f(\boldsymbol{x})](\boldsymbol{\omega})\cos(\sigma\|\boldsymbol{\omega}\|t) + \mathfrak{F}[g(\boldsymbol{x})](\boldsymbol{\omega})\frac{\sin(\sigma\|\boldsymbol{\omega}\|t)}{\sigma\|\boldsymbol{\omega}\|}. \qquad (4.4.2)$$

Now suppose there are functions $W_1(\boldsymbol{x}, t)$ and $W_2(\boldsymbol{x}, t)$ so that $\mathfrak{F}[W_1(\boldsymbol{x}, t)](\boldsymbol{\omega}) = \cos(\sigma\|\boldsymbol{\omega}\|\,t)$ and $\mathfrak{F}[W_2(\boldsymbol{x}, t)](\boldsymbol{\omega}) = \frac{\sin(\sigma\|\boldsymbol{\omega}\|t)}{\sigma\|\boldsymbol{\omega}\|}$. That is, $W_1(\boldsymbol{x}, t) = \mathfrak{F}^{-1}[\cos(\sigma\|\boldsymbol{\omega}\|\,t)](\boldsymbol{x})$ and $W_2(\boldsymbol{x}, t) = \mathfrak{F}^{-1}\left[\frac{\sin(\sigma\|\boldsymbol{\omega}\|t)}{\sigma\|\boldsymbol{\omega}\|}\right](\boldsymbol{x})$. Notice that $\frac{\partial}{\partial t}\left(\frac{\sin(\sigma\|\boldsymbol{\omega}\|t)}{\sigma\|\boldsymbol{\omega}\|}\right) = \cos(\sigma\|\boldsymbol{\omega}\|t)$ so that by property 4.1.7 of the Fourier transform, $\frac{\partial}{\partial t}\mathfrak{F}[W_2(\boldsymbol{x},t)](\boldsymbol{\omega}) = \mathfrak{F}\left[\frac{\partial}{\partial t}W_2(\boldsymbol{x},t)\right](\boldsymbol{\omega}) = \mathfrak{F}[W_1(\boldsymbol{x},t)](\boldsymbol{\omega})$ or $\frac{\partial}{\partial t}W_2(\boldsymbol{x},t) = W_1(\boldsymbol{x}, t)$. With these considerations in mind, (4.4.2) can now be written as

$$\mathfrak{F}[u(\boldsymbol{x},t)](\boldsymbol{\omega}) = \mathfrak{F}[f(\boldsymbol{x})](\boldsymbol{\omega})\cdot\mathfrak{F}[W_1(\boldsymbol{x},t)](\boldsymbol{\omega}) + \mathfrak{F}[g(\boldsymbol{x})](\boldsymbol{\omega})\cdot\mathfrak{F}[W_2(\boldsymbol{x},t)](\boldsymbol{\omega})$$

or by the convolution property 4.1.6

$$\mathfrak{F}[u(\boldsymbol{x},t)](\boldsymbol{\omega}) = \mathfrak{F}[(f \odot W_1)(\boldsymbol{x},t)](\boldsymbol{\omega}) + \mathfrak{F}[(g \odot W_2)(\boldsymbol{x},t)](\boldsymbol{\omega}).$$

That is, $u(\boldsymbol{x},t)=\frac{1}{(\sqrt{2\pi})^n}\left[(f\odot W_1)(\boldsymbol{x},t)+(g\odot W_2)(\boldsymbol{x},t)\right]$ with $\frac{\partial}{\partial t}W_2(\boldsymbol{x},t)=W_1(\boldsymbol{x},t)$.

Equivalently,

$$\begin{aligned}
&u(\boldsymbol{x},t)=\frac{1}{(\sqrt{2\pi})^n}\int_{\mathbb{R}^n}\left[f(\xi)\cdot W_1(\boldsymbol{x}-\xi,t)+g(\xi)\cdot W_2(\boldsymbol{x}-\xi,t)\right]d\xi\\
&\frac{\partial}{\partial t}W_2(\boldsymbol{x},t)=W_1(\boldsymbol{x},t)\\
&W_2(\boldsymbol{x},t)=\mathfrak{F}^{-1}\left[\frac{\sin(\sigma\|\boldsymbol{\omega}\|t)}{\sigma\|\boldsymbol{\omega}\|}\right](\boldsymbol{x})
\end{aligned}\tag{4.4.3}$$

is the general solution of the n–dimensional wave equation (4.4.1). Naturally, the question to ask is: Can $\mathfrak{F}^{-1}\left[\frac{\sin(\sigma\|\boldsymbol{\omega}\|t)}{\sigma\|\boldsymbol{\omega}\|}\right](\boldsymbol{x})$ be computed in a closed analytic form? The case of $n\geq 2$ is deferred to Chapter 5. The case of $n=1$ yields a result with a special form known as *d'Alembert's solution*.

§4.5. d'Alembert's Solution of the Wave Equation

Consider the $n=1$ dimensional version of the wave equation (4.4.1). That is,

$$\left.\begin{aligned}
&\frac{\partial^2 u}{\partial t^2}(x,t)=\sigma^2\frac{\partial^2 u}{\partial x^2}(x,t), && \text{for } x\in\mathbb{R},\ t>0\\
&u(x,0)=f(x), && \text{for } x\in\mathbb{R}\\
&\frac{\partial u}{\partial t}(x)=g(x), && \text{for } x\in\mathbb{R}
\end{aligned}\right\}\tag{4.5.1}$$

for $f\in C^2(\mathbb{R})$ and $g\in C^1(\mathbb{R})$. Then the solution (4.4.3) still applies but the frequency vector $\boldsymbol{\omega}$ is now merely the one–dimensional variable ω and the norm or length of the vector $\boldsymbol{\omega}$ is $\|\boldsymbol{\omega}\| = |\omega|$. Consider the function $\frac{\sin(\sigma\|\boldsymbol{\omega}\|t)}{\sigma\|\boldsymbol{\omega}\|}=\frac{\sin(\sigma|\omega|t)}{\sigma|\omega|}$ from (4.4.3). If $|\omega|=-\omega$, then $\frac{\sin(\sigma|\omega|t)}{\sigma|\omega|}=\frac{\sin(-\sigma\omega t)}{-\sigma\omega}=\frac{\sin(\sigma\omega t)}{\sigma\omega}$ since *sine* is an odd function. Similarly, $|\omega|=\omega$ means that $\frac{\sin(\sigma|\omega|t)}{\sigma|\omega|}=\frac{\sin(\sigma\omega t)}{\sigma\omega}$. In either case, it must be that $W_2(x,t)=\mathfrak{F}^{-1}\left[\frac{\sin(\sigma\omega t)}{\sigma\omega}\right](x)=t\cdot\mathfrak{F}^{-1}\left[\frac{\sin(\sigma t\omega)}{\sigma t\omega}\right](x)$.

By Table 4.1, $t\cdot\mathfrak{F}^{-1}\left[\frac{\sin(\sigma t\omega)}{\sigma t\omega}\right](x)=t\cdot\sqrt{\frac{\pi}{2}}\frac{1}{\sigma t}\mathfrak{I}_{[-\sigma t,\sigma t]}(x)=\sqrt{\frac{\pi}{2}}\frac{1}{\sigma}\mathfrak{I}_{[-\sigma t,\sigma t]}(x)=W_2(x,t)$. It is seen further that $\mathfrak{F}\left[\boldsymbol{\delta}(\boldsymbol{x}-\boldsymbol{x}_o)+\boldsymbol{\delta}(\boldsymbol{x}+\boldsymbol{x}_o)\right](\boldsymbol{\omega}) = \frac{1}{\left(\sqrt{2\pi}\right)^n}\left[e^{i(\boldsymbol{x}_o\bullet\boldsymbol{\omega})}+e^{-i(\boldsymbol{x}_o\bullet\boldsymbol{\omega})}\right] = \frac{2}{\left(\sqrt{2\pi}\right)^n}\cdot\frac{1}{2}\left[e^{i(\boldsymbol{x}_o\bullet\boldsymbol{\omega})}+e^{-i(\boldsymbol{x}_o\bullet\boldsymbol{\omega})}\right]$

$= \frac{2}{\left(\sqrt{2\pi}\right)^n}\cdot\cos(\boldsymbol{x}_o\bullet\boldsymbol{\omega})$. For $n = 1$, $\frac{2}{\left(\sqrt{2\pi}\right)} = \sqrt{\frac{2}{\pi}}$ and $\boldsymbol{x}_o\bullet\boldsymbol{\omega} = x_o\cdot\omega$ and therefore $\cos(\sigma t\omega) = \sqrt{\frac{\pi}{2}}\,\mathfrak{F}[\delta(x-\sigma t)$ $+\,\delta(x+\sigma t)](\omega)$ or $W_1(x, t) = \sqrt{\frac{\pi}{2}}\,[\delta(x-\sigma t)+\delta(x+\sigma t)]$. Returning to (4.4.3) produces the result

$$u(x,t) = \frac{1}{\sqrt{2\pi}}\sqrt{\frac{\pi}{2}}\int_{-\infty}^{\infty} f(\xi)\left[\delta(x-\sigma t-\xi)+\delta(x+\sigma t-\xi)\right]d\xi$$

$$+\frac{1}{\sqrt{2\pi}}\sqrt{\frac{\pi}{2}}\int_{-\infty}^{\infty} g(\xi)\cdot\mathcal{S}_{[-\sigma t,\sigma t]}(x-\xi)\,d\xi$$

$$= \frac{1}{2}\left[f(x-\sigma t)+f(x+\sigma t)+\frac{1}{\sigma}\int_{x-\sigma t}^{x+\sigma t} g(\xi)\,d\xi\right].$$

By defining the function $G(x) = \frac{1}{\sigma}\int_0^x g(\xi)\,d\xi$, then the solution of the one–dimensional wave equation can be written as *d'Alembert's traveling wave solution* (4.5.2).

$$u(x,t) = \frac{1}{2}\left[f(x+\sigma t)+f(x-\sigma t)+G(x+\sigma t)-G(x-\sigma t)\right] \tag{4.5.2}$$

Equation (4.5.2) can be interpreted geometrically as the ripple effect that dropping a stone into still water produces. The initial wave displacement f is averaged across the two perpendicular rays $x - \sigma t = 0$ and $x + \sigma t = 0$. The wave then moves away from the origin of the displacement by the difference in the initial velocity integral G. This geometry is illustrated in Figure 4.1 with $f(x) = e^{-x^2}$ and $g(x) = -\tanh(x)\cdot\text{sech}(x)$. In this case, $G(x) = \text{sech}(x)$ and *d'Alembert's* solution is $u(x, t)$ $= ½\left[e^{-(x+\sigma t)^2}+e^{-(x-\sigma t)^2}+\frac{1}{\sigma}\left(\text{sech}(x+\sigma t)-\text{sech}(x-\sigma t)\right)\right]$.

From Figure 4.1, it is seen that the wave front separates and moves away from each side with increasing distance as time progresses. Temporal cross-sections of the wave are illustrated in Figure 4.2. The arrows indicate the direction of the wave front and show that the wave is dispersing.

Naturally, different Cauchy data $f(x)$ and initial wave velocity $g(x)$ will generate different solution surfaces. The exercises suggest additional initial conditions which may provide insight into wave behavior.

Next, the question arises as to what happens when an additional force is introduced to the model. That is, how is d'Alembert's solution altered if a forcing function ρ is added to the one–dimensional inhomogeneous wave equation resulting in (4.5.3)?

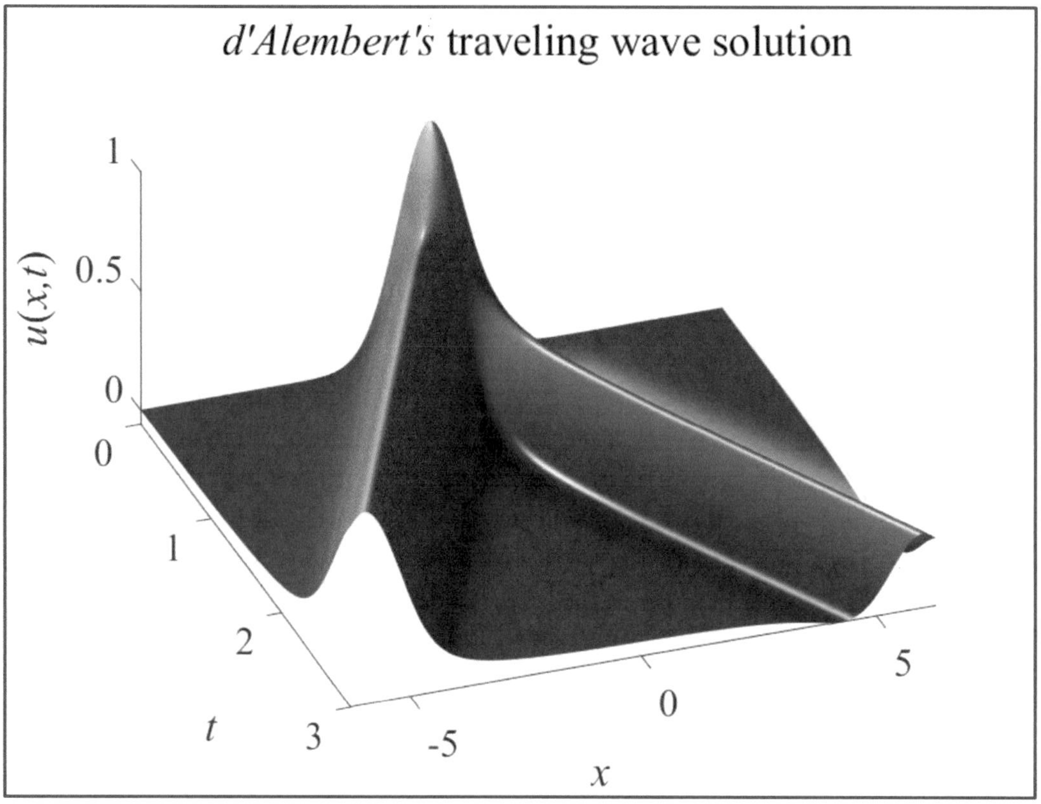

Figure 4.1. *d'Alembert's* solution for $f(x) = e^{-x^2}$, $g(x) = -\tanh(x)\cdot\text{sech}(x)$, and $\sigma = 2$.

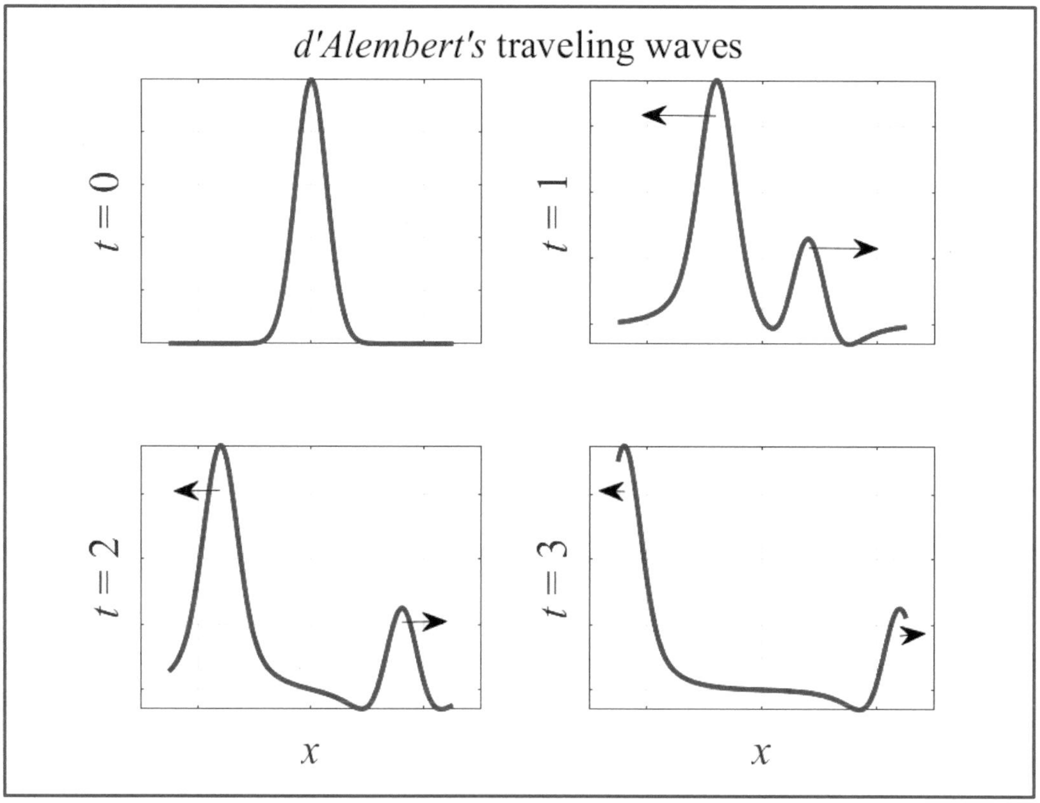

Figure 4.2. Traveling waves dispersing.

$$\left.\begin{aligned} &\frac{\partial^2 u}{\partial t^2}(x,t) = \sigma^2 \frac{\partial^2 u}{\partial x^2}(x,t) + \rho(x,t), && \text{for } x \in \mathbb{R},\, t > 0 \\ &u(x,0) = f(x), && \text{for } x \in \mathbb{R} \\ &\frac{\partial u}{\partial t}(x) = g(x), && \text{for } x \in \mathbb{R} \end{aligned}\right\} \tag{4.5.3}$$

As with the homogeneous wave equation, using the Fourier transform is the proposed method of solution. Applying $\mathfrak{F}$ to the top equation in (4.5.3) produces

$$\frac{\partial^2}{\partial t^2}\mathfrak{F}\left[u(x,t)\right](\omega) = -(\sigma\omega)^2\mathfrak{F}\left[u(x,t)\right](\omega) + \mathfrak{F}\left[\rho(x,t)\right](\omega). \tag{4.5.4}$$

It is well known from methods in ordinary differential equations that solutions of the inhomogeneous second order linear equation $\frac{d^2 y}{dt^2} + p(t)\frac{dy}{dt} + q(t)y = r(t)$ take the form $y(t) = y_p(t) + y_c(t)$. In this setting, $y_p(t)$ is referred to as the *particular solution* while $y_c(t)$ is the *complementary solution*. See, for example, Boyce and DiPrima [12]. If $y_1(t)$ and $y_2(t)$ are the linearly independent solutions of the homogeneous equation $\frac{d^2 y}{dt^2} + p(t)\frac{dy}{dt} + q(t)y = 0$, then the particular solution of the inhomogeneous second order linear *ODE* is $y_p(t) = \int_{t_o}^{t} \frac{y_1(\tau)y_2(t) - y_1(t)y_2(\tau)}{y_1(\tau)y_2'(\tau) - y_1'(\tau)y_2(\tau)} r(\tau)\,d\tau$. The complementary solution, conversely, is a linear combination of the linearly independent solutions $y_c(t) = c_1 \cdot y_1(t) + c_2 \cdot y_2(t)$. Casting (4.5.4) as in inhomogeneous *ODE* with $y(t) = \mathfrak{F}[u(x, t)](\omega)$, $p(t) = 0$, $q(t) = -(\sigma\omega)^2$, and $r(t) = \mathfrak{F}[\rho(x, t)](\omega)$, it is seen that the homogeneous *ODE* associated with this equation is $\frac{d^2 y}{dt^2} - (\sigma\omega)^2 y = 0$. The independent solutions of this equation are $y_1(t) = \cos(\sigma\omega t)$ and $y_2(t) = \sin(\sigma\omega t)$ so that the complementary solution is $y_c(t) = c_1 \cdot \cos(\sigma\omega t) + c_2 \cdot \sin(\sigma\omega t)$. Moreover, $y_1(\tau)\, y_2(t) - y_1(t)\, y_2(\tau) = \cos(\sigma\omega\tau)\, \sin(\sigma\omega t) - \cos(\sigma\omega t)\, \sin(\sigma\omega\tau) = \sin(\sigma\omega[t-\tau])$, while $y_1(\tau)\cdot y_2'(\tau) - y_1'(\tau)\cdot y_2(\tau) = \sigma\omega\cdot\cos^2(\sigma\omega\tau) - (-\sigma\omega\cdot\sin^2(\sigma\omega\tau)) = \sigma\omega$. Thus,

$$y_p(t) = \int_{t_o}^{t} \frac{\sin(\sigma\omega[t-\tau])}{\sigma\omega} r(\tau)\,d\tau = \int_{t_o}^{t} \frac{\sin(\sigma\omega[t-\tau])}{\sigma\omega}\mathfrak{F}\left[\rho(x,\tau)\right](\omega)\,d\tau.$$

Since $t_o = 0$, it is seen that the complementary solution of (4.5.4) is $\mathfrak{F}[u(x, t)](\omega) = y_c(t) = c_1 \cdot \cos(\sigma\omega t) + c_2 \cdot \sin(\sigma\omega t)$ which implies that $\mathfrak{F}[f(x)](\omega) = \mathfrak{F}[u(x, 0)](\omega) = c_1$. Now $\mathfrak{F}[g(x)](\omega) = \mathfrak{F}\left[\frac{\partial u}{\partial t}(x,0)\right](\omega) = \frac{\partial}{\partial t}\mathfrak{F}\left[u(x,t)\right](\omega)\Big|_{t=0} = \frac{\partial}{\partial t} y_c(0) = \sigma\omega\cdot c_2 \cos(\sigma\omega t)\,|_{t=0} = \sigma\omega\cdot c_2$ or

$c_2 = \dfrac{1}{\sigma\omega}\,\mathfrak{F}[g(x)](\omega)$. Thus, the solution of (4.5.4) is $y(t) = y_c(t) + y_p(t) \Rightarrow \mathfrak{F}[u(x,t)](\omega) = c_1 \cdot y_1(t) + c_2 \cdot y_2(t) + y_p(t) = \mathfrak{F}[f(x)](\omega)\cdot\cos(\sigma\omega t) + \mathfrak{F}[g(x)](\omega)\cdot\dfrac{\sin(\sigma\omega t)}{\sigma\omega} + \displaystyle\int_{t_o}^{t}\dfrac{\sin(\sigma\omega[t-\tau])}{\sigma\omega}\mathfrak{F}\left[\rho(x,\tau)\right](\omega)\,d\tau$

$$= \sqrt{\tfrac{\pi}{2}}\mathfrak{F}\left[f \odot \left(\delta(x+\sigma t) + \delta(x-\sigma t)\right)\right](\omega) + \tfrac{1}{\sigma}\sqrt{\tfrac{\pi}{2}}\mathfrak{F}\left[g \odot \mathfrak{I}_{[-\sigma t,\sigma t]}(x)\right](\omega)$$

$$+ \tfrac{1}{\sigma}\sqrt{\tfrac{\pi}{2}}\int_0^t \mathfrak{F}\left[\rho \odot \mathfrak{I}_{[-\sigma[t-\tau],\sigma[t-\tau]]}(x)\right](\omega).$$

As the last integral from $[0, t]$ is taken over a compact interval, the inverse Fourier transform can be applied inside of $\int_0^t \mathfrak{F}\left[\rho \odot \mathfrak{I}_{[-\sigma[t-\tau],\sigma[t-\tau]]}(x)\right](\omega)$ so that the solution of the inhomogeneous wave equation (4.5.3) is given by (4.5.5).

$$\boxed{\begin{aligned} u(x,t) &= \frac{1}{2}\Big[f(x+\sigma t) + f(x-\sigma t) + G(x+\sigma t) - G(x-\sigma t) \\ &\quad + \int_0^t \left(R(x+\sigma[t-\tau],\tau) - R(x-\sigma[t-\tau],\tau)\right)d\tau\Big] \\ G(x) &= \frac{1}{\sigma}\int_0^x g(\xi)\,d\xi, \quad R(x,\tau) = \frac{1}{\sigma}\int_0^x \rho(\xi,\tau)\,d\xi \end{aligned}} \tag{4.5.5}$$

The equation (4.5.5) indicates that the solution of the one–dimensional inhomogeneous wave equation is a traveling wave.

Before discussing the general n–dimensional solution of the inhomogeneous wave equation and hence a solution of Maxwell's equations, a return to the classical methods and problems in wave propagation is required. The classical solutions of the vibrating string (piano, violin, guitar) and the vibrating membrane (drums) problems are now explored.

§4.6. Examples of the Wave Equation: Finite Domains

As mentioned at the end of §4.5, there are (at least) two important physical examples of the wave equation restricted to finite domains: the string equation and the vibrating drum.

Example 4.6.1. *The vibrating string.* The physical setting is that a metal (or nylon) wire is stretched taut over a finite length. It is assumed, at either end of the string length, there are no vibrations. The string is then "plucked" or excited with force $f(x)$ at time 0 with a particular velocity $g(x)$. The vibration of the string is modeled by the one (spatial) dimensional wave equation over a length L as equation (4.6.1).

$$\left.\begin{array}{ll} \frac{\partial^2 u}{\partial t^2}(x,t) = \sigma^2 \frac{\partial^2 u}{\partial x^2}(x,t) & \text{for } x \in [0,L] \text{ and } t > 0 \\ u(x,0) = f(x) & \text{for } x \in [0,L] \\ \frac{\partial u}{\partial x}(x,0) = g(x) & \text{for } x \in [0,L] \\ u(0,t) = 0 = u(L,t) & \text{for } t > 0 \end{array}\right\} \quad (4.6.1)$$

Solution: The approach will be to solve (4.6.1) for the case of general initial string displacement $f(x)$ and velocity $g(x)$. The constant σ is known as the *wave speed* or velocity of propagation. As indicated above, the usual classical method of separation of variables is utilized. Consequently, the assumption that $u(x, t) = X(x)\cdot T(t)$ results in $\frac{X''(x)}{X(x)} = \frac{T''(t)}{\sigma^2 \cdot T(t)} \equiv \lambda$ (constant). The boundary conditions $u(0, t) = 0 = u(L, t)$ mean that $0 = X(0)\cdot T(t)$ and $0 = X(L)\cdot T(t)$. The only way to avoid the trivial solution $u(x, t) \equiv 0$ is to have $X(0) = 0 = X(L)$. This produces the *ODE* $X''(x) - \lambda X(x) = 0$ and $X(0) = 0 = X(L)$. This is just (*ii*) of Example 3.5.1 which has eigenvalues $\lambda_k = -(\pi k/L)^2$ and solutions $X_k(x) = \sin\left(\frac{\pi k}{L} x\right)$. The second *ODE* $T''(t) = \lambda\sigma^2 T(t) \Rightarrow T''(t) = \left(\frac{\pi k\sigma}{L}\right)^2 T(t)$ or $T_k(t) = A_k\cdot \sin\left(\frac{\pi k\sigma}{L} t\right) + B_k\cdot \cos\left(\frac{\pi k\sigma}{L} t\right)$. This means that $u_k(x, t) = \sin\left(\frac{\pi k}{L} x\right) \cdot\left[A_k\cdot \sin\left(\frac{\pi k\sigma}{L} t\right) + B_k\cdot \cos\left(\frac{\pi k\sigma}{L} t\right)\right]$ form a sequence of solutions to the top equation in (4.6.1). By the superposition principle,

$$u(x, t) = \sum_{k=1}^{\infty} \sin\left(\tfrac{\pi k}{L} x\right)\cdot\left[A_k \cdot \sin\left(\tfrac{\pi k\sigma}{L} t\right) + B_k \cdot \cos\left(\tfrac{\pi k\sigma}{L} t\right)\right]$$

is a solution of the one–dimensional, finite domain wave equation. The initial condition $f(x) = u(x, 0)$ implies that $f(x) = \sum_{k=1}^{\infty} B_k \cdot \sin\left(\frac{\pi k}{L} x\right)$ while $g(x) = \frac{\partial u}{\partial x}(x,0) = \sum_{k=1}^{\infty} A_k \cdot \frac{\pi k\sigma}{L} \cdot \sin\left(\frac{\pi k}{L} x\right)$. Using equation (3.1.1ℓ) on Fourier series, it must be the case that $B_k = \frac{2}{L}\int_0^L f(x)\sin\left(\frac{\pi k}{L} x\right) dx$ for $k \in \mathbb{Z}^+$ as ℓ is replaced by L/σ. Similarly, $A_k = \frac{L}{\pi k\sigma}\frac{2}{L}\int_0^L g(x)\sin\left(\frac{\pi k}{L} x\right) dx = \frac{2}{\pi k\sigma}\int_0^L g(x)\sin\left(\frac{\pi k}{L} x\right) dx$ for $k \in \mathbb{Z}^+$. Therefore, (4.6.2) solves (4.6.1).

$$\left.\begin{array}{l} u(x,t) = \sum_{k=1}^{\infty} u_k(x,t) \\ u_k(x,t) = \sin\left(\frac{\pi k}{L} x\right)\cdot\left[A_k \cdot \sin\left(\frac{\pi k\sigma}{L} t\right) + B_k \cdot \cos\left(\frac{\pi k\sigma}{L} t\right)\right] \\ A_k = \frac{2}{\pi k\sigma}\int_0^L g(x)\sin\left(\frac{\pi k}{L} x\right) dx \text{ and } B_k = \frac{2}{L}\int_0^L f(x)\sin\left(\frac{\pi k}{L} x\right) dx \end{array}\right\} \quad (4.6.2)$$

The component solutions $u_k(x, t)$ are known as the kth *harmonics* or modes of vibration of the string.

Special case: Consider plucking a guitar string directly above the sound hole of the instrument. That is, displace the string by a height h approximately $1/5^{th}$ of the string length from the bridge (the origin of the string). Assume zero initial velocity. In this special case, $g(x) = 0$ while $f(x) = \begin{cases} \frac{5h}{L}x & \text{for } 0 \le x \le \frac{1}{5}L \\ \frac{5h}{4L}(L-x) & \text{for } \frac{1}{5}L < x \le L \end{cases}$, $h << L$. Figure 4.3 illustrates this general notion of plucking a string at a height considerably smaller than its overall length.

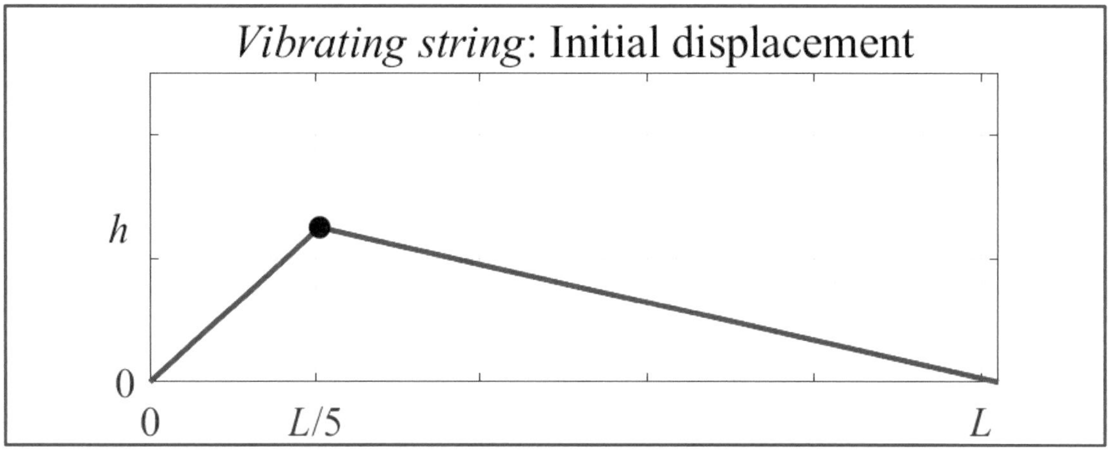

Figure 4.3. The initial displacement of the "plucked" string.

More specifically, suppose the string length $L = 25.5''$ and the string is raised $0.25''$ at $5.1''$ from the base of the instrument (i.e., over the approximate center of the sound hole). The propagation speed σ of the string is a function of the string tension τ along with the string density ρ: $\sigma = \sqrt{\tau/\rho}$. According to the D'Addario sponsored site [21], an "A" string on a guitar with a $0.042''$ diameter has a density of 0.00036722 *lb/in* and a frequency $\omega = 110$ *Hz*. Therefore, its tension is $T = [\rho(2L\omega)^2]/c_o$ where c_o is the non–dimensional constant 386.4. This means that $T = 29.91$ *lb-in/sec*^{2} and the propagation speed is $\sigma = 285.3936$ *in/sec*. Zachmanoglou and Thoe [87] provide a more detailed discussion of the string propagation speed. With these assumptions in mind, it is immediate from (4.6.2) that $A_k = 0$ for all k as $g(x) \equiv 0$. A bit of work can show that $B_k = \frac{10h}{(\pi k)^2}\left(5\sin^3\left(\frac{\pi k}{5}\right) - 4\sin^5\left(\frac{\pi k}{5}\right)\right)$. Using the trigonometric identity $\sin(a)\cdot\cos(b) = ½\,(\sin(a-b) + \sin(a+b))$ in (4.6.2) produces the harmonics $\boxed{u_k(x, t) = ½\, B_k\cdot\left(\sin\left(\frac{\pi k}{L}(x-\sigma t)\right) + \sin\left(\frac{\pi k}{L}(x+\sigma t)\right)\right)}$ which are traveling wave solutions. Figure 4.4 illustrates the solution surface for the sum of the first nine harmonics $u(x, t) \approx \sum_{k=1}^{9} u_k(x,t)$, while Figure 4.5 depicts the string wave dispersion as time progresses over the interval [0, 1]. Note that the wave moves from left ($x = 0$) to right ($x = L = 25.5''$).

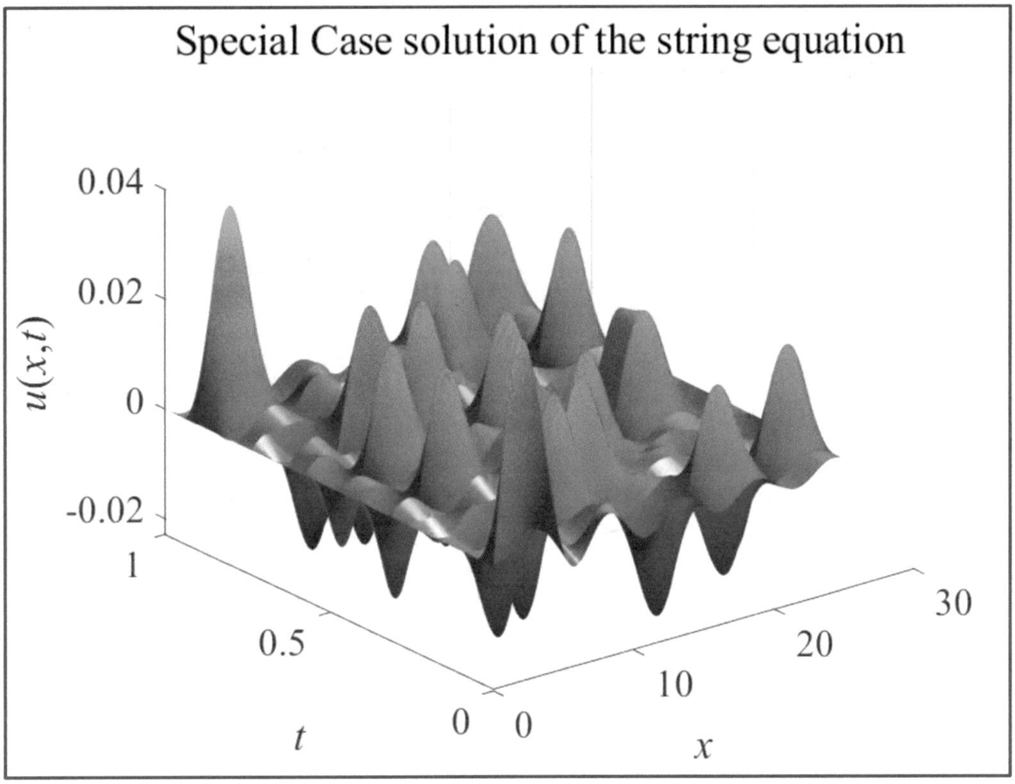

Figure 4.4. The plucked guitar string with initial displacement in Figure 4.3.

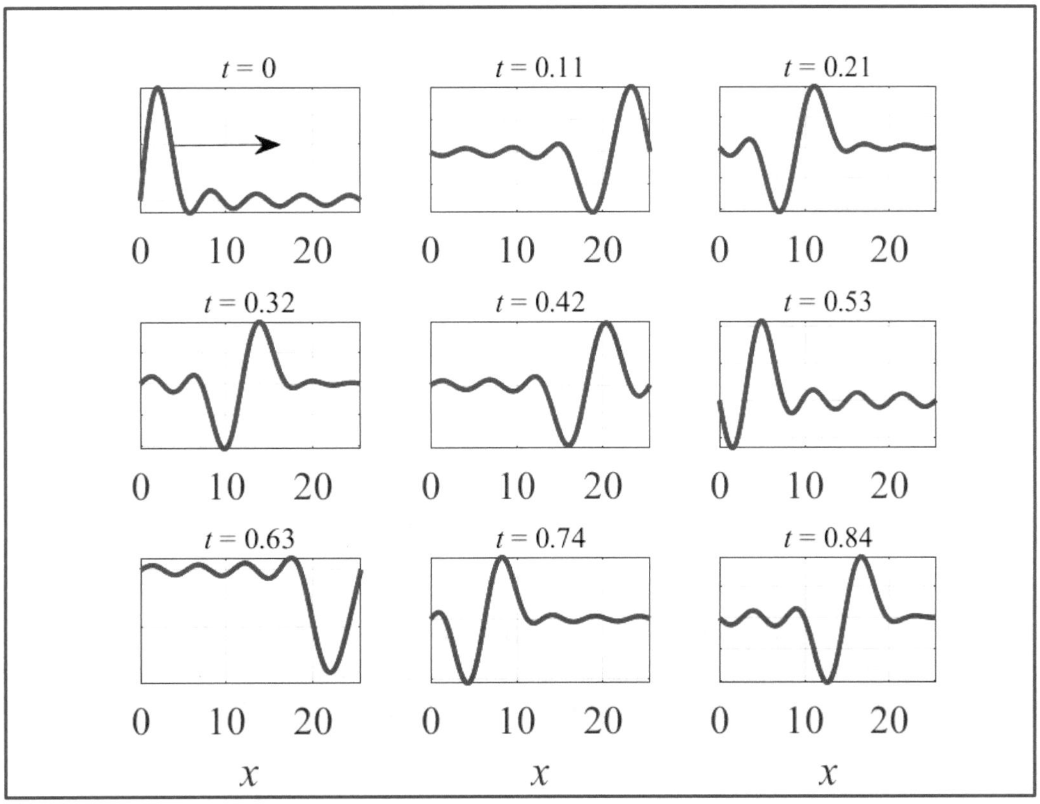

Figure 4.5. Plucked string as a traveling wave (first nine harmonics).

Example 4.6.2. *The vibrating membrane.* Rather than being restricted to a single spatial dimension, consider the wave equation in two spatial dimensions $\boldsymbol{x} = (x, y)$ over a rectangle of length L_1 by L_2.

$$\left.\begin{array}{ll}
\dfrac{\partial^2 u}{\partial t^2}(\boldsymbol{x},t) = \sigma^2\left(\dfrac{\partial^2 u}{\partial x^2}(\boldsymbol{x},t) + \dfrac{\partial^2 u}{\partial y^2}(\boldsymbol{x},t)\right) & \text{for } x \in [0, L_1], y \in [0, L_2], \text{ and } t > 0 \\
u(\boldsymbol{x},0) = f(\boldsymbol{x}) & \text{for } x \in \Omega = [0, L_1] \times [0, L_2] \\
\dfrac{\partial u}{\partial x}(\boldsymbol{x},0) = g(\boldsymbol{x}) & \text{for } x \in \Omega = [0, L_1] \times [0, L_2] \\
u(0, y, t) = 0 = u(L_1, y, t) & \text{for } t > 0 \\
u(x, 0, t) = 0 = u(x, L_2, t) & \text{for } t > 0
\end{array}\right\} \quad (4.6.3)$$

Solution: It is a straightforward exercise in the method of separation of variables to determine that the solution modes are

$$\boxed{u_{nm}(\boldsymbol{x},t) = \left[a_{nm}\cos\left(\sigma\sqrt{|\lambda_{nm}|}\,t\right) + b_{nm}\sin\left(\sigma\sqrt{|\lambda_{nm}|}\,t\right)\right]\sin\left(\tfrac{n\pi}{L_1}x\right)\sin\left(\tfrac{m\pi}{L_2}y\right)}$$

with associated eigenvalues $\lambda_{nm} = -\left[\left(\dfrac{n\pi}{L_1}\right)^2 + \left(\dfrac{m\pi}{L_2}\right)^2\right]$. By the superposition principle, $u(\boldsymbol{x}, t) = \sum_{m=1}^{\infty}\sum_{n=1}^{\infty} u_{nm}(\boldsymbol{x},t)$ satisfies the top equation of (4.6.3). The initial condition $f(\boldsymbol{x}) = u(\boldsymbol{x}, 0) = \sum_{m=1}^{\infty}\sum_{n=1}^{\infty} a_{nm}\sin\left(\tfrac{n\pi}{L_1}x\right)\sin\left(\tfrac{n\pi}{L_2}y\right)$, and the Fourier formulae of (3.2.2ℓ) yields

$$a_{mn} = \tfrac{4}{L_1L_2}\int_0^{L_2}\int_0^{L_1} f(x,y)\sin\left(\tfrac{n\pi}{L_1}x\right)\sin\left(\tfrac{n\pi}{L_2}y\right)dx\,dy\text{, while}$$

$$g(\boldsymbol{x}) = \frac{\partial u}{\partial t}(x,0) = \sum_{m=1}^{\infty}\sum_{n=1}^{\infty}\sigma b_{nm}\sqrt{|\lambda_{nm}|}\sin\left(\tfrac{n\pi}{L_1}x\right)\sin\left(\tfrac{n\pi}{L_2}y\right) \text{ implies that}$$

$$b_{mn} = \frac{4}{\sigma\sqrt{|\lambda_{nm}|}L_1L_2}\int_0^{L_2}\int_0^{L_1} g(x,y)\sin\left(\tfrac{n\pi}{L_1}x\right)\sin\left(\tfrac{n\pi}{L_2}y\right)dx\,dy. \qquad \blacksquare$$

Example 4.6.3. *The circular drum.* No discussion of vibrating membranes is realistic without an examination of the circular drum. In this scenario, a flexible skin is stretched across a rigid circular frame of radius r_o and clamped across the boundary. As with the heat equation on the cylinder

(Example 3.5.4), the polar coordinate version of the Laplacian is applied and $\Delta u = \frac{1}{r}\frac{\partial}{\partial r}\left(r\frac{\partial u}{\partial r}\right) + \frac{1}{r^2}\frac{\partial^2 u}{\partial \theta^2}$. The wave equation is now written as

$$\left.\begin{array}{l} \frac{\partial^2 u}{\partial t^2}(r,\theta,t) = \sigma^2\left[\frac{1}{r}\frac{\partial}{\partial r}\left(r\frac{\partial u}{\partial r}(r,\theta,t)\right) + \frac{1}{r^2}\frac{\partial^2 u}{\partial \theta^2}(r,\theta,t)\right] \\ u(r,\theta,0) = \phi(r,\theta) \\ \frac{\partial u}{\partial r}(r,\theta,0) = \psi(r,\theta) \\ u(r_o,\theta,t) = 0 \text{ and } u(r,\theta,t) = u(r,\theta+2\pi,t) \quad \text{for all } r,t \\ 0 \le r \le r_o, -\pi \le \theta \le \pi, t > 0 \end{array}\right\}. \tag{4.6.4}$$

Solution: The separation of variables substitution $u(r, \theta, t) = R(r)\cdot\Theta(\theta)\cdot T(t)$ and the circular d'Alembertian operator $\square_o u = \frac{\partial^2 u}{\partial t^2} - \sigma^2\left(\frac{1}{r}\frac{\partial}{\partial r}\left(r\frac{\partial u}{\partial r}\right) - \frac{1}{r^2}\frac{\partial^2 u}{\partial \theta^2}\right)$ combine to produce

$$0 = \square_o u = \tfrac{1}{\sigma^2}T''(t)\cdot\Theta(\theta)\cdot R(r) - T(t)\cdot\Theta(\theta)\cdot R''(r) - \frac{1}{r}T(t)\cdot\Theta(\theta)\cdot R'(r) - \frac{1}{r^2}T(t)\cdot\Theta''(\theta)\cdot R(r).$$

Dividing through by the non–trivial product $R(r)\cdot\Theta(\theta)\cdot T(t)$ and simplifying yields the differential equation

$$\frac{T''(t)}{\sigma^2 T(t)} = \frac{R''(r)}{R(r)} + \frac{R'(r)}{r\cdot R(r)} + \frac{\Theta''(\theta)}{r^2\cdot\Theta(\theta)} \equiv C_o.$$

For those reading this text linearly, this type of expression is by now markedly familiar. The left-hand side is a function purely of t while the right-hand side is a function of the independent variables r and θ. Therefore, these ratios must be equal to a constant, $C_o = -\lambda$. For technical reasons, the constant is assumed to be negative as this assumption makes the subsequent analysis simpler. The Dirichlet boundary condition $u(r_o, \theta, t) = 0$ and the periodicity requirement $u(r, \theta, t) = u(r, \theta+2\pi, t)$ imply that $R(r_o) = 0$ and $\Theta(\theta) = \Theta(\theta+2\pi)$. In particular, $\Theta(0) = \Theta(2\pi)$. These conditions result in the set of *ODE*s

$$(i)\;\; T''(t) + \lambda\sigma^2 T(t) = 0 \quad (ii)\begin{cases} r^2 R''(r) + r\,R'(r) + \left(\lambda r^2 - \mu\right)R(r) = 0 \\ R(r_o) = 0 \end{cases} \quad (iii)\begin{cases} \Theta''(\theta) + \mu\,\Theta(\theta) = 0 \\ \Theta(0) = \Theta(2\pi) \end{cases}.$$

The *ODE* (*iii*) is readily seen to require $\mu = n^2$ which produces the solution $\Theta(\theta) = c_1 \cos(n\theta) + c_2 \sin(n\theta)$. If the substitution $x = \sqrt{\lambda}\, r$ is made, then the *ODE* (*ii*) becomes the *Bessel equation of order n*, first seen in Example 3.5.4 and equation (3.5.5) of Chapter 3.

$$\xi^2 \frac{d^2R}{d\xi^2} + \xi \frac{dR}{d\xi} + (\xi^2 - n^2)R = 0 \tag{4.6.5}$$

As detailed in Chapter 3, the solutions of (4.6.5) are *Bessel functions of order n* identified by the series $J_n(\xi) = \sum_{k=0}^{\infty} \frac{(-1)^k}{2^{2k+n} k!(n+k)!} \xi^{k+n}$. Hence, $R_n(r) = d_o J_n(\sqrt{\lambda}\, r)$ solves the top equation of *ODE* (*ii*). To satisfy the initial condition, $R(r_o) = 0$, it is required that $J_n(\sqrt{\lambda}\, r_o) = 0$ or $\lambda_{n,k} = \left(\frac{\zeta_{n,k}}{r_o}\right)^2$ where $\zeta_{n,1} < \zeta_{n,2} < \ldots < \zeta_{n,k} < \ldots$ are the ascending zeros of the Bessel function J_n. Thus, $R_{n,k}(r) = J_n\left(\frac{\zeta_{n,k}}{r_o} r\right)$ are the eigenfunctions of (*ii*) with eigenvalues $\lambda_{n,k}$. Setting $v_{n.k}(r, \theta) = R_{n,k}(r) \cdot (\cos(n\theta) + \sin(n\theta))$ it is seen that $v_{n.k}(r,\theta)$ solves $\frac{1}{r}\frac{\partial}{\partial r}\left(r\frac{\partial v}{\partial r}\right) + \frac{1}{r^2}\frac{\partial^2 v}{\partial \theta^2} + \lambda v = 0$ over $(0, r_o)\times[-\pi, \pi]$ with initial condition $v(r_o,\theta) = 0$. Hence, $v_{n.k}(r, \theta)$ are the eigenfunctions of this *PDE* with eigenvalues $\lambda_{n,k} = \left(\zeta_{n,k}/r_o\right)^2$. But it is a well-known result that any function $v(r,\theta) \in C^2(f: [0, r_o]\times[-\pi, \pi] \to \mathbb{R} : f(r_o, \theta) = 0 \text{ for } -\pi \le \theta \le \pi)$ can be written as the absolutely and uniformly convergent series (4.6.6). See, for example, Tikhonov and Samarskii [79] and Zachmanoglou and Thoe [87] among other sources. Therefore, if $T(t)$ is the fundamental solutions of (*i*), then $u_{n,k}(r, q, t) = v_{n.k}(r, \theta)\cdot T(t)$ are the eigenfunctions of the wave equation (4.6.4).

$$\boxed{\begin{aligned} v(r,\theta) &= \sum_{n=0}^{\infty}\sum_{k=1}^{\infty}\left[\alpha_{n,k}\cos(n\theta) + \beta_{n,k}\sin(n\theta)\right] J_n\left(\tfrac{\zeta_{n,k}}{r_o} r\right) \\ \alpha_{n,k} &= \frac{1}{\pi r_o^2 \left[J_{n+1}(\zeta_{n,k})\right]} \int_{-\pi}^{\pi}\int_0^{r_o} v(r,\theta)\cos(n\theta) J_n\left(\tfrac{\zeta_{n,k}}{r_o} r\right) r\, dr\, d\theta \\ \beta_{n,k} &= \frac{1}{\pi r_o^2 \left[J_{n+1}(\zeta_{n,k})\right]} \int_{-\pi}^{\pi}\int_0^{r_o} v(r,\theta)\sin(n\theta) J_n\left(\tfrac{\zeta_{n,k}}{r_o} r\right) r\, dr\, d\theta \end{aligned}}, \; n \in \mathbb{Z}_o^+, \; k \in \mathbb{Z}^+ \tag{4.6.6}$$

As $\lambda > 0$, the solution of (*i*) is $T_{n.k}(t) = e_1 \cos\left(\sqrt{\lambda_{n,k}}\sigma t\right) + e_2 \sin\left(\sqrt{\lambda_{n,k}}\sigma t\right)$. Hence,

$$u_{n,k}(r,\theta,t) = A_{n,k} v_{n,k}(r,\theta)\cos\left(\sqrt{\lambda_{n,k}}\sigma t\right) + B_{n,k} v_{n,k}(r,\theta)\sin\left(\sqrt{\lambda_{n,k}}\sigma t\right)$$

are the fundamental solutions of (4.6.4). The Cauchy condition $\phi(r, \theta) = u(r, \theta, 0)$ means that the fundamental solutions must satisfy $\phi(r, \theta) = u_{n,k}(r, \theta, 0) = A_{n,k} v_{n,k}(r, \theta)$. Similarly, $\psi(r, \theta) = \frac{\partial u}{\partial t}(r,\theta,0)$ implies $\psi(r, \theta) = \sqrt{\lambda_{n.k}}\, B_{n,k} v_{n,k}(r, \theta)$. Combining these calculations with (4.6.6), it is seen that

$$
\begin{aligned}
u(r,\theta,t) &= \sum_{n=0}^{\infty}\sum_{k=1}^{\infty} u_{n,k}(r,\theta,t) \\
&= \sum_{n=0}^{\infty}\sum_{k=1}^{\infty} J_n\left(\tfrac{\zeta_{n,k}}{r_o} r\right)\Big((\cos(n\theta)\Big[A_{n,k}\cos\left(\tfrac{\zeta_{n,k}}{r_o}\sigma t\right) + B_{n,k}\sin\left(\tfrac{\zeta_{n,k}}{r_o}\sigma t\right)\Big] \\
&\quad + \sin(n\theta)\Big[C_{n,k}\cos\left(\tfrac{\zeta_{n,k}}{r_o}\sigma t\right) + D_{n,k}\sin\left(\tfrac{\zeta_{n,k}}{r_o}\sigma t\right)\Big]\Big)
\end{aligned}
\tag{4.6.7a}
$$

and

$$
\begin{aligned}
A_{n,k} &= \frac{2}{\varepsilon_n \cdot \pi \left[r_o \cdot J_n'(\zeta_{n,k})\right]^2} \int_{-\pi}^{\pi}\int_0^{r_o} \phi(r,\theta) J_n\left(\tfrac{\zeta_{n,k}}{r_o} r\right)\cos(n\theta)\, r\, dr\, d\theta \\
B_{n,k} &= \frac{2}{\varepsilon_n \cdot \pi \left[r_o \cdot J_n'(\zeta_{n,k})\right]^2} \int_{-\pi}^{\pi}\int_0^{r_o} \phi(r,\theta) J_n\left(\tfrac{\zeta_{n,k}}{r_o} r\right)\sin(n\theta)\, r\, dr\, d\theta \\
C_{n,k} &= \frac{2}{\varepsilon_n \cdot \pi \cdot \sigma \frac{\zeta_{n,k}}{r_o}\left[r_o \cdot J_n'(\zeta_{n,k})\right]^2} \int_{-\pi}^{\pi}\int_0^{r_o} \psi(r,\theta) J_n\left(\tfrac{\zeta_{n,k}}{r_o} r\right)\cos(n\theta)\, r\, dr\, d\theta \\
D_{n,k} &= \frac{2}{\varepsilon_n \cdot \pi \cdot \sigma \frac{\zeta_{n,k}}{r_o}\left[r_o \cdot J_n'(\zeta_{n,k})\right]^2} \int_{-\pi}^{\pi}\int_0^{r_o} \psi(r,\theta) J_n\left(\tfrac{\zeta_{n,k}}{r_o} r\right)\sin(n\theta)\, r\, dr\, d\theta \\
\varepsilon_n &= \begin{cases} 2 & \text{for } n = 0 \\ 1 & \text{for } n \neq 0 \end{cases}
\end{aligned}
\quad . \tag{4.6.7b}
$$

Note that the derivative of the Bessel function J_n satisfies the recursion relationship $J_n'(x) = J_{n-1}(x) - \frac{n}{x} J_n(x)$, see, e.g., Watson [81] and Olver et al. [57]. Therefore, the one portion of denominators in (4.6.7*b*) can be replaced by $J_n'(\zeta_{n,k}) = J_{n-1}(\zeta_{n,k}) - \frac{n}{\zeta_{n,k}} J_n(\zeta_{n,k}) = J_{n-1}(\zeta_{n,k})$ as $\zeta_{n,k}$ is a zero of J_n. It is evident from Figures 3.10–3.11 of Chapter 3 that $J_{n-1}(0) = 0$ for $n \in \mathbb{Z}^+$. Therefore, all instances in which $\zeta_{n,k} = 0$ must be excluded from the calculations of $A_{n,k}$, $B_{n,k}$, $C_{n,k}$, and $D_{n,k}$. The coefficient ε_n is required as the Bessel functions J_n are orthogonal but not orthonormal. In particular, as per Tikhonov and Samarskii [79],

$$
\int_0^{r_o} J_m\left(\tfrac{\zeta_{n,k}}{r_o} r\right) J_n\left(\tfrac{\zeta_{n,j}}{r_o} r\right) r\, dr = \begin{cases} 0 & \text{for } m \neq n \text{ or } k \neq j \\ \frac{\pi r_o^2}{2}\left[J_n'(\zeta_{n,k})\right]^2 & \text{for } m = n \text{ and } j = k \\ \pi r_o^2 \left[J_0'(\zeta_{o,k})\right]^2 & \text{for } m = n = 0 \text{ and } j = k \end{cases} .
$$

The family of functions $u_{n,k}(r, \theta, t) = J_n(\xi_{n,k} r) \cdot \left(\cos(n\theta - \xi_{n,k} t) + \sin(n\theta - \xi_{n,k} t)\right)$, $\xi_{n,k} = \zeta_{n,k}/r_o$, help define *fundamental nodes of the circular drum*: $U_n(r, \theta, t) = \sum_{k=1}^{N_k} u_{n,k}(r,\theta,t)$ and N_k are the number of zeros $\zeta_{n,k}$ of $J_n(z)$ for $z \in [0, r_o]$. That is, each node $u_{n,k}(r, \theta, t)$ determines a vibration pattern at a particular frequency. When all of the nodes[14] are combined, then all tones (or harmonics) are activated on the drumhead. Select nodes are illustrated in Figure 4.6.

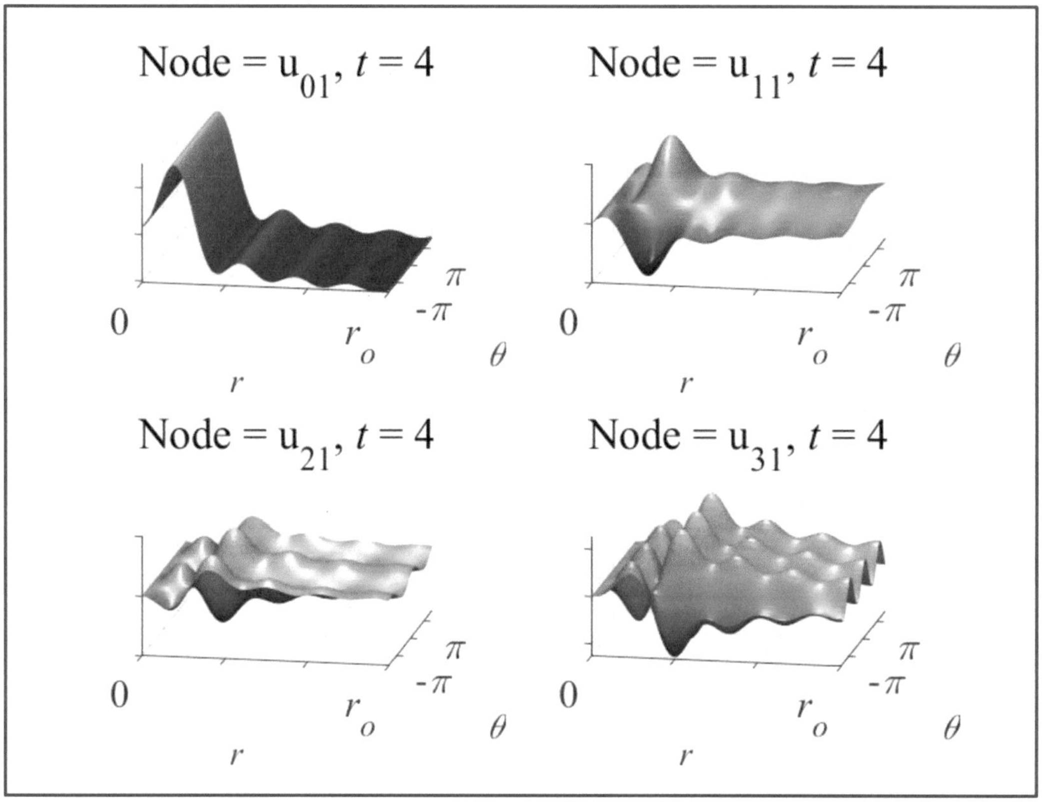

Figure 4.6. Nodes of the vibrating circular drum.

To produce a numerical representation of the solution of the vibrating circular membrane (4.6.4) via (4.6.7*a*)–(4.6.7*b*), a MATLAB program (`circdrum.m`) has been written and applied to the initial conditions $u(r, \theta, t) = \phi(r, \theta) = \exp(-[r^2+\theta^2])$ and $\frac{\partial u}{\partial t}(r,\theta,t) = \psi(r, \theta) = 1$. The wave speed is set to unity $\sigma = 1$. This approximates a drum strike with a velocity of 1 unit. Figure 4.7 illustrates the idea. For this example, the drum radius is selected to be 30 *cm* (approximately a 12-*inch* radius).

[14] In this context, some sources refer to nodes as vibration *modes*. See, for example, the Wikipedia article https://en.wikipedia.org/wiki/Vibrations_of_a_circular_membrane .

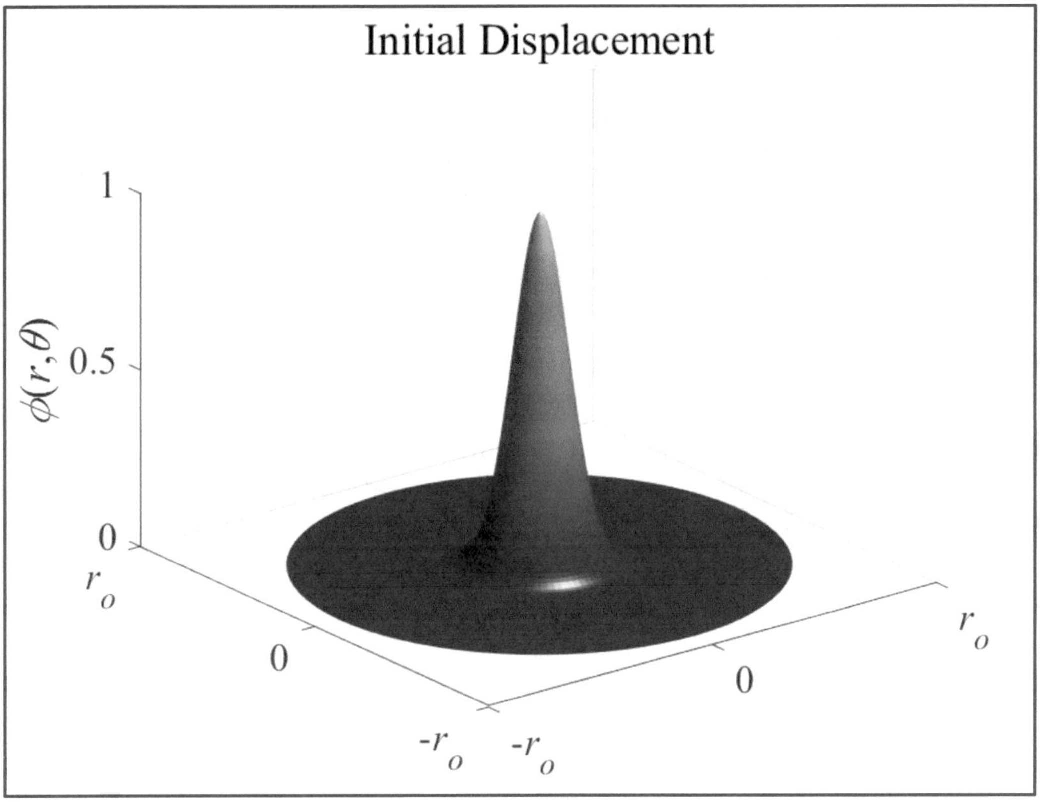

Figure 4.7. Initial displacement of the circular membrane (drum).

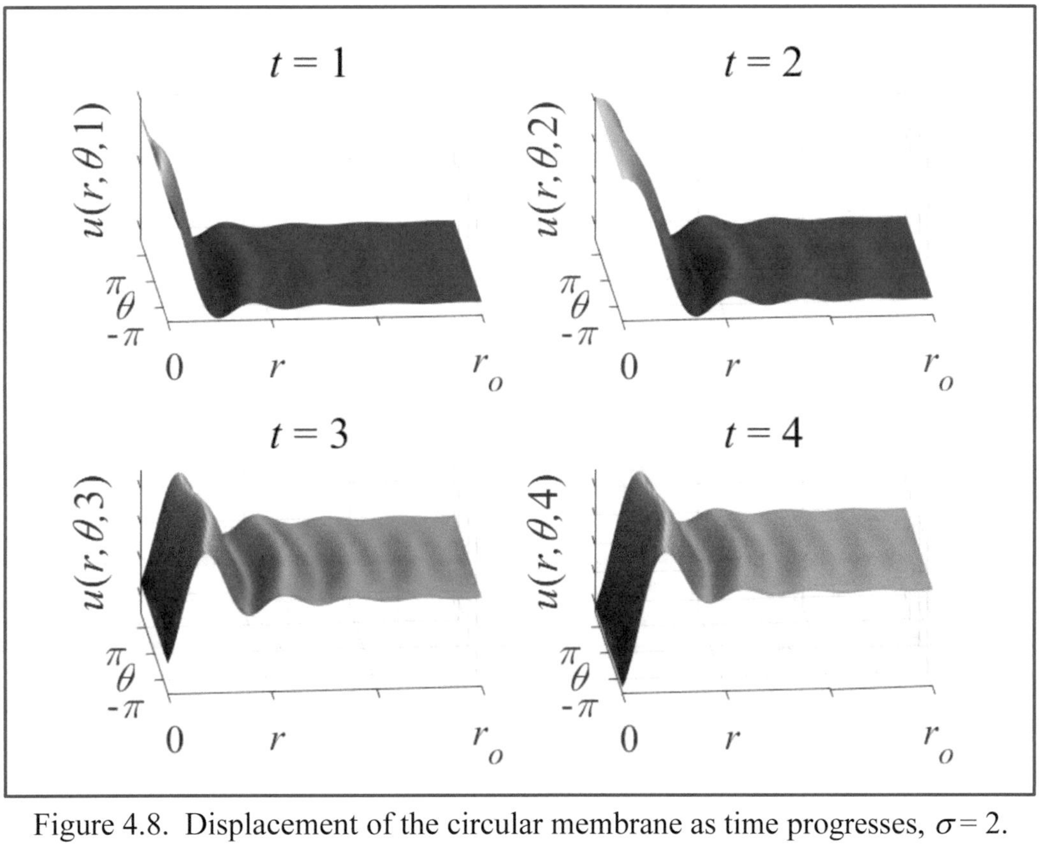

Figure 4.8. Displacement of the circular membrane as time progresses, $\sigma = 2$.

It is noted that a more realistic model for $\phi(r, \theta)$ could be the *negative* of the exponential $-e^{-2(r^2+\theta^2)}$ which mimics the force *downward* on the drum top by a drumstick. The mathematics remains the same regardless of the sign of the initial displacement. Since the nodes (Figure 4.6) are selected *independent* of initial displacement, they have a positive orientation. Therefore, the positive exponential is used in this example. It can be thought of as a local upward "tug" on the drumskin followed by a release. As Figure 4.8 indicates, an increase in time results in the wave moving out from the center of the drum (i.e., $r = 0$) toward the exterior edge. The reader is encouraged to use the M–file `circdrum.m` to determine how drums with different radii r react to the initial displacement $\phi(r,\theta) = e^{-2(r^2+\theta^2)}$ and (initial) constant velocity $\psi(r,\theta) = 1$.

As stated in Remark 3.1 of Chapter 3, the techniques above cannot be applied to the vibrating sphere. Corresponding to the work done on the heat equation, spherical coordinates are required. The ambitious reader is welcome to follow the methodology presented in Remark 3.1 to see how the separation of variables transform fails to bring about an analytic solution for the vibrating sphere. See Exercise 4.7 below for details.

Example 4.6.4. *The vibrating volume.* The final consideration in this tour of the homogeneous wave equation over finite domains is the vibrating parallelepiped. In this setting, the domain is a rigid parallelepiped filled with gelatin. The three–dimensional wave equation that governs the motion of the gelatin within the volume is

$$\left.\begin{array}{l} \frac{\partial^2 u}{\partial t^2}(\boldsymbol{x},t) = \sigma^2\left(\frac{\partial^2 u}{\partial x^2}(\boldsymbol{x},t) + \frac{\partial^2 u}{\partial y^2}(\boldsymbol{x},t) + \frac{\partial^2 u}{\partial z^2}(\boldsymbol{x},t)\right) \\ \boldsymbol{x} = (x,y,z) \in \Omega = [0,L_1]\times[0,L_2]\times[0,L_3] \text{ and } t > 0 \\ u(\boldsymbol{x},0) = f(\boldsymbol{x}),\ \frac{\partial u}{\partial t}(\boldsymbol{x},0) = g(\boldsymbol{x}),\ \boldsymbol{x}\in\Omega \\ u(\boldsymbol{x},t) = 0 \text{ for } \boldsymbol{x}\in\partial\Omega \end{array}\right\}. \tag{4.6.8}$$

The Dirichlet condition $u(\boldsymbol{x}, t) = 0$ for $\boldsymbol{x} \in \partial\Omega$ means that $u(0, y, z, t) = u(L_1, y, z, t) = 0$, $u(x, 0, z, t) = u(x, L_2, z, t) = 0$, and $u(x, y, 0, t) = u(x, y, L_3, t) = 0$ for all $t > 0$.

Solution: As with Example 4.6.2, the separation of variables transform $u(x, y, z, t) = X(x)\cdot Y(y)\cdot Z(z)\cdot T(t)$ produces the fundamental solutions (eigen solutions)

$$\boxed{u_{n,m,k}(\boldsymbol{x}, t) = \left[a_{n,m,k}\cos\left(\sqrt{|\lambda_{n,m,k}|}\,\sigma t\right) + b_{n,m,k}\sin\left(\sqrt{|\lambda_{n,m,k}|}\,\sigma t\right)\right]\sin\left(\tfrac{n\pi}{L_1}x\right)\sin\left(\tfrac{m\pi}{L_2}y\right)\sin\left(\tfrac{k\pi}{L_3}z\right)}$$

with the corresponding eigenvalues $\lambda_{n,m,k} = -\left[\left(\frac{n\pi}{L_1}\right)^2 + \left(\frac{m\pi}{L_2}\right)^2 + \left(\frac{k\pi}{L_3}\right)^2\right]$. By the superposition principle, $u(\boldsymbol{x}, t) = \sum_{n=1}^{\infty}\sum_{m=1}^{\infty}\sum_{k=1}^{\infty} u_{n,m,k}(\boldsymbol{x},t)$ solves $\frac{\partial^2 u}{\partial t^2} = \sigma\,\Delta u$. The initial conditions $u(\boldsymbol{x}, 0) = f(\boldsymbol{x})$ and $\frac{\partial u}{\partial t}(\boldsymbol{x},t) = g(\boldsymbol{x})$ require that

$$f(\boldsymbol{x}) = \sum_{n=1}^{\infty}\sum_{m=1}^{\infty}\sum_{k=1}^{\infty} a_{n,m,k} \sin\left(\tfrac{n\pi}{L_1}x\right)\sin\left(\tfrac{m\pi}{L_2}y\right)\sin\left(\tfrac{k\pi}{L_3}z\right) \text{ with}$$

$$a_{n,m,k} = \tfrac{8}{L_1 L_2 L_3}\int_0^{L_3}\int_0^{L_2}\int_0^{L_1} f(\boldsymbol{x})\sin\left(\tfrac{n\pi}{L_1}x\right)\sin\left(\tfrac{m\pi}{L_2}y\right)\sin\left(\tfrac{k\pi}{L_3}z\right)dx\,dy\,dz \text{ and}$$

$$g(\boldsymbol{x}) = \sum_{n=1}^{\infty}\sum_{m=1}^{\infty}\sum_{k=1}^{\infty} b_{n,m,k}\sqrt{|\lambda_{n,m,k}|}\sin\left(\tfrac{n\pi}{L_1}x\right)\sin\left(\tfrac{m\pi}{L_2}y\right)\sin\left(\tfrac{k\pi}{L_3}z\right) \text{ with}$$

$$b_{n,m,k} = \frac{8}{\sqrt{|\lambda_{n,m,k}|}L_1 L_2 L_3}\int_0^{L_3}\int_0^{L_2}\int_0^{L_1} g(\boldsymbol{x})\sin\left(\tfrac{n\pi}{L_1}x\right)\sin\left(\tfrac{m\pi}{L_2}y\right)\sin\left(\tfrac{k\pi}{L_3}z\right)dx\,dy\,dz\,.$$

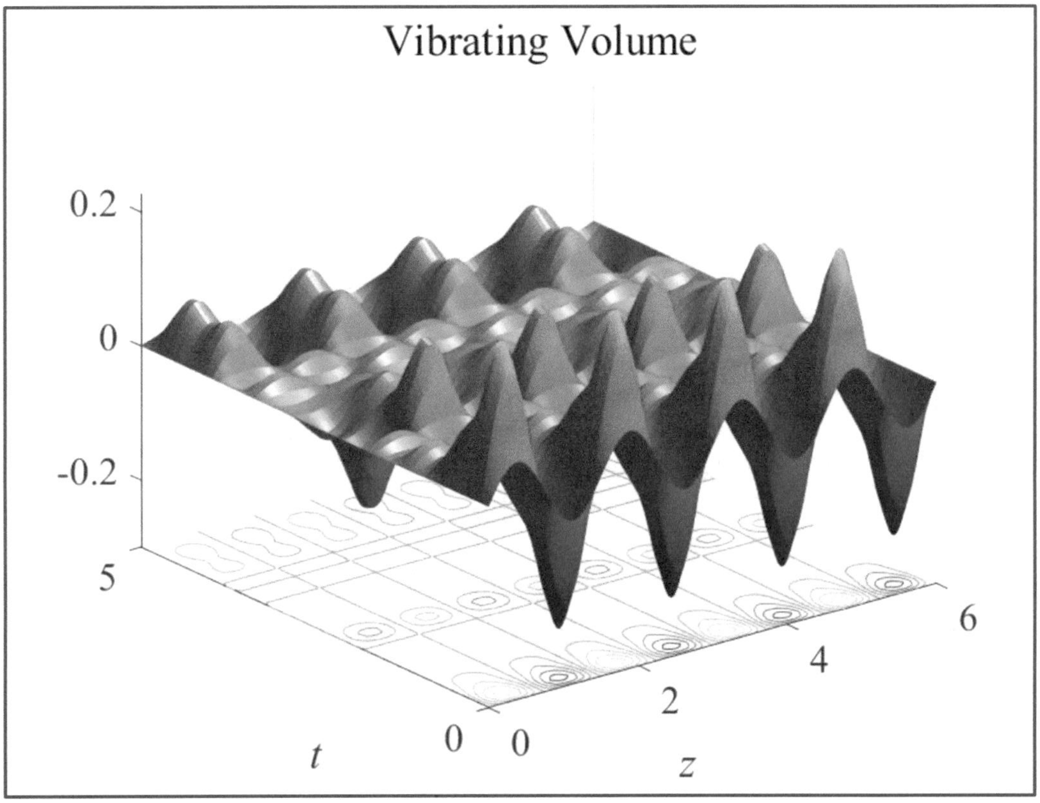

Figure 4.9. A two–dimensional cut of the vibrating volume ($x = 1, y = 2$).

The contour plot in Figure 4.9 illustrates a cross section of the four–dimensional solution of (4.6.8) with $f(\boldsymbol{x}) = \exp(-(x+y+z))$ on $\Omega = [0, L_1]\text{x}[0, L_2]\text{x}[0, L_3]$, $g(\boldsymbol{x}) \equiv 1$ on $\partial\Omega$, $\sigma = 1$, $L_1 = 2$, $L_2 = 4$, $L_3 = 6$, $t \in [0, 5]$, and $n, m, k = 1, 2, \ldots, 8$. Since the formulae for the eigen functions $u_{n,m,k}$ and Fourier coefficients $a_{n,m,k}$, $b_{n,m,k}$ are symmetric with respect to the spatial variables, then a two–dimensional "slice" of $u(\boldsymbol{x}, t)$ is given at $x = L_1/2 = 1$ and $y = L_2/2 = 2$. Notice that as t moves forward from 0, the amplitude of the waves decreases. ■

In this chapter, the power of the Fourier transform to reduce multi–[spatial] dimensional *PDE*s to single variable *ODE*s is demonstrated with respect to the heat and wave equations. The vibrating guitar string and vibrating circular drum models are explored and analytic solutions are determined. D'Alembert's solution of the one–spatial dimensional wave equation is produced via the Fourier transform for both the homogeneous and inhomogeneous forms. This technique sets the stage for the higher dimensional problems. Solutions of the n–dimensional homogeneous wave equation (4.4.1) and its inhomogeneous counterpart (4.6.9) are quite complicated for $n \geq 2$. Analytic techniques used to solve these equations and consequently Maxwell's equations are presented in the next chapter.

$$\left.\begin{aligned} &\frac{\partial^2 u}{\partial t^2}(\boldsymbol{x},t) = \sigma^2 \Delta u(\boldsymbol{x},t) + \rho(\boldsymbol{x},t), && \text{for } \boldsymbol{x} \in \mathbb{R}^n, t > 0 \\ &u(\boldsymbol{x},0) = f(\boldsymbol{x}), \frac{\partial u}{\partial t}(\boldsymbol{x}) = g(\boldsymbol{x}), && \text{for } \boldsymbol{x} \in \mathbb{R}^n \end{aligned}\right\} \tag{4.6.9}$$

Exercises

4.1 Validate properties 4.1.1 *Linearity*, 4.1.3 *Scaling*, and 4.1.4 *Translation* of the Fourier transform.

4.2 Find the Fourier transform of the function $f(x) = e^{-\alpha|x|}$ for $\alpha > 0$.

4.3 Using the Fourier transform, show that the system of coupled equations

$$\begin{aligned} &\frac{\partial u}{\partial t}(x,t) = \frac{\partial^2 u}{\partial x^2}(x,t) - v(x,t) \\ &\frac{\partial v}{\partial t}(x,t) = \frac{\partial^2 v}{\partial x^2}(x,t) + u(x,t) \\ &u(x,0) = f(x), v(x,0) = g(x) \end{aligned} \tag{4.E.1}$$

has the solution

$$\begin{aligned} u(x,t) &= \sqrt{\tfrac{2\pi}{2t}}\cos(t)\big(f \odot H\big)(x,t) - \sqrt{\tfrac{2\pi}{2t}}\sin(t)\big(g \odot H\big)(x,t) \\ v(x,t) &= \sqrt{\tfrac{2\pi}{2t}}\sin(t)\big(f \odot H\big)(x,t) + \sqrt{\tfrac{2\pi}{2t}}\cos(t)\big(g \odot H\big)(x,t) \end{aligned} \tag{4.E.2}$$

where H is the heat kernel with dispersion constant $\sigma = 1$ and $\odot$ is convolution. More specifically, $(f \odot H)(x,t) = \frac{1}{\sqrt{2\pi}} \int_{-\infty}^{\infty} f(\xi) \cdot H(x-\xi, t)\, d\xi$. If $f(x) = \exp(-x^2)$ and $g(x) = f(x)/10$, find the analytic form of the solutions $u(x, t)$ and $v(x, t)$. Use MATLAB and/or the Symbolic Math Toolbox to compute and plot the solution surfaces.

Hint: $(f \odot H)(x,t) = \dfrac{\sqrt{t}}{2\sqrt{\pi(1+t)}} \exp\left(-\dfrac{x^2}{1+t}\right)$ when $\sigma = 1$ and $f(x) = \exp(-x^2)$.

4.4 For $\sigma = 1$, $f(x) = \text{sech}(x)$, and $g(x) = \cos(x)$, find d'Alembert's solution of the one–dimensional wave equation (4.5.1) via (4.5.2). What happens for $\sigma > 1$ and $0 < \sigma < 1$?

4.5 For $\sigma = 1$, $f(x) = \exp(-x^2)$, $g(x) = -\tanh(x)\cdot\text{sech}(x)$, and $\rho(x, t) = \cos(\pi x)\cdot\sin(\pi x/2)$, use (4.5.5) to find the solution of the one–dimensional inhomogeneous wave equation (4.5.3). What happens for $\sigma > 1$ and $0 < \sigma < 1$?

4.6 Show that, in polar coordinates, the d'Alembertian operator $\Box u = \dfrac{\partial^2 u}{\partial t^2} - \sigma^2\left(\dfrac{\partial^2 u}{\partial x^2} + \dfrac{\partial^2 u}{\partial y^2}\right)$ becomes $\Box_{\circ} u = \dfrac{\partial^2 u}{\partial t^2} - \sigma^2\left(\dfrac{1}{r}\dfrac{\partial}{\partial r}\left(r\dfrac{\partial u}{\partial r}\right) - \dfrac{1}{r^2}\dfrac{\partial^2 u}{\partial \theta^2}\right)$. *Hint*: Let $x = r\cos(\theta)$ and $y = r\sin(\theta)$.

4.7. Consider the wave equation for the vibrating sphere.

$$
\begin{aligned}
\frac{\partial^2 u}{\partial t^2}(r,\theta,\omega,t) &= \sigma^2\Big[\frac{\partial^2 u}{\partial r^2}(r,\theta,\omega,t) + \frac{2}{r}\frac{\partial u}{\partial r}(r,\theta,\omega,t) \\
&\quad + \frac{1}{r^2}\left(\frac{\partial^2 u}{\partial \theta^2}(r,\theta,\omega,t) + \cot(\theta)\frac{\partial u}{\partial \theta}(r,\theta,\omega,t) + \csc^2(\theta)\frac{\partial^2 u}{\partial \omega^2}(r,\theta,\omega,t)\right)\Big] \\
u(r_o,\theta,\omega,0) &= \phi(\theta,\omega) \\
\frac{\partial u}{\partial r}(r_o,\theta,\omega,0) &= \psi(\theta,\omega) \\
u(r_o,\theta,\omega,t) &= 0,\ u(r,\theta+\pi,\omega,t) = u(r,\theta,\omega,t), \\
u(r,\theta,\omega,t) &= u(r,\theta,\omega+2\pi,t) \quad \text{for all } r,t \\
0 \le r \le r_o,\ &0 \le \theta \le \pi,\ 0 < \omega < 2\pi,\ t > 0
\end{aligned}
$$

Use the methods developed in Remark 3.1 of Chapter 3 to see how the separation of variables transform *fails* to produce an analytic solution to the vibrating sphere equation above.

Suggested approach:

(*i*) The transform for spherical coordinates is $x = r \cdot \sin(\theta) \cdot \cos(\omega)$, $y = r \cdot \sin(\theta) \cdot \sin(\omega)$, $y = r \cdot \cos(\theta)$. Use this transform and multivariate calculus to derive the Laplacian in spherical coordinates. That is, demonstrate

$$\Delta u = \frac{\partial^2 u}{\partial x^2} + \frac{\partial^2 u}{\partial y^2} + \frac{\partial^2 u}{\partial z^2}$$

$$= \frac{\partial^2 u}{\partial r^2} + \frac{2}{r}\frac{\partial u}{\partial r} + \frac{1}{r^2}\left(\frac{\partial^2 u}{\partial \theta^2} + \cot(\theta)\frac{\partial u}{\partial \theta} + \csc^2(\theta)\frac{\partial^2 u}{\partial \omega^2}\right) \equiv \Delta_{\mathbb{S}} u.$$

(*ii*) The separation of variables transform is $u(r, \theta, \omega, t) = T(t) \cdot U(r, \theta, \omega)$. Apply the d'Alembertian operator in spherical coordinates $\Delta_{\mathbb{S}}$ to the wave equation and divide through the resulting *ODE* by $T(t) \cdot U(r, \theta, \omega)$.

(*iii*) Obtain $\dfrac{T''}{\sigma^2 T} = -\lambda = \dfrac{\Delta_{\mathbb{S}} U(r,\theta,\omega)}{U(r,\theta,\omega)}$. Show that the resulting *ODE* for the radial function $R(r)$ either leads to a non-constant value for λ or a complex–valued function for $R(r)$.

(*iv*) See Remark 3.1 of Chapter 3 above and/or Myint–U with Debnath [52] for details on solving Laplace's equation in spherical coordinates $\Delta_{\mathbb{S}} u(r,\theta,\omega) = 0$.

Chapter 5

The Three–Dimensional Wave Equation

As noted in the first chapter, the solution of Maxwell's equations can be achieved by solving the inhomogeneous wave equation (4.6.9) of Chapter 4. That will be the aim herein. For the reader's convenience, Maxwell's equations (1.4.5) of Chapter 1, are repeated below.

$$\boxed{\begin{array}{ll} \nabla\bullet \boldsymbol{D}(\boldsymbol{x},t) = 4\pi\,\rho(\boldsymbol{x},t) & \text{Coulomb's law} \\ \nabla\times \boldsymbol{E}(\boldsymbol{x},t) = -\dfrac{1}{c}\dfrac{\partial \boldsymbol{B}}{\partial t}(\boldsymbol{x},t) & \text{Faraday's law} \\ \nabla\times \boldsymbol{H}(\boldsymbol{x},t) = \dfrac{1}{c}\dfrac{\partial \boldsymbol{D}}{\partial t}(\boldsymbol{x},t) + \dfrac{4\pi}{c}\boldsymbol{J}(\boldsymbol{x},t) & \text{Ampere's law} \\ \nabla\bullet \boldsymbol{B}(\boldsymbol{x},t) = 0 & \text{no magnetic charges} \end{array}} \tag{5.0.1}$$

Throughout this chapter, the underlying assumption is that the vector $\boldsymbol{x} = (x, y, z) \in \mathbb{R}^3$ and all vector fields operate over three–dimensional Euclidean space and the positive real line $\mathbb{R}^+$ in time.

This form of Maxwell's equations requires that the inhomogeneous term $\rho(\boldsymbol{x}, t)$ in Coulomb's law and the current density $\boldsymbol{J}(\boldsymbol{x}, t)$ in Ampere's law satisfy the continuity equation

$$\boxed{\frac{\partial \rho}{\partial t}(\boldsymbol{x},t) + \varepsilon\,\nabla\bullet \boldsymbol{J}(\boldsymbol{x},t) = 0}\,. \tag{5.0.2}$$

Equation (5.0.2) is known as the *conservation of charge*. Since the magnetic field $\boldsymbol{B}$ is solenoidal (that is, $\nabla\bullet\boldsymbol{B} = 0$), then by the Solenoidal Vector Field Theorem[15] there is a C^1–vector field $\boldsymbol{A}$ whose curl is equal to $\boldsymbol{B}$. Assume that the electric field $\boldsymbol{E}$ is the sum of a scalar gradient field u and a velocity vector field $\boldsymbol{A}$ which satisfies the continuity equation (5.0.2). More precisely, the following assumptions hold throughout this chapter.

(i) $\boldsymbol{B}(\boldsymbol{x}, t) = \nabla \times \boldsymbol{A}(\boldsymbol{x}, t)$

[15] See Chapter 1.

$$(ii)\ \mathbf{E}(\boldsymbol{x},t) = -\left(\nabla u(\boldsymbol{x},t) + c\frac{\partial \boldsymbol{A}}{\partial t}(\boldsymbol{x},t)\right)$$

$$(iii)\ \frac{\partial u}{\partial t}(\boldsymbol{x},t) - c\nabla\bullet\boldsymbol{A}(\boldsymbol{x},t) = 0$$

With this in mind, it is seen that the vector field $\boldsymbol{A}(\boldsymbol{x}, t) = (A_1(\boldsymbol{x}, t), A_2(\boldsymbol{x}, t), A_3(\boldsymbol{x}, t))$ and the scalar field u satisfy the inhomogeneous three–dimensional wave equations (5.0.3*a*)–(5.0.3*b*).

$$\frac{\partial^2 \boldsymbol{A}}{\partial t^2}(\boldsymbol{x},t) = c^2\Delta\boldsymbol{A}(\boldsymbol{x},t) + \tfrac{4\pi\mu}{c\varepsilon}\boldsymbol{J}(\boldsymbol{x},t) \tag{5.0.3a}$$

$$\frac{\partial^2 u}{\partial t^2}(\boldsymbol{x},t) = \Delta u(\boldsymbol{x},t) + \tfrac{4\pi}{\varepsilon}\rho(\boldsymbol{x},t) \tag{5.0.3b}$$

Thus, if a solution of the inhomogeneous wave equation

$$\left.\begin{aligned} &\frac{\partial^2 A}{\partial t^2}(\boldsymbol{x},t) = c^2\Delta A(\boldsymbol{x},t) + p(\boldsymbol{x},t), \qquad \boldsymbol{x}\in\mathbb{R}^3, t>0 \\ &A(\boldsymbol{x},0) = \phi(\boldsymbol{x}) \\ &\frac{\partial A}{\partial t}(\boldsymbol{x},0) = \psi(\boldsymbol{x}) \end{aligned}\right\} \tag{5.0.4}$$

can be determined for the scalar field A, then the solution of (5.0.3*a*)–(5.0.3*b*), along with suitable initial conditions, will be achieved.

§5.1. Fourier Transform Solution

Recall the properties of the n–dimensional Fourier transform $\mathfrak{F}$ from §4.1 of Chapter 4. Note that for $\boldsymbol{x} \in \mathbb{R}^3$, $\Delta A = \dfrac{\partial^2 A}{\partial x^2} + \dfrac{\partial^2 A}{\partial y^2} + \dfrac{\partial^2 A}{\partial z^2}$. Use properties 4.1.4 and 4.1.7 and apply $\mathfrak{F}$ to the top equation of (5.0.4). Utilize the notation $Q(\boldsymbol{\omega},t) = \mathfrak{F}\left[A(\boldsymbol{x},t)\right](\boldsymbol{\omega})$ and $P(\boldsymbol{\omega},t) = \mathfrak{F}\left[p(\boldsymbol{x},t)\right](\boldsymbol{\omega})$ to obtain $\dfrac{\partial^2 Q}{\partial t^2}(\boldsymbol{\omega},t) = -c^2\left\|\boldsymbol{\omega}\right\|^2 Q(\boldsymbol{\omega},t) + P(\boldsymbol{\omega},t)$ where $\|\boldsymbol{\omega}\|^2 = \omega_1^2 + \omega_2^2 + \omega_3^2$. Since c is the speed of light and the norm $\|\boldsymbol{\omega}\|$ are both positive numbers, then $\alpha = c\cdot\|\boldsymbol{\omega}\| > 0$. The Fourier transform of the top equation in (5.0.4) takes the more conventional form

$$\frac{d^2 Q}{dt^2}(\boldsymbol{\omega},t) = -\alpha^2 Q(\boldsymbol{\omega},t) + P(\boldsymbol{\omega},t).$$

This, however, is precisely the equation first encountered in §4.5 of Chapter 4 which has solution $Q(\boldsymbol{\omega}, t) = q_c(\boldsymbol{\omega}, t) + q_p(\boldsymbol{\omega}, t)$, where $q_c(\boldsymbol{\omega}, t) = c_1 \cos(\alpha t) + c_2 \sin(\alpha t) = c_1 \cos(c\|\boldsymbol{\omega}\|t) + c_2 \sin(c\|\boldsymbol{\omega}\|t)$ and $q_p(\boldsymbol{\omega}, t) = \int_0^t \frac{\sin\left(c\|\omega\|[t-\tau]\right)}{c\|\omega\|} P(\omega,\tau)\,d\tau$. This means

$$Q(\omega,t) = c_1 \cos(c\|\omega\|t) + c_2 \sin(c\|\omega\|t) + \int_0^t \frac{\sin\left(c\|\omega\|[t-\tau]\right)}{c\|\omega\|} P(\omega,\tau)\,d\tau . \tag{5.1.1}$$

The first of the initial conditions in (5.0.4) requires via (5.1.1) that $\mathfrak{F}\left[\phi(\boldsymbol{x})\right](\omega) = A(\omega,0) = c_1$. Moreover, by Leibniz's Rule[16],

$$\frac{\partial Q}{\partial t}(\omega,t) = -c\|\omega\|c_1 \sin(c\|\omega\|t) + c\|\omega\|c_2 \cos(c\|\omega\|t) + \cancel{\frac{\sin\left(c\|\omega\|[t-t]\right)}{c\|\omega\|}} P(\omega,t)$$
$$+\int_0^t \cos\left(c\|\omega\|[t-\tau]\right) P(\omega,\tau)\,d\tau.$$

By property 4.1.7 of the Fourier transform, $\mathfrak{F}\left[\psi(\boldsymbol{x})\right](\omega) = \mathfrak{F}\left[\frac{\partial A}{\partial t}(\boldsymbol{x},0)\right](\omega) = \frac{\partial}{\partial t}\mathfrak{F}\left[A(\boldsymbol{x},0)\right](\omega) =$ $\frac{\partial Q}{\partial t}(\omega,0) = c\|\boldsymbol{\omega}\|\, c_2$ or $c_2 = \frac{1}{c\|\omega\|}\mathfrak{F}\left[\psi(\boldsymbol{x})\right](\omega)$. Combining these values for c_1 and c_2 with (5.1.1) produces the transform equation (5.1.2).

$$\boxed{\begin{aligned}\mathfrak{F}\left[A(\boldsymbol{x},t)\right](\omega) &= \mathfrak{F}\left[\phi(\boldsymbol{x})\right](\omega)\cdot\cos\left(c\|\omega\|t\right) + \mathfrak{F}\left[\psi(\boldsymbol{x})\right](\omega)\frac{\sin\left(c\|\omega\|t\right)}{c\|\omega\|} \\ &\quad+\int_0^t \frac{\sin\left(c\|\omega\|[t-\tau]\right)}{c\|\omega\|}\mathfrak{F}\left[p(\boldsymbol{x},\tau)\right](\omega)\,d\tau\end{aligned}} \tag{5.1.2}$$

If there exists functions $K_1(\boldsymbol{x}, t)$ and $K_2(\boldsymbol{x}, t)$ so that $\mathfrak{F}[K_1(\boldsymbol{x}, t)](\boldsymbol{\omega}) = \cos(c\|\boldsymbol{\omega}\|t)$ and $\mathfrak{F}[K_2(\boldsymbol{x}, t)](\boldsymbol{\omega})$ $= \frac{\sin\left(c\|\omega\|t\right)}{c\|\omega\|}$, then (5.1.2) can be written as

[16] If f, $\frac{\partial f}{\partial t} \in C^1(\Omega \text{x}[a, b])$, $\alpha(t)$ and $\beta(t)$ differentiable on $[a, b]$, and $F(t) = \int_{\alpha(t)}^{\beta(t)} f(t,s)\,ds$, then F is differentiable with respect to t and $F'(t) = f\left(\beta(t),t\right)\cdot\beta'(t) - f\left(\alpha(t),t\right)\cdot\alpha'(t) + \int_{\alpha(t)}^{\beta(t)}\frac{\partial f}{\partial t}(t,s)\,ds$.

$$\mathfrak{F}[A(\boldsymbol{x}, t)](\boldsymbol{\omega}) = \mathfrak{F}[\phi(\boldsymbol{x})](\boldsymbol{\omega})\cdot\mathfrak{F}[K_1(\boldsymbol{x}, t)](\boldsymbol{\omega}) + \mathfrak{F}[\psi(\boldsymbol{x})](\boldsymbol{\omega})\cdot\mathfrak{F}[K_2(\boldsymbol{x}, t)](\boldsymbol{\omega})$$

$$+\int_0^t \mathfrak{F}\left[K_2(\boldsymbol{x},t-\tau)\right](\boldsymbol{\omega})\cdot\mathfrak{F}\left[p(x,\tau)\right](\boldsymbol{\omega})\,d\tau$$

$$=\frac{1}{\left(\sqrt{2\pi}\right)^3}\mathfrak{F}\left[\left(\phi\odot K_1\right)(\boldsymbol{x},t)+\left(\psi\odot K_2\right)(\boldsymbol{x},t)+\int_0^t\left(p\odot K_2\right)(\boldsymbol{x},t-\tau)\,d\tau\right](\boldsymbol{\omega}).$$

If $\boldsymbol{\xi} = (\xi_1, \xi_2, \xi_3) \in \mathbb{R}^3$, then the equation above is equivalent to

$$\boxed{\begin{aligned} A(\boldsymbol{x},t) &= \frac{1}{\left(\sqrt{2\pi}\right)^3}\int_{\mathbb{R}^3}\left(\phi(\xi)\cdot K_1(\boldsymbol{x}-\xi,t)+\psi(\xi)\cdot K_2(\boldsymbol{x}-\xi,t)\right)d\xi \\ &+\frac{1}{\left(\sqrt{2\pi}\right)^3}\int_0^t\int_{\mathbb{R}^3} p(\xi,\tau)\cdot K_2(\boldsymbol{x}-\xi,t-\tau)\,d\xi\,d\tau \end{aligned}}\;. \qquad (5.1.3)$$

It remains to construct the functions K_1 and K_2 that are referred to as the *wave kernels*. This task is accomplished in the next section.

§5.2. The Three–Dimensional Wave Kernels

To construct the wave kernels K_1 and K_2 of (5.1.3), define the radial indicator function on the sphere of radius t and "thickness" ε. That is, $\mathfrak{I}_{\mathbb{B}(\boldsymbol{0},t+\varepsilon)\setminus\mathbb{B}(\boldsymbol{0},t)}(\boldsymbol{x}) = \begin{cases}1 & \text{for } x\in\mathbb{B}(\boldsymbol{0},t+\varepsilon)\setminus\mathbb{B}(\boldsymbol{0},t)\\ 0 & \text{otherwise}\end{cases}$

where $\mathbb{B}(\boldsymbol{0}, t) = \{\boldsymbol{x} \in \mathbb{R}^3 : \|\boldsymbol{x}\|^2 = x^2 + y^2 + z^2 \le t^2\}$ is the three–dimensional ball of radius t centered at the origin $\boldsymbol{0}$. The solid $\mathbb{B}(\boldsymbol{0}, t+\varepsilon)\setminus\mathbb{B}(\boldsymbol{0}, t) = \{\boldsymbol{x} \in \mathbb{R}^3 : t^2 \le \|\boldsymbol{x}\|^2 \le (t+\varepsilon)^2\}$ is the three–dimensional ball whose thickness is ε. Figure 5.1 illustrates a cross–section of this "spherical shell" as per Champeney [16] and Dettman [22].

Define $f_\varepsilon(\boldsymbol{x}, t) = \frac{1}{ct\varepsilon}\mathfrak{I}_{\mathbb{B}(\boldsymbol{0},t+\varepsilon)\setminus\mathbb{B}(\boldsymbol{0},t)}(\boldsymbol{x})$. The challenge is to determine the Fourier transform of $f_\varepsilon(\boldsymbol{x}, t)$. It is evident from the geometry that spherical coordinates $x = r\cdot\sin(\beta)\cdot\cos(\theta)$, $y = r\cdot\sin(\beta)\cdot\sin(\theta)$, $z = r\cdot\sin(\beta)$, $0 \le \beta \le \pi$, $0 \le \theta \le 2\pi$, $t \le r \le t+\varepsilon$ should be utilized. Before attempting to compute the Fourier transform, the Jacobian matrix reflecting the change of variables from Euclidean (x, y, z) to spherical (r, β, θ) coordinates must be calculated. That is,

$$J_{r\beta\theta}^{xyz} = \begin{bmatrix} \frac{dx}{dr} & \frac{dx}{d\beta} & \frac{dx}{d\theta} \\ \frac{dy}{dr} & \frac{dy}{d\beta} & \frac{dy}{d\theta} \\ \frac{dz}{dr} & \frac{dz}{d\beta} & \frac{dz}{d\theta} \end{bmatrix} = \begin{bmatrix} \sin(\beta)\cos(\theta) & r\cos(\beta)\cos(\theta) & -r\sin(\beta)\sin(\theta) \\ \sin(\beta)\sin(\theta) & r\cos(\beta)\sin(\theta) & r\sin(\beta)\cos(\theta) \\ \cos(\beta) & -r\sin(\beta) & 0 \end{bmatrix}.$$

A direct calculation shows that $\det\left[J_{r\beta\theta}^{xyz}\right] = r^2\sin(\beta)$.

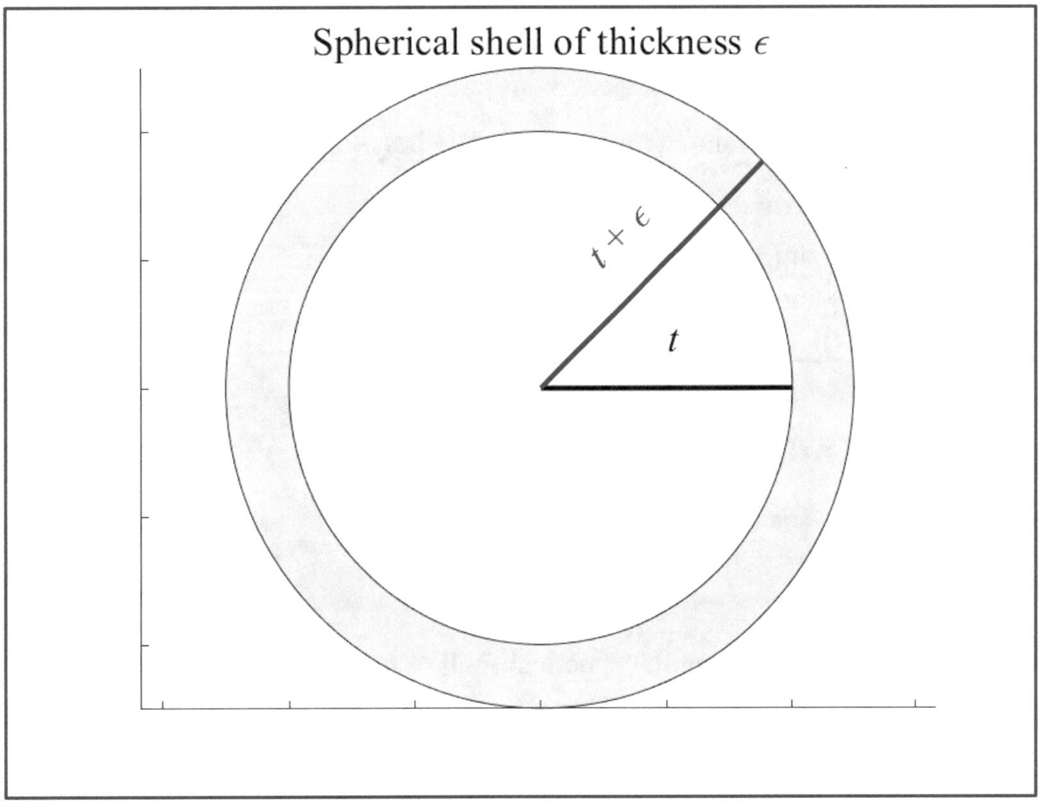

Figure 5.1. Spherical shell of thickness ε.

Recall from vector calculus that the dot product of two unit vectors is the cosine of the angle subtended by the vectors. That is, $\boldsymbol{u}\bullet\boldsymbol{v} = \cos(\beta)$. Setting $\boldsymbol{u}_x = \boldsymbol{x}/\|\boldsymbol{x}\|$ and $\boldsymbol{u}_\omega = \boldsymbol{\omega}/\|\boldsymbol{\omega}\|$ produces the result $\boldsymbol{u}_x\bullet\boldsymbol{u}_\omega = \cos(\beta) \Leftrightarrow \boldsymbol{x}\bullet\boldsymbol{\omega} = \|\boldsymbol{x}\|\cdot\|\boldsymbol{\omega}\|\cdot\cos(\beta)$. Finally, let $r = \|\boldsymbol{x}\| = \sqrt{x^2+y^2+z^2}$ and $\omega = \|\boldsymbol{\omega}\| = \sqrt{\omega_1^2+\omega_2^2+\omega_3^2}$ so that $\boldsymbol{x}\bullet\boldsymbol{\omega} = r\cdot\omega\cdot\cos(\beta)$. With all of these observations in mind, the Fourier transform of $f_\varepsilon(\boldsymbol{x}, t)$ is now computed.

$$
\begin{aligned}
\mathfrak{F}[f_\varepsilon(\boldsymbol{x},t)](\boldsymbol{\omega}) &= \tfrac{1}{\left(\sqrt{2\pi}\right)^3}\frac{1}{ct\varepsilon}\int_{\mathbb{R}^3}\mathcal{I}_{\mathbb{B}(\boldsymbol{0},t+\varepsilon)\setminus\mathbb{B}(\boldsymbol{0},t)}(\boldsymbol{x})\,e^{i(\boldsymbol{x}\bullet\boldsymbol{\omega})}\,d\boldsymbol{x} \\
&= \tfrac{1}{\left(\sqrt{2\pi}\right)^3}\frac{1}{ct\varepsilon}\int_t^{t+\varepsilon}\int_0^{\pi}\int_0^{2\pi} 1\cdot e^{ir\cdot\omega\cdot\cos(\beta)}r^2\sin(\beta)\,d\theta\,d\beta\,dr \\
&= \tfrac{1}{\left(\sqrt{2\pi}\right)^3}\frac{1}{ct\varepsilon}\int_t^{t+\varepsilon}\int_0^{\pi} 2\pi\cdot e^{ir\cdot\omega\cdot\cos(\beta)}r^2\sin(\beta)\,d\beta\,dr \\
&= \frac{1}{\sqrt{2\pi}ct\varepsilon}\int_t^{t+\varepsilon}\tfrac{i}{r\cdot\omega}e^{ir\cdot\omega\cdot\cos(\beta)}\Big|_{\beta=0}^{\pi} r^2\,dr = \frac{1}{\sqrt{2\pi}ct\varepsilon}\int_t^{t+\varepsilon} 2\frac{\sin(r\cdot\omega)}{r\cdot\omega}r^2\,dr \\
&= \frac{2}{\sqrt{2\pi}\omega^3 ct\varepsilon}\Big(\left[\sin(\omega(t+\varepsilon))-\sin(\omega t)\right]-\left[\omega(t+\varepsilon)\cos(\omega(t+\varepsilon))-\omega t\cos(\omega t)\right]\Big)
\end{aligned}
$$

By L'Hopital's rule, $\lim\limits_{\varepsilon\to 0}\dfrac{\sin(\omega(t+\varepsilon))-\sin(\omega t)}{\omega^3 t\varepsilon} = \dfrac{\cos(\omega t)}{\omega^2}$ while $\lim\limits_{\varepsilon\to 0}\dfrac{\omega(t+\varepsilon)\cos(\omega(t+\varepsilon))-\omega t\cos(\omega t)}{\omega^3 t\varepsilon} = -\dfrac{\cos(\omega t)}{\omega^2 t}+\dfrac{\sin(\omega t)}{\omega}$.

Therefore, $\lim\limits_{\varepsilon\to 0}\mathfrak{F}\left[f_\varepsilon(\boldsymbol{x},t)\right](\boldsymbol{\omega}) = \sqrt{\tfrac{2}{\pi}}\dfrac{\sin(\omega t)}{c\omega} = \sqrt{\tfrac{2}{\pi}}\dfrac{\sin(\|\boldsymbol{\omega}\|t)}{c\|\boldsymbol{\omega}\|}$ so that $\lim\limits_{\varepsilon\to 0}\mathfrak{F}\left[f_\varepsilon(\boldsymbol{x},ct)\right](\boldsymbol{\omega}) = \sqrt{\tfrac{2}{\pi}}\dfrac{\sin(tc\|\boldsymbol{\omega}\|)}{c\|\boldsymbol{\omega}\|}$.

But as ε approaches the origin, then the spherical shell of thickness ε converges to the sphere of radius t, $\mathbb{S}(\boldsymbol{0},t)=\left\{\boldsymbol{x}\in\mathbb{R}^3 : \|\boldsymbol{x}\|^2 = t^2\right\}$. That is, $\lim\limits_{\varepsilon\to 0}\mathbb{B}(\boldsymbol{0},t+\varepsilon)\setminus\mathbb{B}(\boldsymbol{0},t)=\mathbb{S}(\boldsymbol{0},t)$. Notice that

$$
\begin{aligned}
\mathfrak{F}\left[\mathcal{I}_{\mathbb{S}(\boldsymbol{0},t)}(\boldsymbol{x})\right](\boldsymbol{\omega}) &= \tfrac{1}{\left(\sqrt{2\pi}\right)^3}\int_{\mathbb{R}^3}\mathcal{I}_{\mathbb{S}(\boldsymbol{0},t)}(\boldsymbol{x})\cdot e^{i(\boldsymbol{x}\bullet\boldsymbol{\omega})}\,d\boldsymbol{x} \\
&= \tfrac{1}{\left(\sqrt{2\pi}\right)^3}\int_0^{\pi}\int_0^{2\pi} 1\cdot e^{i(t\cdot\omega)\cos(\beta)}t^2\sin(\beta)\,d\theta\,d\beta = \tfrac{t^2}{\sqrt{2\pi}}\int_0^{\pi} e^{i(t\cdot\omega)\cos(\beta)}\sin(\beta)\,d\beta \\
&= \tfrac{t^2}{i(t\cdot\omega)\sqrt{2\pi}}e^{i(t\cdot\omega)\cos(\beta)}\Big|_{\beta=0}^{\pi} = \sqrt{\tfrac{2}{\pi}}\,t\,\frac{\sin(t\cdot\omega)}{\omega}.
\end{aligned}
$$

Now since t plays the role of a scalar in this computation, $\mathfrak{F}\left[\tfrac{1}{t}\cdot\mathcal{I}_{\mathbb{S}(\boldsymbol{0},t)}(\boldsymbol{x})\right](\boldsymbol{\omega}) = \tfrac{1}{t}\cdot\mathfrak{F}\left[\mathcal{I}_{\mathbb{S}(\boldsymbol{0},t)}(\boldsymbol{x})\right](\boldsymbol{\omega}) = \sqrt{\tfrac{2}{\pi}}\dfrac{\sin(t\cdot\|\boldsymbol{\omega}\|)}{\|\boldsymbol{\omega}\|}$. This means that $\mathfrak{F}\left[\tfrac{1}{ct}\cdot\mathcal{I}_{\mathbb{S}(\boldsymbol{0},ct)}(\boldsymbol{x})\right](\boldsymbol{\omega}) = \sqrt{\tfrac{2}{\pi}}\dfrac{\sin(t\cdot c\|\boldsymbol{\omega}\|)}{c\|\boldsymbol{\omega}\|}$.

Now as the Fourier transform is a continuous operator, the limit can be moved inside of the integral. That is, $\sqrt{\tfrac{2}{\pi}}\dfrac{\sin(tc\|\boldsymbol{\omega}\|)}{c\|\boldsymbol{\omega}\|} = \lim\limits_{\varepsilon\to 0}\mathfrak{F}\left[f_\varepsilon(\boldsymbol{x},ct)\right](\boldsymbol{\omega}) = \mathfrak{F}\left[\lim\limits_{\varepsilon\to 0}f_\varepsilon(\boldsymbol{x},ct)\right](\boldsymbol{\omega}) = \mathfrak{F}\left[\tfrac{1}{ct}\mathcal{I}_{\mathbb{S}(\boldsymbol{0},ct)}(\boldsymbol{x})\right](\boldsymbol{\omega})$.

By the invertibility of the Fourier transform, $\lim_{\varepsilon\to 0} f_\varepsilon(\boldsymbol{x},ct) = \frac{1}{ct}\cdot \mathcal{G}_{\mathbb{S}(\boldsymbol{0},ct)}(\boldsymbol{x})$. This result parallels the last formula in Table 4.1 of Chapter 4. Consequently, the Fourier pair below is established.

$$f(\boldsymbol{x}, t) = \tfrac{1}{ct}\cdot \mathcal{G}_{\mathbb{S}(\boldsymbol{0},ct)}(\boldsymbol{x}) \leftrightarrow F(\boldsymbol{\omega},t) = \sqrt{\tfrac{2}{\pi}}\,\frac{\sin(t\cdot c\|\boldsymbol{\omega}\|)}{c\|\boldsymbol{\omega}\|}$$

The second wave kernel is therefore $K_2(\boldsymbol{x}, t) = \sqrt{\frac{\pi}{2}}\,\frac{1}{ct}\cdot \mathcal{G}_{\mathbb{S}(\boldsymbol{0},ct)}(\boldsymbol{x})$. Property 4.1.7 of the Fourier transform yields $\cos(tc\|\boldsymbol{\omega}\|) = \frac{\partial}{\partial t}\left(\frac{\sin(tc\|\boldsymbol{\omega}\|)}{c\|\boldsymbol{\omega}\|}\right) = \frac{\partial}{\partial t}\mathfrak{F}\left[K_2(\boldsymbol{x},t)\right](\boldsymbol{\omega}) = \mathfrak{F}\left[\frac{\partial}{\partial t}K_2(\boldsymbol{x},t)\right](\boldsymbol{\omega})$. Since the derivative of the second wave kernel has the desired Fourier transform, then the first wave kernel must be $K_1(\boldsymbol{x}, t) = \frac{\partial}{\partial t}K_2(\boldsymbol{x},t)$. This is summarized in Table 5.1.

In the next section, the wave kernels in Table 5.1 are applied to the inhomogeneous wave solution (5.1.3).

Kernel $K(\boldsymbol{x}, t)$	Functional Value	Fourier Transform $K(\boldsymbol{\omega},t)$
$K_2(\boldsymbol{x}, t)$	$\sqrt{\frac{\pi}{2}}\,\frac{1}{ct}\cdot \mathcal{G}_{\mathbb{S}(\boldsymbol{0},ct)}(\boldsymbol{x})$	$\frac{\sin(tc\|\boldsymbol{\omega}\|)}{c\|\boldsymbol{\omega}\|}$
$K_1(\boldsymbol{x}, t)$	$\frac{\partial}{\partial t}K_2(\boldsymbol{x},t)$	$\cos(tc\|\boldsymbol{\omega}\|)$

Table 5.1. Wave kernels

§5.3. Huygen's Principle and Duhamel's Principle

This section relies heavily upon the work of Zachmanoglou and Thoe [87]. Since convolution is symmetric with respect to the function in translation, then (5.1.3) can be written as

$$A(\boldsymbol{x},t) = \frac{1}{\left(\sqrt{2\pi}\right)^3}\int_{\mathbb{R}^3}\left(\phi(\boldsymbol{x}-\xi)\cdot K_1(\xi,t) + \psi(\boldsymbol{x}-\xi)\cdot K_2(\xi,t)\right)d\xi$$
$$+\frac{1}{\left(\sqrt{2\pi}\right)^3}\int_0^t\int_{\mathbb{R}^3} p(\boldsymbol{x}-\xi,\tau)\cdot K_2(\xi,t-\tau)\,d\xi\,d\tau\,.$$

Utilizing the results of §5.2 and Table 5.1, shifting once again to spherical coordinates over the sphere of radius t/α, selecting $\boldsymbol{\upsilon} = (\sin(\beta)\cdot\cos(\theta), \sin(\beta)\cdot\sin(\theta), \cos(\beta))$ as the position vector on the unit sphere, and setting $\boldsymbol{\xi} = (t/\alpha)\cdot\boldsymbol{\upsilon}$ to be the position vector on $\mathbb{S}(\boldsymbol{0}, t/\alpha)$, results in

$$\int_{\mathbb{R}^3} \psi(\boldsymbol{x}-\xi)\cdot K_2(\xi,t)\,d\xi = \int_0^{2\pi}\int_0^{\pi} \psi(\boldsymbol{x}-ct\cdot\boldsymbol{\upsilon})\cdot K_2(\boldsymbol{\upsilon},t)\sin(\beta)\,d\beta\,d\theta$$
$$= \int_{\mathbb{S}_1} \sqrt{\tfrac{\pi}{2}}\tfrac{1}{ct}\cdot\psi\left(\boldsymbol{x}-ct\cdot\boldsymbol{\upsilon}\right)d\boldsymbol{\upsilon}$$

where $d\boldsymbol{\upsilon} = \sin(\beta)\, d\beta\, d\theta$ is the element of surface area on the unit sphere $\mathbb{S}_1 = \left\{\boldsymbol{x}\in\mathbb{R}^3 : \|\boldsymbol{x}\|^2 = 1\right\}$. Using the relationship between $K_1(\boldsymbol{x}, t)$ and $K_2(\boldsymbol{x}, t)$, it is similarly seen that

$$\int_{\mathbb{R}^3} \phi(\boldsymbol{x}-\xi)\cdot K_1(\xi,t)\,d\xi = \int_0^{2\pi}\int_0^{\pi} \phi(\boldsymbol{x}-ct\cdot\boldsymbol{\upsilon})\cdot\frac{\partial}{\partial t}K_2(\boldsymbol{\upsilon},t)\sin(\beta)\,d\beta\,d\theta$$
$$= \frac{\partial}{\partial t}\int_0^{2\pi}\int_0^{\pi} \phi(\boldsymbol{x}-ct\cdot\boldsymbol{\upsilon})\cdot K_2(\boldsymbol{\upsilon},t)\sin(\beta)\,d\beta\,d\theta$$
$$= \frac{\partial}{\partial t}\int_{\mathbb{S}_1} \sqrt{\tfrac{\pi}{2}}\tfrac{1}{ct}\cdot\phi\left(\boldsymbol{x}-ct\cdot\boldsymbol{\upsilon}\right)d\boldsymbol{\upsilon}$$

and

$$\int_{\mathbb{R}^3} p(\boldsymbol{x}-\xi,\tau)\cdot K_2(\xi,t-\tau)\,d\xi = \int_0^{2\pi}\int_0^{\pi} p(\boldsymbol{x}-ct\cdot\boldsymbol{\upsilon},\tau)\cdot K_2(\boldsymbol{\upsilon},t-\tau)\sin(\beta)\,d\beta\,d\theta$$
$$= \int_{\mathbb{S}_1} \sqrt{\tfrac{\pi}{2}}\tfrac{1}{c(t-\tau)}\cdot p\left(\boldsymbol{x}-ct\cdot\boldsymbol{\upsilon},c(t-\tau)\right)d\boldsymbol{\upsilon}.$$

Therefore the solution of the inhomogeneous wave equation (5.0.4) is given as (5.3.1).

$$\boxed{\begin{aligned} A(\boldsymbol{x},t) &= \frac{1}{4\pi}\frac{\partial}{\partial t}\int_{\mathbb{S}_1}\tfrac{1}{ct}\cdot\phi(\boldsymbol{x}-ct\cdot\boldsymbol{\upsilon})\,d\boldsymbol{\upsilon} + \frac{1}{4\pi}\int_{\mathbb{S}_1}\tfrac{1}{ct}\cdot\psi(\boldsymbol{x}-ct\cdot\boldsymbol{\upsilon})\,d\boldsymbol{\upsilon} \\ &+ \frac{1}{4\pi}\int_0^t \tfrac{1}{c(t-\tau)}\int_{\mathbb{S}_1} p\left(\boldsymbol{x}-ct\cdot\boldsymbol{\upsilon},c(t-\tau)\right)d\boldsymbol{\upsilon}\,d\tau \end{aligned}} \qquad (5.3.1)$$

Observe that the solution $A(\boldsymbol{x}, t)$ depends on the integrals of the Cauchy data ϕ and ψ and the forcing function p over the unit sphere $\mathbb{S}_1$ with respect to the argument $\boldsymbol{x} - ct\cdot\boldsymbol{\upsilon}$. This mimics the d'Alembert wave solution (4.5.5) for the one–spatial dimensional wave equation. Equation (5.3.1) is referred to as *Huygen's Principle* which states that the solution of the three–spatial dimensional wave equation depends only on the spherical means of the Cauchy data and inhomogeneous term.

Definition 5.1: The *spherical mean* of a function f over a sphere of radius r and center $\boldsymbol{0}$ is

$$\boxed{M[f,r]=\tfrac{1}{\omega_{n-1}}\int_{\mathbb{S}_1} f(\boldsymbol{x}+r\cdot\boldsymbol{\upsilon})\,d\boldsymbol{\upsilon}}$$

where ω_{n-1} is the area of the sphere in $\mathbb{R}^n$, $\boldsymbol{\upsilon}$ is the position vector on the unit sphere $\mathbb{S}_1$, and $d\boldsymbol{\upsilon}$ is the element of surface area. For $n = 3$, $\boldsymbol{\upsilon} = (\sin(\beta)\cos(\theta), \sin(\beta)\sin(\theta), \text{con}(\beta))$, $d\boldsymbol{\upsilon} = \sin(\beta)\, d\beta\, d\theta$, and $\omega_{n-1} = \omega_2 = 4\pi$. See John [44] for greater detail and formula (2.2.3) of Chapter 2 to obtain other values of ω_{n-1}.

As equation (5.3.1) indicates, the solution of the inhomogeneous three–dimensional wave equation is constructed via spherical surface integrals. Using this definition, it is seen that (5.3.1) can be written as

$$A(\boldsymbol{x},t)=\frac{\partial}{\partial t}\left(\tfrac{1}{ct}\cdot M[\phi,-ct]\right)+\tfrac{1}{ct}\cdot M[\psi,-ct]+\int_0^t \tfrac{1}{c(t-\tau)}\cdot M[p,-c(t-\tau)]\,d\tau \tag{5.3.2}$$

and leads directly to *Huygen's Principle*.

Huygen's Principle: The solution of the inhomogeneous wave equation (5.0.4) depends only upon the integrals of the Cauchy data ϕ and ψ and the forcing function p over the unit sphere $\mathbb{S}_1$ in $\mathbb{R}^3$.

Note that, because of the manner in which the spherical mean is defined, there is a negative sign on the time argument: $M[f,-ct]=\tfrac{1}{4\pi}\int_{\mathbb{S}_1} f(\boldsymbol{x}-ct\cdot\boldsymbol{\upsilon})\,d\boldsymbol{\upsilon}$. In the case of $p(\boldsymbol{x}, t) \equiv 0$, the inhomogeneous equation (5.0.4) becomes the homogeneous equation (5.3.3).

$$\left.\begin{aligned}
&\frac{\partial^2 A}{\partial t^2}(\boldsymbol{x},t)=c^2\Delta A(\boldsymbol{x},t), \qquad x\in\mathbb{R}^3,\ t>0\\
&A(\boldsymbol{x},0)=\phi(\boldsymbol{x})\\
&\frac{\partial A}{\partial t}(\boldsymbol{x},0)=\psi(\boldsymbol{x})
\end{aligned}\right\} \tag{5.3.3}$$ ■

Using the method of spherical means and (5.3.2), it is seen that the homogeneous equation above has the solution

$$A(\boldsymbol{x},t)=\frac{\partial}{\partial t}\left(\tfrac{1}{ct}\cdot M[\phi,-ct]\right)+\tfrac{1}{ct}\cdot M[\psi,-ct]. \tag{5.3.4}$$

This is known as *Duhamel's Principle.*

Duhamel's Principle: Let $w(\boldsymbol{x}, t) = \frac{1}{ct} M[\phi, -ct]$ be a solution of the homogeneous wave equation

$$\frac{\partial^2 w}{\partial t^2}(\boldsymbol{x},t) = c^2 \Delta w(\boldsymbol{x},t),\ w(\boldsymbol{x},0) = 0, \text{ and } \frac{\partial w}{\partial t}(\boldsymbol{x},0) = \phi(\boldsymbol{x}) \quad \text{for } \boldsymbol{x} \in \mathbb{R}^3. \tag{5.3.5}$$

Then $u(\boldsymbol{x}, t) = \dfrac{\partial}{\partial t} w(\boldsymbol{x},t) + \frac{1}{ct} M[\psi, -ct]$ solves

$$\frac{\partial^2 u}{\partial t^2}(\boldsymbol{x},t) = c^2 \Delta u(\boldsymbol{x},t),\ u(\boldsymbol{x},0) = \phi(\boldsymbol{x}), \text{ and } \frac{\partial u}{\partial t}(\boldsymbol{x},0) = \psi(\boldsymbol{x}) \quad \text{for } \boldsymbol{x} \in \mathbb{R}^3. \tag{5.3.6}$$

The solution $w(\boldsymbol{x}, t) = \frac{1}{ct} M[\phi, -ct] = \frac{1}{4\pi ct} \int_{\mathbb{S}_1} \phi(\boldsymbol{x} - ct \cdot \boldsymbol{\upsilon})\, d\boldsymbol{\upsilon}$ is called *Kirchhoff's formula*, while the sum $u(\boldsymbol{x}, t) = \dfrac{\partial}{\partial t}\left(\frac{1}{ct} M[\phi, -ct]\right) + \frac{1}{ct} M[\psi, -ct]$ is known as *Poisson's solution.* ■

John [44] and Garabedian [26] provide a more detailed discussion.

The development of the solution of the inhomogeneous wave equation via spherical means can be used to produce a solution of the homogeneous and inhomogeneous wave equations in $n = 2$ spatial dimensions. This is the focus of the next section.

§5.4. The Method of Descent

The previous section employed surface integrals and spherical means over the three–dimensional unit sphere $\mathbb{S}_1 = \{\boldsymbol{x} \in \mathbb{R}^3 : \|\boldsymbol{x}\|^2 = x^2 + y^2 + z^2 = 1\}$ to produce the solutions of the homogeneous and inhomogeneous wave equations in $n = 3$ spatial dimensions. In the case of $n = 2$ spatial dimensions, a technique known as *the method of descent* is employed.

To discuss this approach, the symbol $\mathbb{S}(\boldsymbol{x}_o, r) = \{\boldsymbol{x} \in \mathbb{R}^3 : \|\boldsymbol{x}\|^2 = (x - x_o)^2 + (y - y_o)^2 + (z - z_o)^2 = r^2\}$ represents the sphere of radius r with center $\boldsymbol{x}_o = (x_o, y_o, z_o)$. Now consider the surface integral $\int_{\mathbb{S}_1} \frac{1}{ct} \cdot f(\boldsymbol{x} - ct \cdot \boldsymbol{\upsilon})\, d\boldsymbol{\upsilon}$ where, as in §5.3, $\boldsymbol{\upsilon} = (\sin(\beta) \cos(\theta), \sin(\beta) \sin(\theta), \text{con}(\beta))$ is the position vector on the three–dimensional unit sphere $\mathbb{S}_1$ and $d\boldsymbol{\upsilon} = \sin(\beta)\, d\beta\, d\theta$ is the element of surface area. For $\boldsymbol{\upsilon} \in \mathbb{S}_1$, $\boldsymbol{\xi} = \boldsymbol{x} - ct \cdot \boldsymbol{\upsilon}$ is a point on the sphere of radius $r = ct$ and center $\boldsymbol{x}$. That is, $\boldsymbol{\xi} \in \mathbb{S}(\boldsymbol{x}, ct)$. The element of surface area on $\mathbb{S}(\boldsymbol{x}, ct)$ is $d\boldsymbol{\upsilon}_{ct} = (ct)^2 \cdot \sin(\beta)\, d\beta\, d\theta = (ct)^2 \cdot d\boldsymbol{\upsilon}$. The change of variables $\boldsymbol{\xi} = \boldsymbol{x} - ct \cdot \boldsymbol{\upsilon}$ yields $d\boldsymbol{\xi} = ct\, d\boldsymbol{\upsilon} = \dfrac{1}{ct} d\boldsymbol{\upsilon}_{ct}$. Thus, $\int_{\mathbb{S}_1} \frac{1}{ct} \cdot f(\boldsymbol{x} - ct \cdot \boldsymbol{\upsilon})\, d\boldsymbol{\upsilon} = \int_{\mathbb{S}(\boldsymbol{x},ct)} f(\xi) \frac{1}{ct} d\boldsymbol{\upsilon}_{ct}$.

The method of descent requires that the three–dimensional solution $A(\boldsymbol{x}, t)$ of (5.3.1) be projected into two–spatial dimensions. That is, $A(x_1, x_2, x_3, t) \to A(x_1, x_2, 0, t)$. Project the element of surface area onto the plane as per Figure 5.2. Indeed, on $\mathbb{S}(\boldsymbol{x}, ct)$, $(ct)^2 = (\xi_1 - x_1)^2 + (\xi_2 - x_2)^2 + (\xi_3 - x_3)^2$ so that $(\xi_3 - x_3) = \sqrt{(ct)^2 - (\xi_1 - x_1)^2 - (\xi_2 - x_2)^2}$. Using the formula from calculus for surface integrals and elements of surface area, $d\boldsymbol{\upsilon}_{ct} = \sqrt{1 + \left(d\xi_3 / d\xi_1\right)^2 + \left(d\xi_3 / d\xi_2\right)^2}\, d\xi_1\, d\xi_2$, a direct calculation shows $d\boldsymbol{\upsilon}_{ct} = \dfrac{(ct)\, d\xi_1\, d\xi_2}{\sqrt{(ct)^2 - (\xi_1 - x_1)^2 - (\xi_2 - x_2)^2}}$. Now integration over the sphere $\mathbb{S}([x_1, x_2, 0], ct)$ means that $(\xi_1 - x_1)^2 + (\xi_2 - x_2)^2 \le (ct)^2$ or twice the integral over the positive half–disk $ct \ge \sqrt{(\xi_1 - x_1)^2 + (\xi_2 - x_2)^2}$. Therefore,

$$\int_{\mathbb{S}([x_1, x_2, 0], ct)} f(\xi_1, \xi_2, 0) \tfrac{1}{ct}\, d\boldsymbol{\upsilon}_{ct} = 2 \int_{\mathbb{B}([x_1, x_2], ct)} f(\xi_1, \xi_2) \frac{d\xi_1\, d\xi_2}{\sqrt{(ct)^2 - (\xi_1 - x_1)^2 - (\xi_2 - x_2)^2}}.$$

Here $\mathbb{B}([x_1, x_2], r) = \{\boldsymbol{x} \in \mathbb{R}^2 : \|\boldsymbol{x}\|^2 = (x - x_1)^2 + (y - x_2)^2 \le r^2\}$ is the ball of radius r centered about the point $\boldsymbol{x}_o = (x_1, x_2)$. Since the ball exists on the plane $\mathbb{R}^2$, it is more accurately described as a disk.

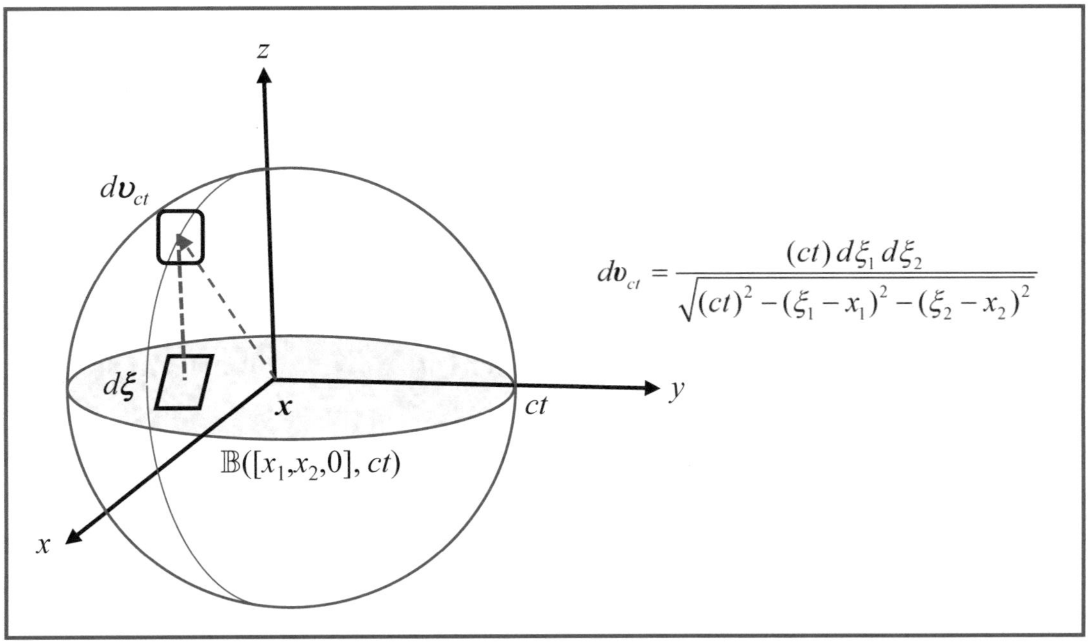

Figure 5.2. Projection of the element of surface area $d\boldsymbol{\upsilon}_{ct}$ onto the planar element $d\boldsymbol{\xi}$.

These calculations help to establish the projection from $n = 3$ to $n = 2$ dimensions via

$$\int_{\mathbb{S}_1} \tfrac{1}{ct}\cdot f(\boldsymbol{x}-ct\cdot\boldsymbol{v})\,d\boldsymbol{v} = \int_{\mathbb{S}_{ct}(\boldsymbol{x})} f(\boldsymbol{\xi})\tfrac{1}{ct}\,d\boldsymbol{v}_{ct} \mapsto 2\int_{\mathbb{B}_{ct}([x_1,x_2])} f(\xi_1,\xi_2)\frac{d\xi_1\,d\xi_2}{\sqrt{(ct)^2-(\xi_1-x_1)^2-(\xi_2-x_2)^2}}. \quad (5.4.1)$$

To solve the two–spatial dimensional wave equation

$$\left.\begin{array}{ll} \dfrac{\partial^2 A}{\partial t^2}(x_1,x_2,t) = c^2\Delta A(x_1,x_2,t)+\rho(x_1,x_2,t), & \boldsymbol{x}\in\mathbb{R}^2,\ t>0 \\ A(x_1,x_2,0)=\varphi(x_1,x_2) & \\ \dfrac{\partial A}{\partial t}(x_1,x_2,0)=\wp(x_1,x_2) & \end{array}\right\}, \quad (5.4.2)$$

it is sufficient to consider the three–dimensional solution (5.3.1) with $x_3 = 0$. That is, combine (5.3.1), (5.4.1), and (5.4.2) to obtain

$$\begin{aligned} A(x_1,x_2,t) = {} & \frac{1}{2\pi}\frac{\partial}{\partial t}\int_{\mathbb{B}([x_1,x_2],ct)} \frac{\varphi(\xi_1,\xi_2)}{\sqrt{(ct)^2-(\xi_1-x_1)^2-(\xi_2-x_2)^2}}\,d\xi_1\,d\xi_2 \\ & +\frac{1}{2\pi}\int_{\mathbb{B}([x_1,x_2],ct)} \frac{\wp(\xi_1,\xi_2)}{\sqrt{(ct)^2-(\xi_1-x_1)^2-(\xi_2-x_2)^2}}\,d\xi_1\,d\xi_2 \\ & +\frac{1}{2\pi}\int_0^t\int_{\mathbb{B}([x_1,x_2],ct)} \frac{\rho(\xi_1,\xi_2,c(t-\tau))}{\sqrt{(ct)^2-(\xi_1-x_1)^2-(\xi_2-x_2)^2}}\,d\xi_1\,d\xi_2\,d\tau. \end{aligned} \quad (5.4.3)$$

In this setting, $\varphi(x_1,x_2)=\phi(x_1,x_2,0)$, $\wp(x_1,x_2)=\psi(x_1,x_2,0)$, and $\rho(x_1,x_2,t)=p(x_1,x_2,0,t)$ are the Cauchy data and forcing functions from (5.0.4) projected into two–spatial dimensions. The function $\pi(\boldsymbol{y}, t) = \dfrac{1}{\sqrt{t^2-y_1^2-y_2^2}}$ is called the *Poisson kernel* with respect to the two–dimensional vector $\boldsymbol{y} = (y_1, y_2)$. If the Poisson kernel is multiplied by the indicator function for the disk of radius t centered at the origin $\boldsymbol{0}$ to produce $\Pi(\boldsymbol{y}, t) = \pi(\boldsymbol{y}, t)\cdot \mathscr{I}_{\mathbb{B}(\boldsymbol{0},t)}(\boldsymbol{y})$, then (5.4.3) can be written as the series of convolutions (5.4.4).

$$A(x_1,x_2,t) = \frac{1}{2\pi}\left(\frac{\partial}{\partial t}(\varphi\odot\Pi)(x_1,x_2,ct)+(\wp\odot\Pi)(x_1,x_2,ct)+\int_0^t(\rho\odot\Pi)(x_1,x_2,c(t-\tau))\,d\tau\right) \quad (5.4.4)$$

In some sense, equations (5.4.3)–(5.4.4) represent a two–dimensional Huygen's Principle in that the solution of the two–dimensional inhomogeneous wave equation depends only on the convolution of the Cauchy data and the forcing function with Poisson's kernel over the disk of radius ct centered at $\boldsymbol{x} = (x_1, x_2) \in \mathbb{R}^2$. Finally, it is a straightforward matter to see that the solution of the two–dimensional homogeneous wave equation can be derived from (5.4.3)–(5.4.4) by setting $\rho(x, t) = 0$.

It is left as an exercise to recover d'Alembert's traveling wave solution (4.5.5) of Chapter 4 for the one–dimensional inhomogeneous wave equation (4.5.3).

The general solutions of the homogeneous and inhomogeneous wave equations are directly related to (5.3.1)–(5.3.2). See Garabedian [26], John [43], Sobolev [71], and Tikhonov and Samarskii [79] for deeper details.

In this chapter, the solution of the homogeneous (4.4.1) and inhomogeneous (4.5.3) wave equations as introduced in Chapter 4, are solved for dimensions $n = 2, 3$. The Fourier transform is used to successfully determine the three–dimensional wave kernels recorded in Table 5.1. Equations (5.3.1)–(5.3.2) are the means through which Maxwell's equations are solved. Specific solutions of Maxwell's equations are determined via Exercises 5.2–5.5. The promise of solutions to Maxwell's equations, first alluded to in Chapter 1, is now fulfilled.

Exercises

5.1 Show, via *Leibniz's Rule*, that with respect to the three–dimensional wave solution

$$\frac{\partial}{\partial t}\left(\tfrac{1}{ct}\cdot M[\phi,-ct]\right) = -\tfrac{1}{ct^2}\int_{\mathbb{S}_1}\phi(x-ct\cdot\boldsymbol{v})\,d\boldsymbol{v} - \tfrac{1}{t}\int_{\mathbb{S}_1}\left(\nabla\bullet\phi(x-ct\cdot\boldsymbol{v})\right)\bullet\boldsymbol{v}\,d\boldsymbol{v}.$$

5.2 Let $\boldsymbol{J}(\boldsymbol{x}, t) = ½\cos(t)\cdot[\|\boldsymbol{x}\|^2, \|\boldsymbol{x}\|^2, \|\boldsymbol{x}\|^2]$ be the current field of an electromagnetic field with $\|\boldsymbol{x}\|^2 = x^2 + y^2 + z^2$, $\boldsymbol{x} = [x, y, z] \in \mathbb{R}^3$. Using the continuity equation $\frac{\partial\rho}{\partial t} + \varepsilon\nabla\bullet\boldsymbol{J} = 0$ with initial condition $\rho(\boldsymbol{x}, 0) = 0$, show that $\rho(\boldsymbol{x}, t) = -\varepsilon\sin(t)\cdot(x + y + z)$.

5.3 For the current density $\rho(\boldsymbol{x}, t)$ of Exercise 5.2 above, solve the inhomogeneous wave equation

$$\left.\begin{aligned} &\frac{\partial^2 u}{\partial t^2}(\boldsymbol{x},t) = c^2\Delta u(\boldsymbol{x},t) + \tfrac{4\pi}{\varepsilon}\rho(\boldsymbol{x},t), \qquad \boldsymbol{x}\in\mathbb{R}^3,\ t>0\\ &u(\boldsymbol{x},0) = \phi(\boldsymbol{x})\\ &\frac{\partial u}{\partial t}(\boldsymbol{x},0) = \psi(\boldsymbol{x}) \end{aligned}\right\} \tag{E5.3}$$

when $\phi(\boldsymbol{x}) = \|\boldsymbol{x}\|^2 = x^2 + y^2 + z^2$ and $\psi(\boldsymbol{x}) = \|\boldsymbol{x}\|^4$. *Hint*: Use formula (5.3.1) and show that

(*a*) $u_1(\boldsymbol{x}, t) = \int_{\mathbb{S}_1}\phi(\boldsymbol{x}-ct\cdot\boldsymbol{v})\,d\boldsymbol{v} = 2\pi^2\,(t^2 + \|\boldsymbol{x}\|^2)$,

(*b*) $u_2(\boldsymbol{x}, t) = \int_{\mathbb{S}_1}\psi(\boldsymbol{x}-ct\cdot\boldsymbol{v})\,d\boldsymbol{v} = 2\pi^2\,(t^2[t^2 + 3\|\boldsymbol{x}\|^2 + z^2] + \|\boldsymbol{x}\|^4)$, and (*c*) $u_3(\boldsymbol{x}, t) = \int_0^t \frac{1}{t-\tau} \int_{\mathbb{S}_1} p(\boldsymbol{x} - ct\cdot\boldsymbol{v}, t-\tau)\, d\boldsymbol{v}\, d\tau = -2\pi^2\, Si(t)\cdot(x + y + z)$.

Here $Si(t)$ is the *sine integral* function $Si(t) = \int_0^t \frac{\sin(\eta)}{\eta}\, d\eta$.

5.4 Given the solution $u(\boldsymbol{x}, t) = \frac{1}{4\pi}\frac{\partial}{\partial t}\left(\frac{1}{t}u_1(\boldsymbol{x},t)\right) + \frac{1}{4\pi}u_2(\boldsymbol{x},t) + \frac{1}{4\pi}u_3(\boldsymbol{x},t)$ of (E5.3) above use (*iii*) $\frac{\partial u}{\partial t}(\boldsymbol{x},t) - c\nabla\bullet\boldsymbol{A}(\boldsymbol{x},t) = 0$ from the beginning of Chapter 5 to find the field $\boldsymbol{A}$. *Hint*: Set $c\frac{\partial A_1}{\partial x} = \frac{1}{4\pi}\frac{\partial}{\partial t}\left(\frac{1}{t}u_1(\boldsymbol{x},t)\right)$, $c\frac{\partial A_2}{\partial y} = u_2(\boldsymbol{x},t)$, and $c\frac{\partial A_3}{\partial z} = u_3(\boldsymbol{x},t)$ with $\boldsymbol{x} = (x, y, z)$ to form $\boldsymbol{A}(\boldsymbol{x}, t) = [A_1(\boldsymbol{x}, t), A_2(\boldsymbol{x}, t), A_3(\boldsymbol{x}, t)]$.

5.5 From the results of Exercises 5.2–5.4, use equations (*i*) and (*ii*) from the beginning of Chapter 5 to find the magnetic field $\boldsymbol{B}$ and electric field $\boldsymbol{E}$ to solve Maxwell's equations.

5.6 Use the *method of descent* to show that the solution of the three–dimensional inhomogeneous equation (5.3.1) reduces to d'Alembert's result (4.5.5) of Chapter 4 by setting $x_2 = 0 = x_3$. *Hint*: Show that $\frac{1}{4\pi ct}\int_{\mathbb{S}_{ct}([x_1,x_2,x_3])} \phi(\xi_1,0,0)\, d\boldsymbol{v}_{ct} = \frac{1}{4\pi ct}\int_{\mathbb{S}_{ct}([x_1,0,0])} \phi(\xi_1,0,0)\, d\boldsymbol{v}_{ct} = \frac{1}{2c}\int_{x_1-ct}^{x_1+ct} f(\xi_1)\, d\xi_1$ where $f(\xi_1) = \phi(\xi_1, 0, 0)$.

Chapter 6

An Introduction to Nonlinear Partial Differential Equations

Through the first five chapters of this book, the focus has been on *linear* partial differential equations and the methods of solutions to Laplace's, Poisson's, Maxwell's, and the (homogeneous and inhomogeneous) heat and wave equations. The mathematical devices used to arrive at these solutions included Green's functions, separation of variables, Fourier series, and Fourier transforms. It is natural ask "How do we proceed in the *nonlinear* case?" Nonlinear partial differential equations are inherently more difficult to solve than their linear counterparts. Fourier series and transforms cannot be directly applied. In this chapter, a few special examples are examined and a glimpse at a general methodology (*transformation of variables*) for a specific class of nonlinear *PDE*s is provided.

A Separable Nonlinear Partial Differential Equation

As seen in Chapter 3, linear *PDE*s subject to boundary conditions over finite regions can be solved via the *separation of variables* method. For example, the Dirichlet problem over the unit cube, $\Delta u(\boldsymbol{x}) = 0$ for $\boldsymbol{x} \in \mathcal{C}_3(1) \equiv [0,1] \times [0,1] \times [0,1]$, can readily be solved by assuming the solution surface is the product of functions of each of the independent variables. That is, $u(\boldsymbol{x}) = X(x)\cdot Y(y)\cdot Z(z)$. This method can also be applied to the heat and wave equations. Such equations are called *separable*. One question arises (i.e., [63]) "Are there separable nonlinear *PDE*s?" This section is able to offer an affirmative answer via the *nonlinear Klein–Gordon* equation.

§6.1. Nonlinear Klein–Gordon Equation

It is established in Chapter 4 that the *inhomogeneous wave equation*

$$\left.\begin{aligned}
&\frac{\partial^2 u}{\partial t^2}(x,t) = \sigma^2 \frac{\partial^2 u}{\partial x^2}(x,t) + \phi(x,t), \text{ for } x \in \mathbb{R} \text{ and } t > 0 \\
&u(x,0) = f(x), \text{ for } x \in \mathbb{R} \\
&\frac{\partial u}{\partial t}(x,0) = g(x), \text{ for } x \in \mathbb{R}
\end{aligned}\right\} \tag{6.1.1}$$

has a traveling wave solution

$$\boxed{\begin{aligned} u(x,t) = \tfrac{1}{2}[&f(x+\sigma t) + f(x-\sigma t) + G(x+\sigma t) - G(x-\sigma t) \\ &+ \int_0^t \left(\Phi(x+\sigma(\tau - t),\tau) - \Phi(x-\sigma(\tau-t),\tau)\right) d\tau] \end{aligned}} \tag{6.1.2a}$$

where

$$G(x) = \tfrac{1}{\sigma}\int_0^x g(\xi)\, d\xi \tag{6.1.2b}$$

and

$$\Phi(x,t) = \tfrac{1}{\sigma}\int_0^x \phi(\xi,t)\, d\xi\,. \tag{6.1.2c}$$

This result (6.1.2*a*)–(6.1.2*c*) is known as *d'Alembert's solution*.

If, instead of the inhomogeneous term $\phi(x, t)$, a nonlinear monomial is substituted and an initial value in time (t) and space (x) is provided, then equation (6.1.1) becomes the *nonlinear Klein–Gordon* (*nKG*) equation (6.1.3).

$$\left.\begin{aligned} &\frac{\partial^2 u}{\partial t^2}(x,t) = \sigma^2 \frac{\partial^2 u}{\partial x^2}(x,t) + \alpha \bullet u^n(x,t), \text{ for } x \in \mathbb{R}, t > 0, \text{ and } n > 1 \\ &u(0,0) = 1 \end{aligned}\right\} \tag{6.1.3}$$

As is seen in the next section, a change of variables will greatly simplify the form of the *nKG*.

§6.2. Change of Variables

Since the inhomogeneous wave equation (6.1.1) admits a traveling wave solution, it seems reasonable to apply a *traveling–wave change of variables* to the *nKG* equation as per Strang [74, 75]. To that end, let

$$\left.\begin{aligned} \xi &= \tfrac{1}{2}(\sigma t + x) \\ \tau &= \tfrac{1}{2}(\sigma t - x) \end{aligned}\right\}. \tag{6.2.1}$$

A direct application of the chain rule yields $\dfrac{\partial u}{\partial t} = \dfrac{\partial u}{\partial \tau}\dfrac{\partial \tau}{\partial t} + \dfrac{\partial u}{\partial \xi}\dfrac{\partial \xi}{\partial t} = \tfrac{1}{2}\sigma\left(\dfrac{\partial u}{\partial \tau} + \dfrac{\partial u}{\partial \xi}\right)$ and, consequently, $\dfrac{\partial^2 u}{\partial t^2} = \tfrac{1}{4}\sigma^2\left(\dfrac{\partial^2 u}{\partial \tau^2} + \dfrac{\partial^2 u}{\partial \tau \partial \xi} + \dfrac{\partial^2 u}{\partial \xi \partial \tau} + \dfrac{\partial^2 u}{\partial \xi^2}\right)$.

In a similar manner, it is seen that $\frac{\partial u}{\partial x} = \tfrac{1}{2}\left(-\frac{\partial u}{\partial \tau} + \frac{\partial u}{\partial \xi} \right)$ and therefore,

$\frac{\partial^2 u}{\partial x^2} = \tfrac{1}{4}\left(\frac{\partial^2 u}{\partial \tau^2} - \frac{\partial^2 u}{\partial \tau \partial \xi} - \frac{\partial^2 u}{\partial \xi \partial \tau} + \frac{\partial^2 u}{\partial \xi^2} \right)$. Assume that the solution of (6.1.3) is sufficiently smooth so that $\frac{\partial^2 u}{\partial \tau \partial \xi} = \frac{\partial^2 u}{\partial \xi \partial \tau}$. That is, *u* is in the space of all twice continuously differentiable functions over the plane $u \in C^2(\mathbb{R}^2)$.

In such a circumstance, the change of variables (6.2.1) produces the simplification

$$\frac{\partial^2 u}{\partial t^2} - \sigma^2 \frac{\partial^2 u}{\partial x^2} = \tfrac{1}{4}\sigma^2 \left(\frac{\partial^2 u}{\partial \tau^2} + \frac{\partial^2 u}{\partial \tau \partial \xi} + \frac{\partial^2 u}{\partial \xi \partial \tau} + \frac{\partial^2 u}{\partial \xi^2} \right) - \tfrac{1}{4}\sigma^2 \left(\frac{\partial^2 u}{\partial \tau^2} - \frac{\partial^2 u}{\partial \tau \partial \xi} - \frac{\partial^2 u}{\partial \xi \partial \tau} + \frac{\partial^2 u}{\partial \xi^2} \right)$$
$$= \tfrac{1}{2}\sigma^2 \left(\frac{\partial^2 u}{\partial \tau \partial \xi} + \frac{\partial^2 u}{\partial \xi \partial \tau} \right) = \sigma^2 \frac{\partial^2 u}{\partial \tau \partial \xi}.$$

By equating $u(x,t) = u\left(\xi - \tau, \tfrac{1}{\sigma}(\xi + \tau)\right) = U(\xi, \tau)$, the *nKG* (6.1.3) becomes

$$\left.\begin{array}{l} \sigma^2 \dfrac{\partial^2 U}{\partial \tau \partial \xi} = \alpha \bullet U^n(\xi, \tau) \text{ for } (\xi, \tau) \in \mathbb{R}^2 \\ U(0,0) = 1 \end{array}\right\}. \tag{6.2.2}$$

Notice by (6.2.1), that $x = 0 = t$ implies that $\xi = 0 = \tau$. Hence, $u(0,0) = 1$ implies $U(0,0) = 1$. As is shown in the next section, (6.2.2) is separable. Since $n \neq 1$, then (6.2.2) is a nonlinear, separable *PDE*.

§6.3. Separation of Variables

As with linear *PDE*s assume that the solution *u* is a product of two independent functions of independent variables: $U(\xi, \tau) = \Xi(\xi) \bullet \mathrm{T}(\tau)$. In this case, $\frac{\partial^2 U}{\partial \tau \partial \xi}(\xi, \tau) = \Xi'(\xi) \bullet \mathrm{T}'(\tau)$ and the *nKG* (6.2.2) becomes $\sigma^2\, \Xi'(\xi) \bullet \mathrm{T}'(\tau) = \alpha \bullet \Xi^n(\xi) \bullet \mathrm{T}^n(\tau)$ or $\sigma^2 \frac{\Xi'(\xi)}{\Xi^n(\xi)} = \alpha \bullet \frac{\mathrm{T}^n(\tau)}{\mathrm{T}'(\tau)} \equiv \lambda$, λ an arbitrary constant. This ratio forms a separable differential equation. Moreover, the initial condition $1 = U(0,0) = \Xi(0) \bullet \mathrm{T}(0)$ combines with the *ODE* above to form the two distinct equations below.

$$\left.\begin{aligned} &\sigma^2 \frac{d\Xi}{d\xi}(\xi) = \lambda\,\Xi^n(\xi) \\ &\Xi(0) = 1 \end{aligned}\right\} \tag{6.3.1a}$$

$$\left.\begin{aligned} &\lambda \frac{d\mathrm{T}}{d\tau}(\tau) = \alpha \bullet \mathrm{T}^n(\tau) \\ &\mathrm{T}(0) = 1 \end{aligned}\right\} \tag{6.3.1b}$$

These are first order differential equations and are readily integrable. After a straightforward calculation, it is seen that

$$\Xi(\xi) = \left(\frac{\lambda(1-n)}{\sigma^2}\xi + 1\right)^{\frac{1}{1-n}} \tag{6.3.2a}$$

and

$$\mathrm{T}(\tau) = \left(\frac{\alpha(1-n)}{\lambda}\tau + 1\right)^{\frac{1}{1-n}} \tag{6.3.2b}$$

are the solutions of (6.3.1a) and (6.3.1b), respectively.

Therefore, $u(x,\ t) = U(\xi,\tau) = \Xi(\xi)\bullet\mathrm{T}(\tau) = \left[\left(\frac{\lambda(1-n)}{\sigma^2}\xi + 1\right)\bullet\left(\frac{\alpha(1-n)}{\lambda}\tau + 1\right)\right]^{\frac{1}{1-n}}$. After some algebraic simplification, the solution of the nKG equation (6.2.1) is the traveling wave solution presented in (6.3.3).

$$\boxed{u(x,t) = \left[\alpha\cdot\left(\frac{1-n}{2\sigma}\right)^2\cdot(\sigma t + x)\bullet(\sigma t - x) + \left(\frac{\alpha(1-n)}{2\lambda}\right)\cdot(\sigma t - x) + \left(\frac{\lambda(1-n)}{2\sigma^2}\right)\cdot(\sigma t + x) + 1\right]^{\frac{1}{1-n}}} \tag{6.3.3}$$

Observe that (6.3.3) can be written as

$$\boxed{u(x,t) = \left[\left(\frac{\lambda(1-n)}{2\sigma^2}(\sigma t + x) + 1\right)\cdot\left(\frac{\alpha(1-n)}{2\lambda}(\sigma t - x) + 1\right)\right]^{\frac{1}{1-n}}}\,. \tag{6.3.4}$$

Plainly, when $n > 1$, the exponent is negative: $\frac{1}{1-n} < 0$. Consequently, whenever either factor of the base is zero, then the solution is singular. Therefore

$$\frac{\lambda(1-n)}{2\sigma^2}(\sigma t + x) + 1 = 0 \Rightarrow x = \frac{2\sigma^2}{\lambda(n-1)} - \sigma t \tag{6.3.5a}$$

$$\frac{\alpha(1-n)}{2\lambda}(\sigma t - x) + 1 = 0 \Rightarrow x = \frac{-2\lambda}{\alpha(n-1)} + \sigma t \tag{6.3.5b}$$

are singularities of (6.3.4).

Figure 6.1 below depicts a solution surface with the wave speed σ =1, nonlinear dispersion coefficient $\alpha = 1$, the separation parameter $\lambda = 1$, and the degree of nonlinearity $n = 2$. Note that along the lines of singularity (6.3.5*a*)–(6.5.3*b*), the solution surface is undefined and flips from concave up to concave down.

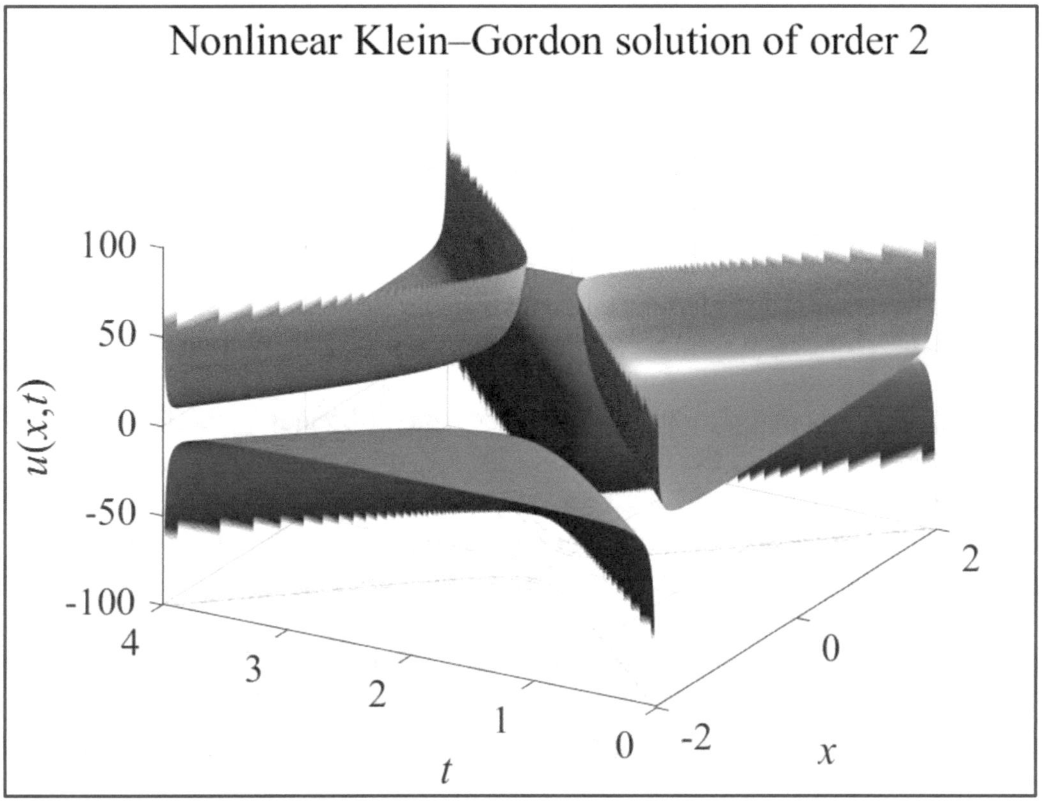

Figure 6.1. Solution of the second order Klein–Gordon equation (6.3.4) with $\sigma = \alpha = \lambda = 1$.

Example 6.3.1: Figure 6.2 illustrates the contour lines of the solution surface in the case of $n = 2$, $\sigma = \alpha = \lambda = 1$. More specifically, for the stated values of these parameters, the solution (6.3.4) becomes singular on the lines $x = 2 - t$ and $x = -2 + t$. These *lines of singularity* are indicated as dashed black (----) in Figure 6.2. The four quadrants over which the solution surface can divided indicate where the surface tends toward $+\infty$ (quadrants I and III) and $-\infty$ (quadrants II and IV). Therefore, the solution surface is not only discontinuous along the lines of singularities but changes concavity and direction.

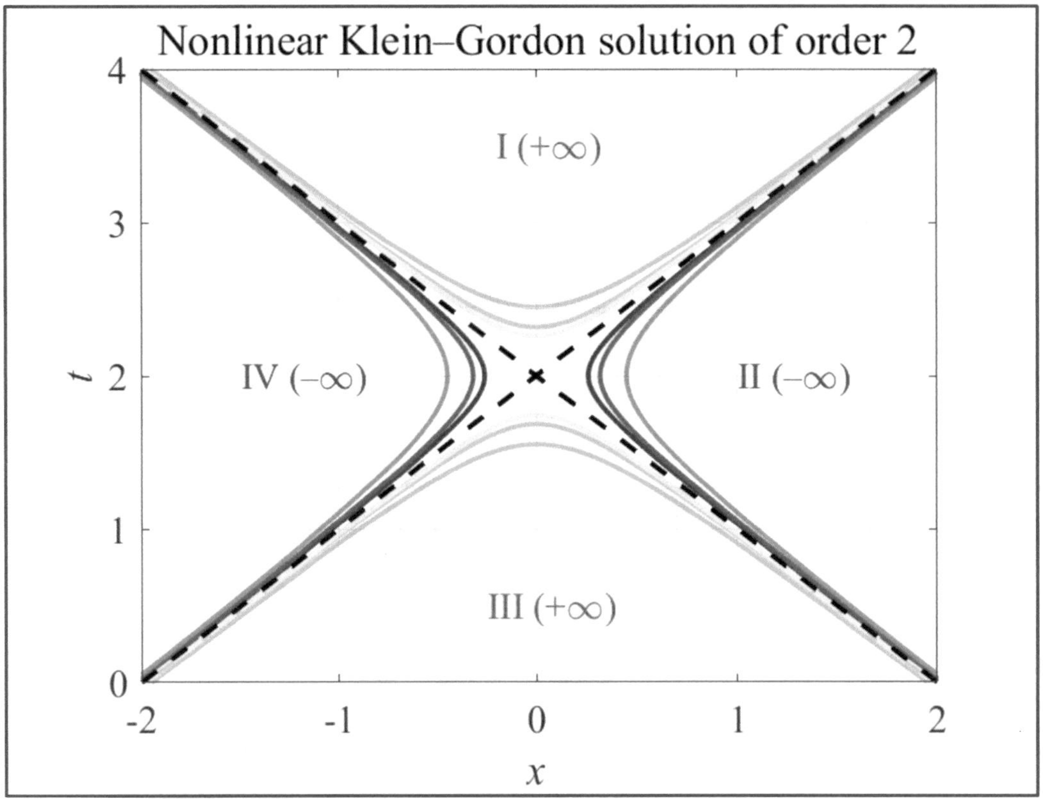

Figure 6.2. Contour and characteristic lines of the second order *nKG* equation.

Remark 6.1: *(i)* It is an open problem to determine a set of boundary conditions that will yield distinct (and defined) values of the constant λ. Indeed, this constitutes a nonlinear eigenvalue problem for the *nKG* equation (6.1.3). For a more detailed discussion see, for example, Berger and Berger [9]. It is left as an exercise to the interested reader to determine the solution surface for varying values of n, σ, α, and λ.

(ii) Rather than use the constant Cauchy data $u(0, 0) = 1$ in the *nKG* equation (6.1.3) consider the general initial condition $u(x, 0) = f(x)$. Under the change of variables (6.2.1) and the identification $u(x, t) = U(\xi, \tau)$ the top equation in (6.2.2) is obtained. Reversing (6.2.1), produces the relations $x = \xi - \tau$ and $t = \frac{1}{\sigma}(\xi + \tau)$. Hence, $u(x, 0) = f(x) = f(\xi - \tau)$. When $t = 0$, however, $\xi = -\tau$ so that $u(x, 0) = U(2\xi, 0)$ as $x = \xi - \tau = 2\xi$. Subsequently, (6.2.2) can now be written as

$$\left.\begin{aligned} &\sigma^2 \frac{\partial^2 U}{\partial \tau \partial \xi}(\xi,\tau) = \alpha \cdot U''(\xi,\tau) \quad \text{for } (\xi,\tau) \in \mathbb{R}^2 \\ &U(2\xi,0) = f(\xi - \tau) \end{aligned}\right\}. \tag{6.3.6}$$

The separation of variables transform $U(\xi, \tau) = \Xi(\xi)\cdot\mathrm{T}(\tau)$ will *not* yield independent *ODE*s in ξ and τ for general Cauchy data $f(x)$. In the special case of $f(x + y) = g(x)\cdot h(y)$, the separation of variables transform produces the set of first order *ODE*s below.

$$\left.\begin{aligned}&\sigma^2\frac{d\Xi}{d\xi}(\xi)=\lambda\cdot\Xi^n(\xi)\\&\Xi(0)=g(0)\end{aligned}\right\}\quad(a)\qquad\qquad \left.\begin{aligned}&\lambda\cdot\frac{d\mathrm{T}}{d\tau}(\tau)=\alpha\cdot\mathrm{T}^n(\tau)\\&\mathrm{T}(0)=h(-\tau)\end{aligned}\right\}\quad(b)$$

Indeed, the initial condition $\Xi(2\xi)\cdot\mathrm{T}(0) = U(2\xi, 0) = f(\xi - \tau) = g(\xi)\cdot h(-\tau)$ implies that $\mathrm{T}(0) = h(-\tau)$, $\Xi(2\xi) = g(\xi)$, and hence $\Xi(0) = g(0)$. The resulting solutions of (*i*) and (*ii*) are combined to form the solution (6.3.6) of the *nKG* with initial condition $U(2\xi, 0) = f(\xi - \tau)$.

$$\boxed{U(\xi,\tau)=\left[\left(\frac{\lambda(1-n)}{\sigma^2}\xi+g(0)^{1-n}\right)\cdot\left(\frac{\alpha(1-n)}{\lambda}\tau+h(-\tau)^{1-n}\right)\right]^{\frac{1}{1-n}}}\qquad(6.3.7)$$

When $g(\xi) = 1 = h(\tau)$ and hence $f(x) = 1$, then (6.3.7) is equivalent to (6.3.4).

(iii) The nonlinear Klein–Gordon equation presents an opportunity to instruct the reader of the importance of units. This section describes not only a mathematical development but also a physical problem. Partial differential equations are used extensively in the mathematical modeling of physical phenomena. In particular, the *nKG* equation (6.1.3) models nonlinear pulses and dispersive waves. Therefore, it is imperative that the solution makes sense physically. One such check is to ensure that the units of the solution are consistent. Assume that the solution $u(x, t)$ of (6.1.3) models wave displacement. It is ascribed the unit of length L. Consequently, the acceleration terms $\frac{\partial^2 u}{\partial t^2}$ and $\frac{\partial^2 u}{\partial x^2}$ are measured in units $\frac{L}{T^2}$ and $\frac{L}{L^2}=\frac{1}{L}$, respectively. To insure that the equation (6.1.3) is consistent across units (that is, consistent *dimensionally*), the constants σ and α must have units $\frac{L}{T}$ and $\frac{L^{1-n}}{T^2}$, respectively. From these assumptions and (6.2.1), the variables ξ and τ are both measured in units of L. To guarantee that the solution (6.3.4) of (6.1.3) is measured in units of L, select λ to have units $\frac{L}{T^2}$. In this case, the first term of (6.3.4) $\left(\frac{\lambda(1-n)}{2\sigma^2}(\sigma t+x)+1\right)$ is dimensionless with units $\frac{L}{T^2}\frac{T^2}{L^2}L=1$. Conversely, the second term $\left(\frac{\alpha(1-n)}{2\lambda}(\sigma t-x)+1\right)$ has units $\frac{L^{1-n}}{T^2}\frac{T^2}{L}L=L^{1-n}$. Since the solution $u(x, t)$ is the product of these two terms raised to the power $\frac{1}{1-n}$, it is apparent that $u(x, t)$ is measured in units of L.

§6.4. Burgers' Equation and a Nonlinear Transformation

Just as the previous section indicates how the inhomogeneous wave equation can be altered to form a nonlinear *PDE*, so too can the heat equation (4.3.4) of Chapter 4. To simplify matters, consider the case of $n = 1$ spatial dimension. Replace the inhomogeneous term $\rho(x, t)$ with the nonlinear expression $-u(x,t)\bullet\frac{\partial u}{\partial x}(x,t)=-\frac{1}{2}\frac{\partial}{\partial x}\left(u(x,t)^2\right)$. In this case, the inhomogeneous heat equation becomes *Burgers' equation* [14].

$$\left.\begin{aligned}&\frac{\partial u}{\partial t}(x,t)=\sigma^2\frac{\partial^2 u}{\partial x^2}(x,t)-\tfrac{1}{2}\frac{\partial}{\partial x}\left(u(x,t)^2\right),\ \text{for } x\in\mathbb{R} \text{ and } t>0\\ &u(x,0)=f(x),\ \text{for } x\in\mathbb{R}\end{aligned}\right\}\tag{6.4.1}$$

Applying the Fourier transform to the *PDE* above yields the nonlinear integro–differential equation below.

$$\frac{d}{dt}\mathscr{F}[u](\omega)=-(\sigma\omega)^2\,\mathscr{F}[u](\omega)-\frac{i\omega}{2}\mathscr{F}[u^2](\omega)\tag{6.4.2}$$

There is no analytic solution of (6.4.2). How can a solution be determined? Since a linear [Fourier] transformation does not produce a manageable simplification of (6.4.1), it seems reasonable to try a *nonlinear transformation*. The transformation which achieves an analytic form of Burgers' equation is described next.

§6.5. Hopf–Cole Transformation

The value of the Fourier transform in solving *linear* partial differential equations is its universality. That is, an application of the Fourier transform to a linear[17] *PDE* produces a linear ordinary differential equation. For nonlinear equations, however, this methodology fails in even the simplest cases. Hence, a different kind of transform is warranted. E. Hopf [40] and J. Cole determined that the following change of variables will linearize Burgers' equation (6.4.1).

More precisely, setting

$$u(x,t)=-2\sigma^2\frac{\partial}{\partial x}\ln\left(v(x,t)\right)=\frac{-2\sigma^2}{v(x,t)}\frac{\partial v}{\partial x}(x,t)\tag{6.5.1}$$

will reduce (6.4.1) to the linear heat equation. The change of variables (6.5.1) is called the *Hopf–Cole transformation*. The consequence of the transform can be evaluated in the following two-step process (as per Myint–U with Debnath [52]). First, unraveling (6.5.1) results in the following representation for $v(x, t)$.

$$v(x,t)=\exp\left(-\tfrac{1}{2\sigma^2}\int_{-\infty}^{x}u(\xi,t)\,d\xi\right)\Leftrightarrow v(x,0)=\exp\left(-\tfrac{1}{2\sigma^2}\int_{-\infty}^{x}f(\xi)\,d\xi\right)\tag{6.5.2}$$

Now, let $u(x, t)=\frac{\partial w}{\partial x}(x,t)$ for a sufficiently smooth function w. Then, using a bit of calculus, the substitution of this transformation into (6.4.1) yields

[17] With constant coefficients

$$\frac{\partial}{\partial x}\left[\frac{\partial w}{\partial t}+\tfrac{1}{2}\left(\frac{\partial w}{\partial x}\right)^2\right]=\sigma^2\frac{\partial}{\partial x}\left[\frac{\partial^2 w}{\partial x^2}\right] \tag{6.5.3a}$$

and after integrating with respect to x gives

$$\frac{\partial w}{\partial t}+\tfrac{1}{2}\left(\frac{\partial w}{\partial x}\right)^2=\sigma^2\frac{\partial^2 w}{\partial x^2}. \tag{6.5.3b}$$

Next, set $w(x, t) = -2\sigma^2 \ln(v(x, t))$ and note that $\frac{\partial w}{\partial t}=-2\sigma^2\frac{1}{v}\frac{\partial v}{\partial t}$, $\frac{\partial w}{\partial x}=-2\sigma^2\frac{1}{v}\frac{\partial v}{\partial x}$, and $\frac{\partial^2 w}{\partial x^2}=$ $-2\sigma^2\frac{1}{v^2}\left(\frac{\partial^2 v}{\partial x^2}-\left[\frac{\partial v}{\partial x}\right]^2\right)$. Inserting these identifications into (6.5.3b) results in $-2\sigma^2\frac{1}{v}\frac{\partial v}{\partial t}$ + $\frac{4\sigma^2}{2}\frac{1}{v^2}\left(\frac{\partial v}{\partial x}\right)^2 = -2\sigma^2\frac{1}{v^2}\left(v\frac{\partial^2 v}{\partial x^2}-\left[\frac{\partial v}{\partial x}\right]^2\right)$. After some simplification, it is observed that v satisfies the heat equation

$$\frac{\partial v}{\partial t}=\sigma^2\frac{\partial^2 v}{\partial x^2}. \tag{6.5.4}$$

Consequently, $u = \frac{\partial}{\partial x}(w) = -2\sigma^2\frac{\partial}{\partial x}\ln(v)$ and $v(x, t)$ satisfies the heat equation with initial condition $v(x, 0)$ of (6.5.2). Therefore, the Hopf–Cole transformation (6.5.1) produces a solution of Burgers' equation (6.4.1). By setting $F(\xi)=\int_{-\infty}^{\xi} f(z)\,dz$ and using formulae (6.5.2) above and (4.3.2) of Chapter 4, it is seen that

$$v(x,t)=\frac{1}{2\sqrt{\pi}\,\sigma t}\int_{-\infty}^{\infty}\exp\left(-\tfrac{1}{2\sigma^2}\left[F(\xi)+\frac{(x-\xi)^2}{2t}\right]\right). \tag{6.5.5}$$

Equations (6.5.1) and (6.5.5) constitute the analytic solution of Burgers' equation via the Hopf–Cole transformation. The process is to apply a nonlinear transformation to a nonlinear *PDE* in order to obtain a linear equation. In the next section, a more sophisticated transform is applied to the much studied *Korteweg de Vries equation*. The, as yet unnamed transform will map the nonlinear *KdV* equation into a linear equation.

§6.6. The Korteweg de Vries Equation and Inverse Scattering

The Korteweg de Vries or *KdV* equation (6.6.1) is one of the primary examples of *integrable* or analytically solvable nonlinear partial differential equations.

$$\frac{\partial u}{\partial t}-6u\frac{\partial u}{\partial x}+\frac{\partial^3 u}{\partial x^3}=0 \tag{6.6.1}$$

In 1895, the Dutch mathematician Diederik Johannes Korteweg and his student Gustav de Vries derived equation (6.6.1) above to model shallow water waves [47, 84]. An immediate calculation confirms that (6.6.1) may be written as

$$\frac{\partial u}{\partial t}+\frac{\partial}{\partial x}\left(-3u^2+\frac{\partial^2 u}{\partial x^2}\right)=0. \tag{6.6.2}$$

Remark 6.2: The *KdV* equation is also frequently written with a positive coefficient in the nonlinear term (i.e., +6 versus –6). A direct calculation shows that if u satisfies (6.6.1), then $v = -u$ satisfies $\frac{\partial v}{\partial t}+6v\frac{\partial v}{\partial x}+\frac{\partial^3 v}{\partial x^3}=0$. Throughout this book, equation (6.6.1) will be used to represent the *KdV* equation.

Traveling Wave Solution

It has been observed by Korteweg and de Vries [47], Kruskal and Zabusky [49], and many others that the *KdV* equation (6.6.1) admits a *traveling wave solution* of the form $u(x, t) = \phi(x - \gamma t)$. Making this assumption and the substitution $\xi = x - \gamma t$, results in

$$\frac{d}{d\xi}\left(3\phi^2+\gamma\phi-\phi''\right)=0. \tag{6.6.3}$$

Equation (6.6.3) is the consequence of the following computations. The identification, $u(x, t) = \phi(x - \gamma t) = \phi(\xi)$ implies $\frac{\partial u}{\partial t}=-\gamma\phi'(\xi)$, $\frac{\partial u}{\partial x}=\phi'(\xi)$, $\frac{\partial^2 u}{\partial x^2}=\phi''(\xi)$, and $\frac{\partial^3 u}{\partial x^3}=\phi'''(\xi)$ which, when substituted into (6.6.1), produces (6.6.3).

To mimic real wave motion, the additional constraint $\phi(\xi)\to 0$ as $|\xi|\to+\infty$ is imposed upon the solution of (6.6.3). Integrating (6.6.3) with respect to ξ and imposing the aforementioned asymptotic boundary condition yields $3\phi^2+\gamma\phi-\phi''=0$. Multiplying this equation through by ϕ' results in $3\phi^2\phi'+\gamma\phi\phi'-\phi''\phi'=0$ or $\frac{d}{d\xi}\left(\phi^3+\tfrac{1}{2}\gamma\phi^2-\tfrac{1}{2}(\phi')^2\right)=0$. Integrating this differential equation and imposing the asymptotic boundary requirement produces $(\phi')^2=\phi^2(\gamma+2\phi)$. That is, $d\xi=\frac{d\phi}{\phi\sqrt{\gamma+2\phi}}$. For $t_o = 0$ and $x(t_o) = x_o$, then $\xi_o = \xi(t_o) = x_o - \gamma t_o = x_o$. Integration of the differential form above from ξ to $\xi_o = x_o$ gives $\frac{1}{\sqrt{\gamma}}\ln\left(\frac{\sqrt{\gamma+2\phi}-\sqrt{\gamma}}{\sqrt{\gamma+2\phi}+\sqrt{\gamma}}\right)=\xi-x_o$, provided $\gamma > 0$. After a considerable amount of algebra, the traveling wave solution (6.6.4) is obtained.

$$u(x,t)=\phi(x-\gamma t)=\tfrac{1}{2}\gamma\,\mathrm{sech}^2\left(\tfrac{\sqrt{\gamma}}{2}\left[x-\gamma\cdot t-x_o\right]\right) \tag{6.6.4}$$

Figure 6.3 illustrates a traveling wave with $\gamma = 2$ and $x_o = 1$.

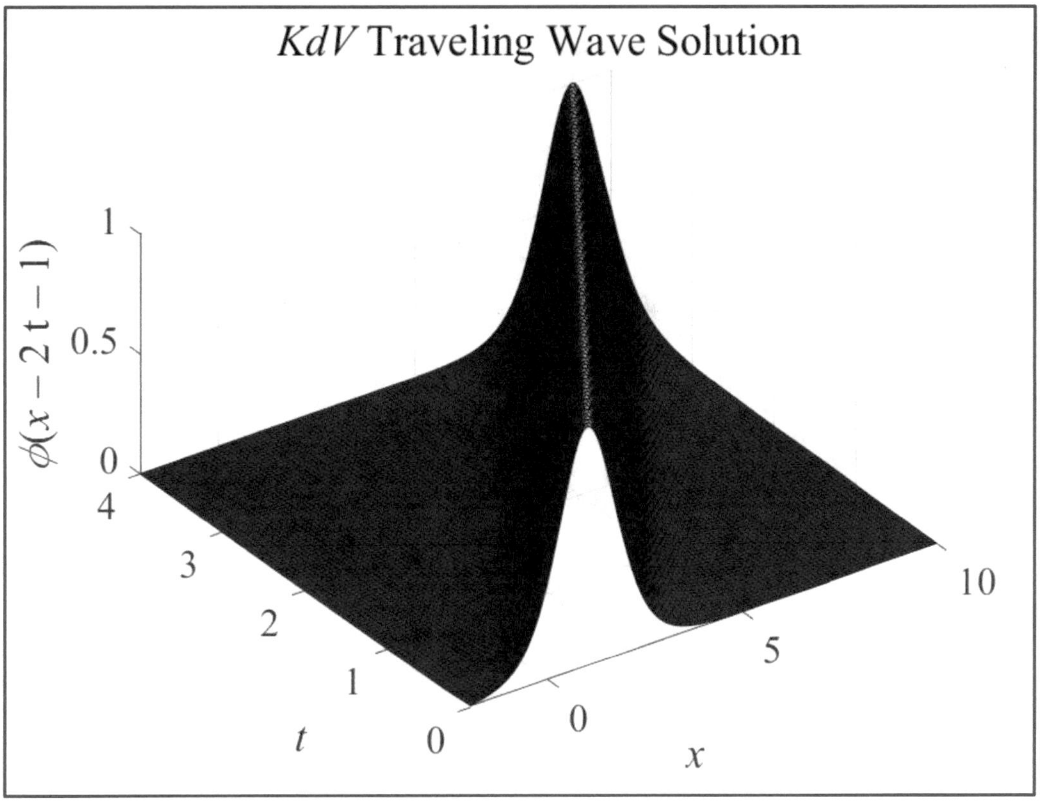

Figure 6.3. Traveling wave solution $u(x, t) = \phi(x - \gamma t - x_o)$, $\gamma = 2$, $x_o = 1$.

Solitons

An examination of Figure 6.3 reveals that the traveling wave solution of the *KdV* equation moves in a single or "solitary" form. The Scottish engineer, John Scott Russell [86], observed such a traveling wave caused by a boat moving through a canal. Russell referred to this as a "wave of translation" and later a solitary wave. Following Russell, physicists adopted the moniker *soliton* to describe solitary waves.

Functions of the form (6.6.4) are called *soliton solutions* of the *KdV* equation. These phenomena have the distinct mathematical characteristic of being invariant to change in speed and amplitude after interaction of two solitons. As Figures 6.4 and 6.5 illustrate, two solitons with different amplitudes and speeds will collide, have a nonlinear interaction at the collision point, and then move past one another unchanged. The nonlinear interaction is depicted in Figure 6.5 as the spike in the soliton of greater amplitude at the two solitons' crossing point. The MATLAB code used to generate these graphs is provided on the website listed below.

http://www.morganclaypoolpublishers.com/Costa/

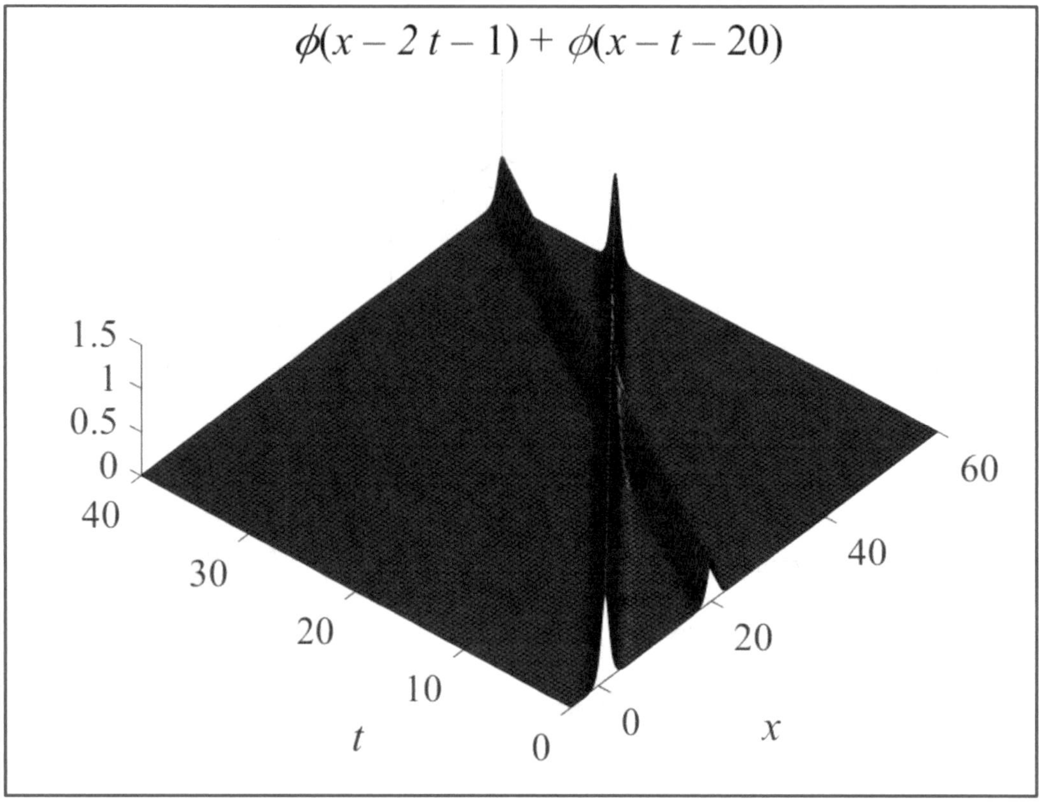

Figure 6.4. Two *solitons* passing through one another.

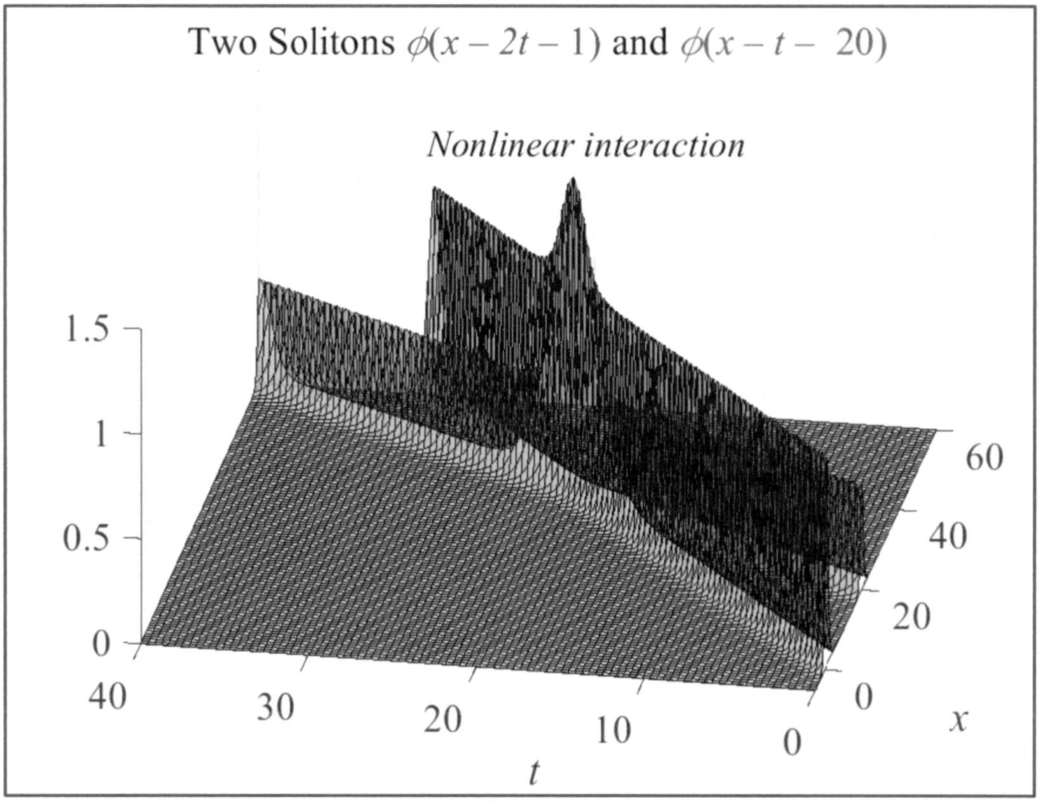

Figure 6.5. Nonlinear interaction of two *solitons*.

The Cauchy Problem and Inverse Scattering

The *KdV* equation (6.6.1) is not governed by initial conditions. As a consequence, the traveling wave/soliton solution (6.6.4) only applies to waves whose initial value $u(x, 0) = \frac{\lambda}{2}\operatorname{sech}^2\left(\frac{\sqrt{\gamma}}{2}[x - x_o]\right)$. What solutions can be formed with respect to the *KdV* equation with general Cauchy data $u(x, 0) = f(x)$ as described by (6.6.5)?

$$\left.\begin{aligned} &\frac{\partial u}{\partial t} - 6u\frac{\partial u}{\partial x} + \frac{\partial^3 u}{\partial x^3} = 0 \\ &u(x,0) = f(x) \end{aligned}\right\} \tag{6.6.5}$$

The Miura transformation [27, 28]

$$u = -\left(v^2 + \frac{\partial v}{\partial x}\right) \tag{6.6.6}$$

results in the following identities

$$\frac{\partial u}{\partial t} = -\left(2v\frac{\partial v}{\partial t} + \frac{\partial^2 v}{\partial t \partial x}\right), \quad \frac{\partial u}{\partial x} = -\left(2v\frac{\partial v}{\partial x} + \frac{\partial^2 v}{\partial x^2}\right), \quad \frac{\partial^2 u}{\partial x^2} = -\left(2v\frac{\partial^2 v}{\partial x^2} + 2\left[\frac{\partial v}{\partial x}\right]^2 + \frac{\partial^3 v}{\partial x^3}\right).$$

Assuming that u is a solution of the *KdV* equation (6.6.5), using the alternate form of the *KdV* (6.6.2), and making the substitutions above yields

$$\begin{aligned} 0 &= -\left(2v\frac{\partial v}{\partial t} + \frac{\partial^2 v}{\partial t \partial x}\right) + \frac{\partial}{\partial x}\left(3\left[-(v^2 + \frac{\partial v}{\partial x})\right]^2 - \left(2v\frac{\partial^2 v}{\partial x^2} + 2\left[\frac{\partial v}{\partial x}\right]^2 + \frac{\partial^3 v}{\partial x^3}\right)\right) \\ &= -\left(2v\frac{\partial v}{\partial t} + \frac{\partial^2 v}{\partial t \partial x}\right) + 12v^3\frac{\partial v}{\partial x} + 6v^2\frac{\partial^2 v}{\partial x^2} + 12v\left(\frac{\partial v}{\partial x}\right)^2 + 2v\frac{\partial^3 v}{\partial x^3} + \frac{\partial^4 v}{\partial x^4} \\ &= -\left(2v + \frac{\partial}{\partial x}\right)\left(\frac{\partial v}{\partial t} + 6v^2\frac{\partial v}{\partial x} + \frac{\partial^3 v}{\partial x^3}\right) \equiv -\left(2v + \frac{\partial}{\partial x}\right)M(v) \end{aligned}$$

with $M(v)$ as defined in (6.6.7).

This is referred to as the *modified KdV equation*

$$M(v) \equiv \frac{\partial v}{\partial t} + 6v^2\frac{\partial v}{\partial x} + \frac{\partial^3 v}{\partial x^3} = 0. \tag{6.6.7}$$

Hence, $\frac{\partial u}{\partial t} + 6u\frac{\partial u}{\partial x} + \frac{\partial^3 u}{\partial x^3} \equiv K(u) = -\left(2u + \frac{\partial}{\partial x}\right)M(u)$ and this can be used to show that the *KdV* equation has an infinite number of conserved quantities, see Ablowitz and Segur [1] for details.

The substitution $v = \frac{1}{\psi}\frac{\partial \psi}{\partial x}$ into the Miura transform (6.6.6) produces the relation $u = -\frac{1}{\psi}\frac{\partial^2 \psi}{\partial x^2}$. That is,

$$\frac{\partial^2 \psi}{\partial x^2} + u\psi = 0. \tag{6.6.8}$$

It is not difficult to show that the solutions of the *KdV equation* are invariant under translation by a constant[18]. More specifically, let $w = u + \lambda$ where u is a solution of the *KdV equation* and λ is a constant. Since $u = w - \lambda$, a direct computation shows $0 = K(u) = \frac{\partial}{\partial t}(w - \lambda) - 6(w - \lambda)\frac{\partial}{\partial x}(w - \lambda) + \frac{\partial^3}{\partial x^3}(w - \lambda)$. Thus, the *KdV* is *Galilean invariant*. Subsequently, the solution u of the *KdV* in (6.6.8) can be replaced by $w = u + \lambda$ or

$$\frac{\partial^2 \psi}{\partial x^2} + (u + \lambda)\psi = 0. \tag{6.6.9}$$

Equation (6.6.9) is the steady–state or *time–independent linear Schrödinger equation*. The phrase "time–independent" is deceptive. While equation (6.6.9) *does not evolve in time t*, the functions u, ψ, and λ nevertheless depend on time. Hence, (6.6.9) can properly be written as $\frac{\partial^2 \psi}{\partial x^2}(x, t) + \big(u(x, t) + \lambda(t)\big)\psi(x, t) = 0$. For now, the explicit dependence on time will be suppressed in the notation. Equation (6.6.9) implies that $u = -\lambda - \frac{1}{\psi}\frac{\partial^2 \psi}{\partial x^2}$. When this value of u is inserted into the *KdV*, as per Infeld and Rowlands [41], then (6.6.9) results in

$$\left.\begin{aligned} \psi^2 \frac{\partial \lambda}{\partial t} &= \frac{\partial}{\partial x}\left(\frac{\partial \psi}{\partial x}\Theta - \psi\frac{\partial \Theta}{\partial x}\right) \\ \Theta &= \frac{\partial \psi}{\partial t} + \frac{\partial^3 \psi}{\partial x^3} - 3(\lambda - u)\frac{\partial \psi}{\partial x} \end{aligned}\right\}. \tag{6.6.10}$$

Next integrate the top equation in (6.6.10) over the real line $\mathbb{R}$ with respect to x and assume that the Schrödinger function $\psi(x, t)$ and all of its derivatives descend to 0 as the spatial variable x approaches either extent of $\mathbb{R}$. That is, assume $\psi(x,t) \to 0$ and $\frac{\partial \psi}{\partial x}(x,t) \to 0$ as $x \to \pm\infty$. Then the integral of the right-hand side of the top equation in (6.6.10) is $\left.\left(\frac{\partial \psi}{\partial x}\Theta - \psi\frac{\partial \Theta}{\partial x}\right)\right|_{x=-\infty}^{\infty}$ which by the assumptions above is 0. Therefore,

[18] That is, Galilean invariant.

$$\frac{\partial \lambda}{\partial t}\int_{-\infty}^{\infty} \psi^2(x,t)\,dx = 0. \tag{6.6.11}$$

The linear Schrödinger equation (6.6.9) has two well-known solutions: *bound states* and *scattering states*. The former (bound) states are bounded by $e^{-|x|}$ as $x \to \pm\infty$. Scattering states, conversely, are asymptotically oscillatory $e^{-i|x|}$ as $x \to \pm\infty$, see Infeld and Rowlands [41], Vvedensky [80], and Ablowitz and Segur [1] for details. It is assumed that there are M bound states $\psi_m(x, t)$ so that the integral $\int_{-\infty}^{\infty} \psi_m^2(x,t)\,dx$ from (6.6.11) exists for $m = 1, 2, \dots, M$. If in addition to these assumptions, $u(x, t)$ is a solution to the *KdV* equation (6.6.1), then the associated eigenvalues λ_m from (6.6.9) are constant. Indeed if (6.6.11) holds, then the derivation of λ with respect to t must be zero: $\frac{\partial \lambda}{\partial t} = 0$. Substituting $\psi_m(x, t)$ for $\psi(x, t)$ in (6.6.11) requires that the associated eigenvalues λ_m are constant. Finally, restrict solutions $u(x, t)$ of the *KdV* equation to those which vanish on the extreme values of the real line so that $u(x, t) \to 0$ as $x \to \pm\infty$. In this case, the eigenfunctions $\psi_m(x, t)$ are asymptotically $C_m(t)\, e^{-k_m x}$ and $\lambda_m = k_m^2$, $m = 1, 2, \dots, M$. Gardner, Greene, Kruskal, and Miura [27, 28] have shown that $C_m(t) = C_m(0)\cdot e^{4k_m^3 t}$ and the associated eigenfunctions $\psi_m(x, t)$ are asymptotically $C_m(0)e^{4k_m^3 t}e^{-k_m x} = C_m(0)e^{k_m(4k_m^2 t - x)}$ which is a traveling wave solution. The *scattering states* have eigenvalues $\lambda = k^2$ for $k \in \mathbb{R}$ with corresponding asymptotic eigenfunctions $\psi(x) \to \begin{cases} e^{-ikx} + b(k)e^{ikx} & \text{as } x \to \infty \\ a(k)e^{-ikx} & \text{as } x \to -\infty \end{cases}$. These forms are reported by Infeld and Rowlands [41], Vvedensky [80], Ablowitz and Segur [1], and Polyanin and Zaitsev [61].

The solution of (6.6.9) under the special case where the potential $u(x, t)$ is replaced by its initial condition $u(x, 0) = f(0)$ is the *scattering problem*. The *inverse scattering problem* is to find the potential function from the eigenfunctions. This is achieved for the *KdV* equation via

$$u(x, t) = 2\,\frac{\partial}{\partial x}K(x,x;t) \tag{6.6.12}$$

where the kernel $K(x, y)$ is a solution of the *Gel'fand–Levitan–Marchenko* (*GLM*) equation

$$K(x,y;t) + B(x+y;t) + \int_x^{\infty} B(y+z;t)K(x,z;t)\,dz = 0\,. \tag{6.6.13}$$

The *basis function B* is defined via (6.6.14)

$$B(x;t) = \sum_{m=1}^{M}\left(C_m(0)\right)^2 e^{8k_m^3 t - k_m x} + \frac{1}{2\pi}\int_{-\infty}^{\infty} b(k)e^{ik(x-8k^2 t)}\,dk\,. \tag{6.6.14}$$

It is observed that (6.6.14) contains both the bound (in the summation) and scattering (integral) states. There are two approaches to solving the *KdV* via inverse scattering. The first is to assume that the *GLM* kernel has the form $K(x, y) = \sum_{m=1}^{M} K_m(x) e^{-k_m y}$. Once a manageable form of the basis function B is at hand, then substituting the sum of exponential for K into (6.6.13) yields M linear algebraic equations for the weighting components $K_m(x)$.

Example 6.6.1: Suppose $B(x; t) = B_o(t)\, e^{-k_o x}$, $B_o(t) = \left(C_o(0)\right)^2 e^{8k_o^3 t}$, and $K(x, y) = K_o(x)\, e^{-k_o y}$ where $k_o > 0$. That is, suppose the function B is comprised of a bound state with scattering weights $b(k) = 0$ for all $k \in \mathbb{R}$. This means that the initial wave displacement $u(x, 0) = f(x)$ is reflectionless. Then the *GLM* equation (6.6.13) becomes

$$\begin{aligned} 0 &= K_o(x)\, e^{-k_o y} + B_o(t)\, e^{-k_o(x+y)} + \int_x^\infty B_o(t) e^{-k_o(y+z)} K_o(x) e^{-k_o z}\, dz \\ &= K_o(x)\, e^{-k_o y} + B_o(t)\, e^{-k_o x}\, e^{-k_o y} + B_o(t)\, e^{-k_o y}\, K_o(x) \int_x^\infty e^{-2k_o z}\, dz \\ &= K_o(x)\, e^{-k_o y} + B_o(t)\, e^{-k_o x}\, e^{-k_o y} + B_o(t)\, e^{-k_o y}\, K_o(x) \frac{1}{2k_o} e^{-2k_o x}\,. \end{aligned}$$

Dividing through by the non-zero quantity $e^{-k_o y}$ results in

$$0 = K_o(x)[1 + B_o(t) \frac{1}{2k_o} e^{-2k_o x}] + B_o(t)\, e^{-k_o x} \text{ or } K_o(x) = \frac{-B_o(t) e^{-k_o x}}{1 + \frac{1}{2k_o} B_o(t) e^{-2k_o x}} = \frac{-2k_o B_o(t)}{2k_o e^{k_o x} + B_o(t) e^{-k_o x}}.$$

Consequently, $K(x, y) = K_o(x)\, e^{-k_o y} = \dfrac{-2k_o B_o(t)\, e^{-k_o y}}{2k_o e^{k_o x} + B_o(t) e^{-k_o x}}$ so that $u(x, t) = 2\dfrac{\partial}{\partial x} K(x, x) = \dfrac{16k_o^3 B_o(t) e^{2k_o x}}{\left(2k_o e^{2k_o x} + B_o(t)\right)^2}$ solves the *KdV* with $f(x) = u(x, 0) = \dfrac{16k_o^3 B_o(0) e^{2k_o x}}{\left(2k_o e^{2k_o x} + B_o(0)\right)^2}$. This solution $u(x, t)$ and initial condition $f(x)$ are displayed in Figures 6.6–6.7 with $k_o = ½$, $C_o(0) = 1$, and $f(x) = \dfrac{2e^x}{\left(e^x + 1\right)^2}$. ■

Note that these assumptions require a special form for the Cauchy data $u(x, 0) = f(x)$. What happens for more general initial functions $f(x)$? The second approach to the solution of the *KdV* equation is achieved via the enhanced work on the inverse scattering method by Gardner, Greene, Kruskal, and Miura [27, 28]. In particular, *Airy's partial differential equation* $\dfrac{\partial u}{\partial t} = -\dfrac{\partial^3 u}{\partial x^3}$ is used to incorporate the *KdV* Cauchy data $f(x)$ into the basis function B via an elliptic *PDE* followed by the *GLM* equation (6.6.13). Then *Airy's* equation is solved to produce a time–dependent basis function $B(x, y; t) \equiv B(x+y; t)$. This basis function is then inserted into the *GLM* equation to produce a solution of the *KdV*. Diagram 6.1 outlines the process.

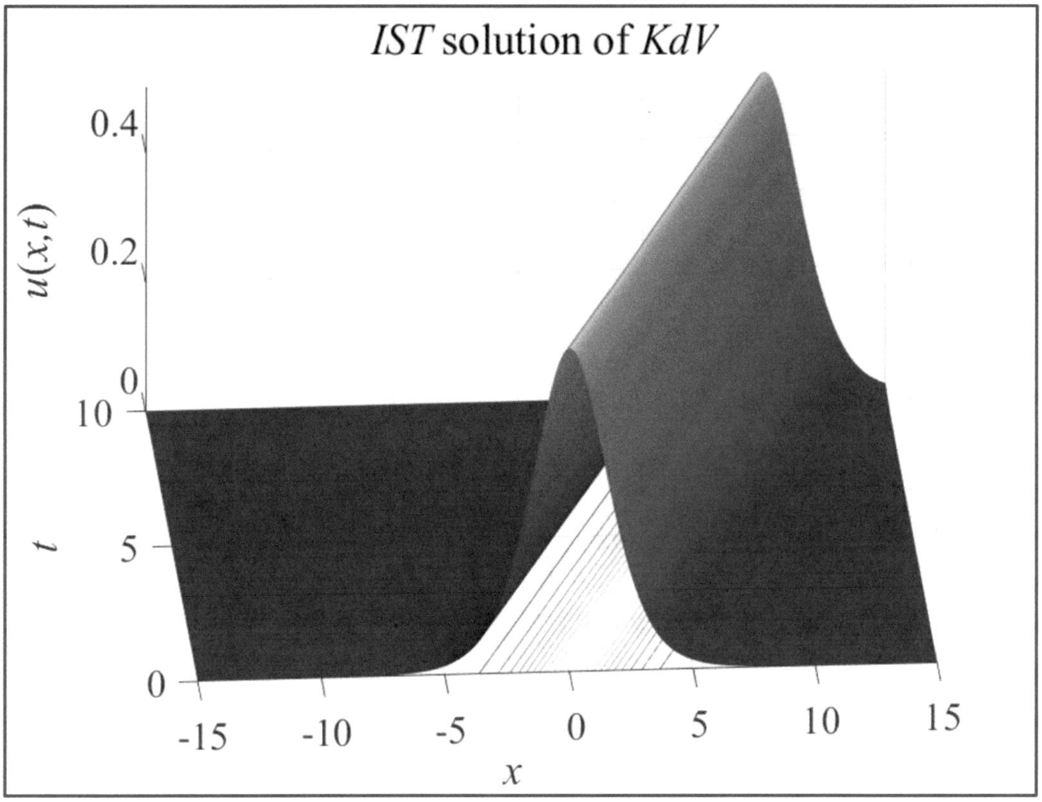

Figure 6.6. The *inverse scattering transform* solution $u(x, t)$ of *KdV*.

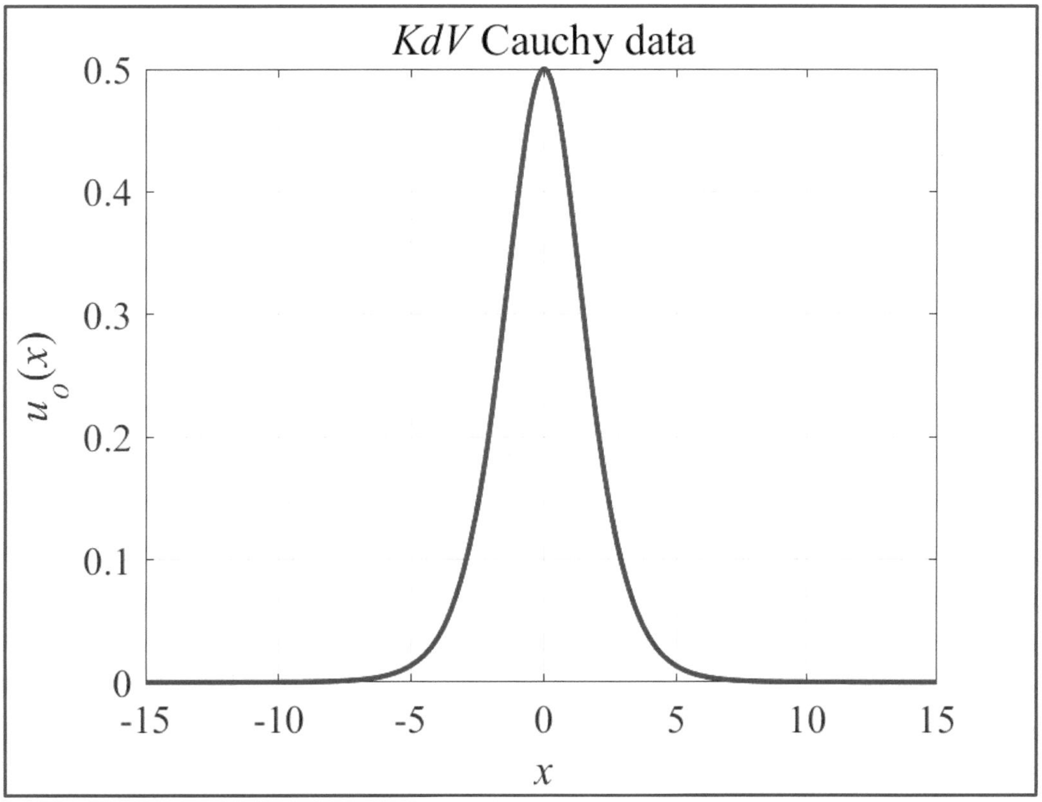

Figure 6.7. The *initial function* $f(x) = 2e^x/(1+e^x)^2$.

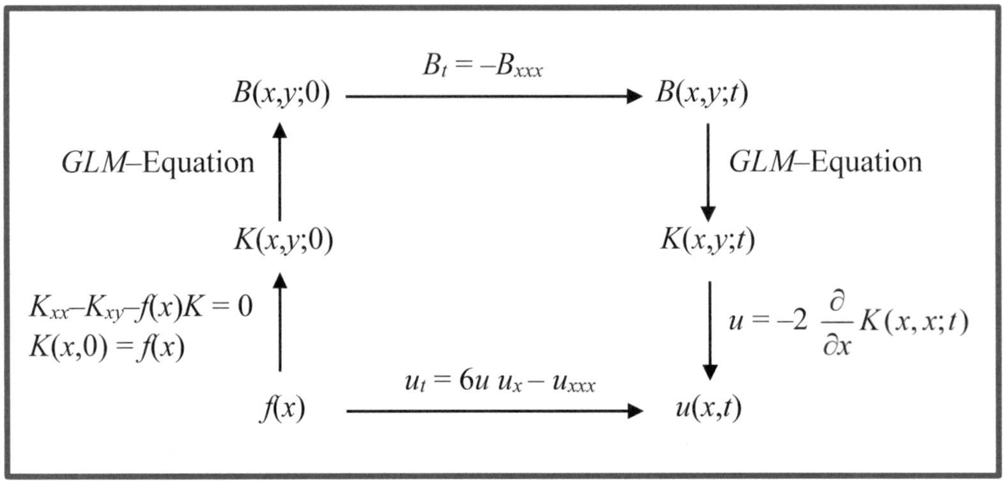

Diagram 6.1. Inverse scattering via Airy's equation.

The description of the *inverse scattering transform* proceeds sequentially.

1. Solve the elliptic *PDE* $\dfrac{\partial^2 K(x,y)}{\partial x^2} - \dfrac{\partial^2 K(x,y)}{\partial y \partial x} - f(x)K(x,y) = 0$ for $K(x, y)$ with Cauchy data $K(x, 0) = f(x)$ and boundary condition $\lim\limits_{x+y\to\infty} K(x,y) = 0$.
2. Insert $K(x, y)$ into the *GLM*–equation and solve for the basis function $B(x, y)$.
3. Solve Airy's *PDE* $\dfrac{\partial B}{\partial t} = -\dfrac{\partial^3 B}{\partial x^3}$ with $B(x, y; t = 0) = B(x, y)$ the basis function obtained in step 2.
4. Insert the time–dependent basis function $B(x+y;\, t)$ into the *GLM* and produce the time–dependent *inverse transform* kernel $K(x, y; t)$.
5. Recover the solution of the *KdV* with Cauchy data $u(x,0) = f(x)$ via $u(x, t) = -2\dfrac{\partial}{\partial x}K(x,x;t)$.

Some comments on this process are in order.

Observe that the time–independent *PDE* for the *GLM*–kernel $K(x, y)$ is separable. That is, assuming $K(x, y) = X(x)\cdot Y(y)$ results in $\dfrac{X''(x) - f(x)\cdot X(x)}{X'(x)} = \dfrac{Y'(y)}{Y(y)}$. Since the left-hand side of this equation is solely a function of x while the right-hand side is purely a function of y, then these ratios must be equal to a constant $\lambda_o \in \mathbb{R}$. Thus, $Y'(y) = \lambda_o \cdot Y(y)$ or $Y(y) = Y(y_o)\cdot\exp(\lambda_o(y - y_o))$. The boundary condition implies that $\lim\limits_{y\to\infty} Y(y) = 0$ so that $\lambda_o = -\mu_o^2$. To satisfy the Cauchy condition, take $y_o = 0$ which means that $Y(y) = Y_o \exp(-\mu_o^2 \cdot y)$ and $Y_o = Y(y_o)$. With this calculation achieved, it remains to solve the second order *ODE* $X''(x) + \mu_o^2 X'(x) - f(x)\cdot X(x) = 0$. This equation resists closed form solution for general $f(x)$. Here a numerical solution of the *ODE* can be carried out. This is illustrated in Figure 6.8.

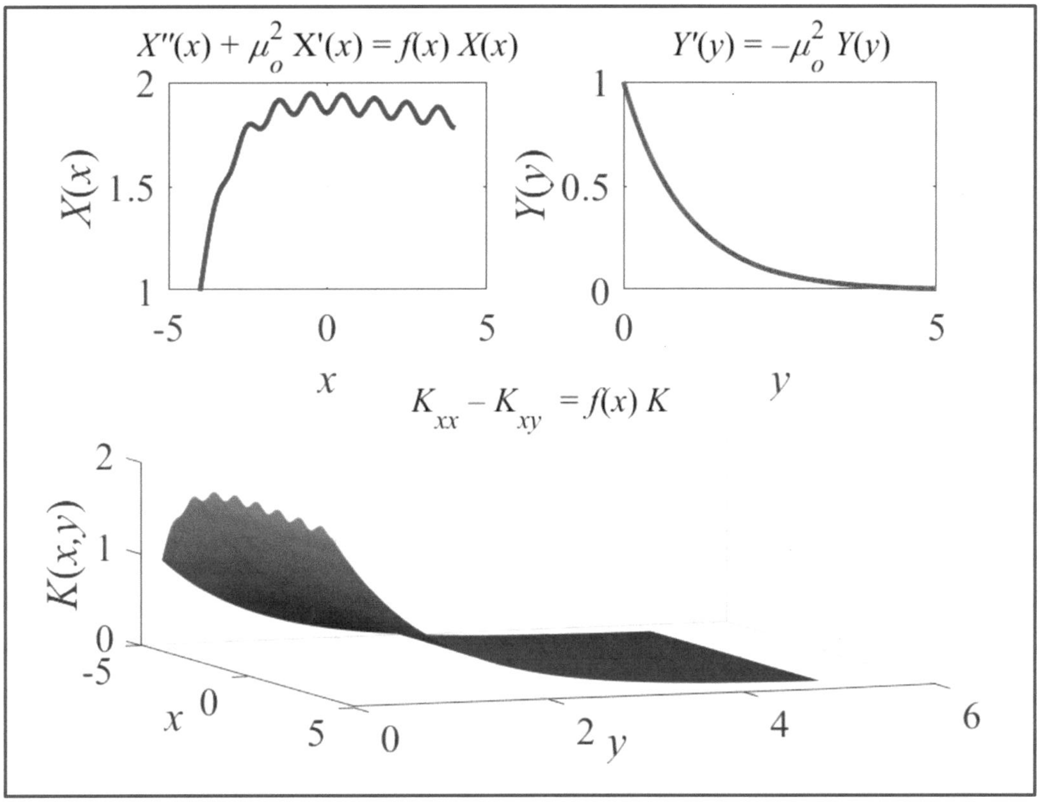

Figure 6.8. Time–independent *GLM* Kernel, $f(x) = \cos(2\pi x)$, $\mu = 1$.

The time–independent *GLM* equation $K(x,y)+B(x+y)+\int_x^{\infty} B(y+z)K(x,z)\,dz=0$ can be discretized as $K(x_\ell,y_k)+B(x_\ell+y_k)+\sum_{j=1}^{N} B(y_k+z_j)K(x_\ell,z_j)\Delta z_j=0$ for $\ell = 1, 2, \ldots, N_x$, $k = 1, 2, \ldots, N_y$, and N is the number of solitons. This produces a matrix equation for the basis function $B(x, y)$ described above in equation (6.6.14). Indeed, let $K_{\ell,k} \equiv K(x_\ell, y_k)$, $B_{\ell,k} \equiv B(x_l + y_k)$, and posit the simplifying assumption $\Delta z_j \equiv \Delta z$ for all j. Then the discretized *GLM* leads to

$$K_{\ell,k}+B_{\ell,k}+\left(B_{\ell,1}K_{\ell,1}+B_{\ell,2}K_{\ell,2}+\cdots+B_{\ell,n}K_{\ell,n}\right)\Delta z=0 \text{ for } \ell = 1, 2, \ldots, N_x \text{ and } k = 1, 2, \ldots, N_y.$$

Finally, introduce the N_x x N_y matrices K_M and B_M as below along with their column truncated partners $K_M(N)$ and $B_M(N)$. That is,

$$K_M = \begin{bmatrix} K_{1,1} & K_{1,2} & \cdots & K_{1,N_y} \\ K_{2,1} & K_{2,2} & \cdots & K_{2,N_y} \\ \vdots & \vdots & \ddots & \vdots \\ K_{N_x,1} & K_{N_x,2} & \cdots & K_{N_x,N_y} \end{bmatrix}, B_M = \begin{bmatrix} B_{1,1} & B_{1,2} & \cdots & B_{1,N_y} \\ B_{2,1} & B_{2,2} & \cdots & B_{2,N_y} \\ \vdots & \vdots & \ddots & \vdots \\ B_{N_x,1} & B_{N_x,2} & \cdots & B_{N_x,N_y} \end{bmatrix} \in \mathcal{M}at_{N_x \times N_y}(\mathbb{R}),$$

$$K_M(N) = \begin{bmatrix} K_{1,1} & K_{1,2} & \cdots & K_{1,N} \\ K_{2,1} & K_{2,2} & \cdots & K_{2,N} \\ \vdots & \vdots & \ddots & \vdots \\ K_{N_x,1} & K_{N_x,2} & \cdots & K_{N_x,N} \end{bmatrix} \in \mathcal{M}at_{N_x \times N}(\mathbb{R}), \text{ and}$$

$$B_M(N) = \begin{bmatrix} B_{1,1} & B_{1,2} & \cdots & B_{1,N} \\ B_{2,1} & B_{2,2} & \cdots & B_{2,N} \\ \vdots & \vdots & \ddots & \vdots \\ B_{N_y,1} & B_{N_y,2} & \cdots & B_{N_y,N} \end{bmatrix} \in \mathcal{M}at_{N_x \times N}(\mathbb{R}).$$

The discretized *GLM* equation for *N*–solitons is

$$K_M + B_M = -\Delta z \cdot K_M(N) \cdot B_M^T(N). \tag{6.6.15}$$

The underlying assumption is that N_x, $N_y >> N$. Once a discrete approximation for the basis function $B(x, y)$ is calculated, then an analytic model can be applied as the data suggest.

Estimates of the time–independent *GLM* kernel *K* and basis function *B* lead to *Airy's equation*

$$\left.\begin{aligned} &\frac{\partial B}{\partial t} = -\frac{\partial^3 B}{\partial x^3} \\ &B(x,0) = B_o(x) \end{aligned}\right\}. \tag{6.6.16}$$

The solution of the Airy *PDE* can be obtained from the Fourier Transform $\mathfrak{F}$. Applying the transform to (6.6.16) and recalling the properties of the transform developed in Chapter 4, it is seen that

$$\frac{\partial}{\partial t}\mathfrak{F}\big[B\big](\omega) = -(i\omega)^3 \mathfrak{F}\big[B\big](\omega) = i\omega^3 \mathfrak{F}\big[B\big](\omega). \tag{6.6.17}$$

Setting $\beta(\omega) = \mathfrak{F}[B(x;t)](\omega)$, $\beta_o(\omega) = \mathfrak{F}[B_o(x)](\omega)$, and solving the resulting *ODE* from (6.6.17), $\frac{d}{dt}\beta(\omega) = i\omega^3\,\beta(\omega)$, produces the solution $\beta(\omega)/\beta_o(\omega) = e^{i\omega^3 t}$ or $\beta(\omega) = \beta_o(\omega)\cdot e^{i\omega^3 t}$. The *Airy function* $Ai(x)$ is defined via (6.6.18)

$$Ai(x) = \frac{1}{\pi}\int_0^\infty \cos\left(\tfrac{1}{3}\xi^3 + x\cdot\xi\right)d\xi \tag{6.6.18}$$

and has the property that its (ordinary frequency) Fourier transform $\mathfrak{F}_1$ yields $\mathfrak{F}_1[Ai(x)](\lambda) \equiv \int_{-\infty}^{\infty} Ai(x)e^{-2\pi i\lambda x}\,dx = \exp\left(\tfrac{i}{3}[2\pi\lambda]^3\right)$ as per Abramowitz and Stegun [2] and Wikipedia [82]. Therefore,

$$\mathfrak{F}_1\left[Ai\left(\tfrac{1}{2\pi}\sqrt[3]{3t}\cdot x\right)\right](\lambda) = \int_{-\infty}^{\infty} A(x,t)e^{-2\pi i x[\sqrt[3]{3t}\cdot\lambda/2\pi]}\,dx = \exp\left(i\lambda^3\right)$$

where $A(x, t) = Ai\left(\tfrac{1}{2\pi}\sqrt[3]{3t}\cdot x\right)$. Consequently, $\mathfrak{F}[A(x;t)](\omega) \equiv \frac{1}{\sqrt{2\pi}}\int_{-\infty}^{\infty} A(x;t)e^{-ix\omega}\,dx = \frac{1}{\sqrt{2\pi}}e^{it\omega^3}$.

It is now seen that, $\mathfrak{F}[B(x;t)](\omega) = \mathfrak{F}[B_o(x)](\omega)\,\mathfrak{F}[\sqrt{2\pi}\,A(x;t)](\omega) = \mathfrak{F}[B_o(x)\odot A(x;t)](\omega)$ or $B(x;t) = B_o(x)\odot A(x;t)$ where $\odot$ is convolution. Note that the factor $\sqrt{2\pi}$ is absent due to the convolution theorem of Chapter 4. Thus,

$$B(x;t) = \int_{-\infty}^{\infty} B_o(\xi)\cdot Ai\left(\sqrt[3]{3t}[x-\xi]\right)d\xi \tag{6.6.19}$$

is the solution of Airy's equation (6.6.16). An illustration of a solution to Airy's equation is provided in Figure 6.9.

Now going from a solution $B(x, t)$ of the *Airy PDE* to a solution $u(x, t)$ of the *KdV* is a matter of solving the *GLM* equation (6.6.13). This can be achieved via the discretization approach discussed above. Finally, taking the derivative of the *GLM* kernel $K(x, x; t)$ with respect to x produces $u(x, t)$. This rather lengthy discussion is presented to demonstrate that the solution to a nonlinear partial differential equation can be determined via a series of linear steps. Indeed, the solution of the (nonlinear) *KdV* is dependent on the solution of the (linear) *Airy's* equation. Similarly, solutions of the (nonlinear) Klein–Gordon and Burgers' equations are achieved either by linear techniques (separation of variables) or a *change of variables* which transform the nonlinear *PDE* into a linear *PDE*.

This is the central idea of this book. Namely, that a change of variables in the form of separation of variables, the Hopf–Cole transformation, or the inverse scattering transform can be determined so that the underlying nonlinear equation is mapped to a linear equation. Lie–Bäcklund transformations can also be used to solve the *KdV* and other nonlinear *PDE*s as per Anderson and Ibragimov [3]. A general theory for the existence of transformations which linearize nonlinear *PDE*s will be undertaken in Chapter 8.

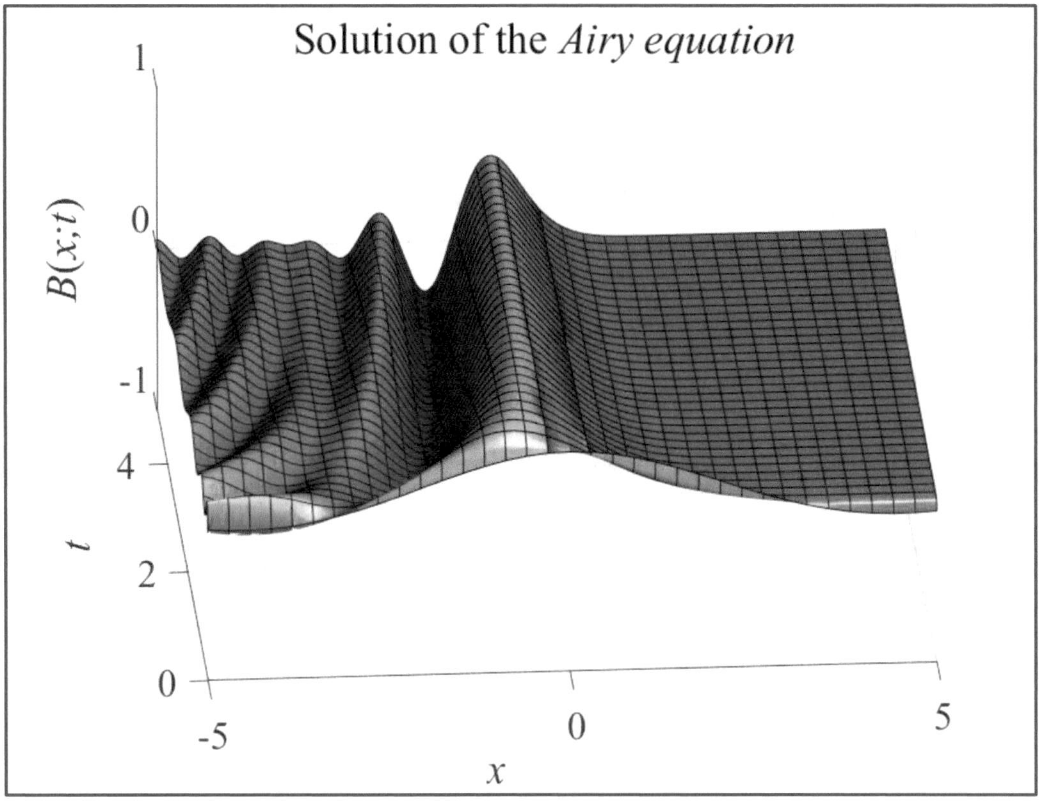

Figure 6.9. Solution of the *Airy equation* with $B_o(x) = \sin(x)/x$.

Exercises

6.1 Plot the solution of the nonlinear Klein–Gordon equation (6.3.4) for $n = 3$. $\lambda = \sigma = a = 1$. What happens if the values of the parameters λ, σ, and/or a are changed?

6.2 Show that the solution of the nonlinear Klein–Gordon equation (6.3.7) is written in the original coordinates as

$$u(x,t) = \left[\left(\frac{\lambda(1-n)}{2\sigma^2}(\sigma t + x) + g(0)^{1-n}\right)\cdot\left(\frac{\alpha(1-n)}{2\lambda}(\sigma t - x) + h(-\tau)^{1-n}\right)\right]^{\frac{1}{1-n}}.$$ If the Cauchy function is $f(x) = e^x$, show $g(x) = e^x = h(x)$. Plot the solution for $n = 2$. $\lambda = \sigma = a = 1$. Determine the region of singularities.

6.3 Using (6.5.4), compute the analytic solution of Burgers' equation (6.4.1) for

$$f(x) = \begin{cases} 1+x, & \text{for } -1 \le x \le 0 \\ 1-x, & \text{for } 0 \le x \le 1 \\ 0, & \text{otherwise} \end{cases}$$ (the triangle function).

Plot a solution for $\sigma = 1$.

Chapter 7

Raman Scattering and Numerical Methods

In this chapter, the mathematical model for *stimulated Raman scattering* is developed. This phenomenon constitutes the interaction of two lasers and results in a system of parabolic partial differential equations with third order nonlinearities. A quasi–implicit finite difference scheme, used to solve the nonlinear system, is presented. It is proved that the system is both stable and convergent under certain restrictions on the "time" step size with respect to the system's Cauchy data. The one–spatial dimensional case is solved via the quasi–implicit scheme. Generalizations and physical interpretation of the results are discussed.

§7.1 The Stimulated Raman Scattering Laser Model

The interaction of two lasers within an electro-magnetic medium can be described by a system of partial differential equations. The method of spectroscopy referred to as *stimulated Raman scattering* can be roughly described as an extension of Maxwell's equations and Figure 7.1. More specifically, let M_1 and M_2 be two mirrors which are coated in such a manner as to be transparent to an input or *pump laser*. The electric field for the pump laser will be denoted as E_p. Moreover, the coatings on the mirrors reflect the energy from a second input or *Stokes laser* within the aforementioned *Raman gain medium*. The electric field on the Stokes laser is represented by E_s. Finally, suppose that the mirror M_1 is 100% reflective of the Stokes laser while M_2 is less than 100% reflective (e.g., 95%). The energy from the pump laser E_p excites photons within the Raman gain medium that, in turn, transfer energy to the Stokes laser E_s. This energy is amplified by the back and forth reflection of the pump laser between the mirrors. With respect to the reflection rates mentioned above, energy from the pump laser (solid lines) is transmitted via the gain medium to the Stokes laser (dashed lines). The Stokes field then transmits energy through the second mirror M_2 at the leakage rate (in this case 5%).

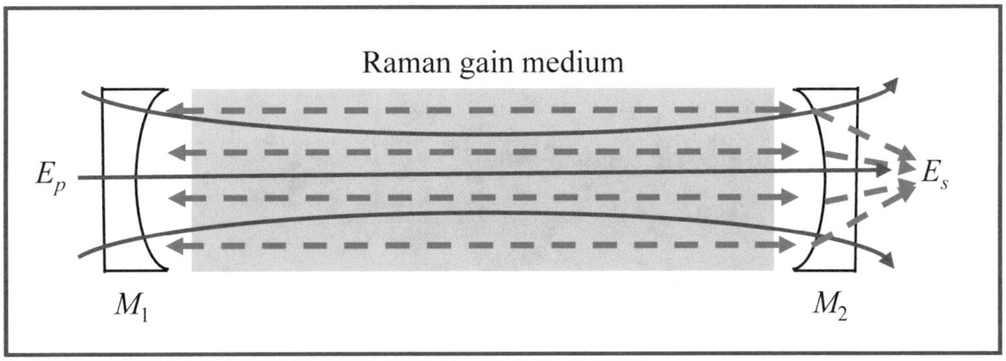

Figure 7.1. Stimulated Raman scattering.

The two lasers undergo a nonlinear interaction within the gain medium that is modeled by the system of *PDE*s (7.1.1), see Newell and Moloney [53], Butcher and Cotter [15], and Penzkopfer, Laubereau, and Kaiser [60] for greater detail. The development presented here follows Costa [20].

$$\left.\begin{aligned} &\left(\frac{\partial}{\partial z}+\frac{1}{c}\frac{\partial}{\partial t}\right)E_p+\frac{i}{2k_p}\Delta_\perp E_p=i\frac{\hbar\,\omega_p}{2\varpi c}\nu(\omega_p)\,q_{ps}\,\sigma_{ps}\,E_s \\ &\left(\frac{\partial}{\partial z}+\frac{1}{c}\frac{\partial}{\partial t}\right)E_s+\frac{i}{2k_s}\Delta_\perp E_s=i\frac{\hbar\,\omega_s}{2\varpi c}\nu(\omega_s)\,q_{ps}\,\overline{\sigma}_{ps}\,E_p \\ &\frac{\partial}{\partial t}\sigma_{ps}+\beta_{ps}\,\sigma_{ps}=i\,q_{ps}\,E_p\overline{E}_s \end{aligned}\right\} \tag{7.1.1}$$

This rather formidable looking set of equations employs the notation listed below.

c is the speed of light (299,792,458 *meters/second*)
$\hbar$ is Planck's constant ($6.62607015\text{x}10^{-34}$ *Joule–seconds*)
k_p, k_s are pump and Stokes wave numbers
ω_p, ω_s are pump and Stokes frequencies
λ_p, λ_s are pump and Stokes wavelengths
q_{ps} is the effective beam coupling parameter
β_{ps} is the laser interaction broadening constant
σ_{ps} and $\overline{\sigma}_{ps}$ are the vibrational amplitude of the interacting lasers and its complex conjugate
ϖ is the dielectric constant within the gain medium
$\nu(\omega_p)$ is the index of refraction of the pump laser in the Raman gain medium
$\nu(\omega_s)$ is the index of refraction of the Stokes laser in the Raman gain medium
E_p and $\sqrt{\frac{\lambda_s}{\lambda_p}}\cdot E_s$ are complex valued electric fields of the pump and Stokes lasers
$\overline{E}_p$ and $\overline{E}_s$ are the complex conjugates of E_p and E_s
$\Delta_\perp=\sum_{j=1}^{n_d}\frac{\partial^2}{\partial x_j^2}$ is the n–dimensional transverse Laplacian operator in which the variables $x_1, x_2, \ldots,$ x_{n_d} have units of length (i.e., meters or centimeters)
$\boldsymbol{x}=[x_1, x_2, \ldots, x_{n_d}]$
$\Omega\subset\mathbb{R}^{n_d}$ is the *support* of the electric fields E_p and E_s
$\text{supp}(E)=\Omega=\{\boldsymbol{x}\in\mathbb{R}^{n_d}: E(\boldsymbol{x})\neq 0\}$
$\partial\Omega$ is the *boundary* of the support Ω

The physical cases are $n_d = 1$ or $n_d = 2$.

The special case of *steady–state scattering* is now considered. That is, assume $\frac{\partial E_p}{\partial t} = \frac{\partial E_s}{\partial t} = \frac{\partial \sigma_{ps}}{\partial t} = 0$. Under this scenario, the third equation of (7.1.1) indicates that $\sigma_{ps} = i\frac{q_{ps}}{\beta_{ps}} E_p \overline{E}_s$. Substituting into the remaining portion of (7.1.1) produces the coupled set of *PDE*s

$$\frac{\partial E_p}{\partial z} = -i\tfrac{1}{k_p}\Delta_\perp E_p - \gamma_p |E_s|^2 E_p$$
$$\frac{\partial E_s}{\partial z} = -i\tfrac{1}{k_s}\Delta_\perp E_s + \gamma_s |E_p|^2 E_s$$

in which $\gamma_p = \frac{\hbar\omega_p \cdot \nu(\omega_p)}{2\varpi c \beta_{ps}} q_{ps}^2$ and $\gamma_s = \frac{\hbar\omega_s \cdot \nu(\omega_s)}{2\varpi c \beta_{ps}} q_{ps}^2$. But the wave numbers are the products of the laser frequency and index of refraction so that $k_p = \omega_p \cdot \nu(\omega_p)$ and $k_s = \omega_s \cdot \nu(\omega_s)$. Further, $g_r = \frac{\hbar\omega_s \cdot \nu(\omega_s)}{\varpi c \beta_{ps}} q_{ps}^2$ is the *Raman gain coefficient*. In addition to these observations, add the requirements that the electric fields evolve with respect to z over a compact set $\Omega \subset \mathbb{R}^{n_d}$. Further, it is required that the electric fields have *compact support* Ω and have non-zero Cauchy data $E_p(\boldsymbol{x};0) \equiv E_p(0)$ and $E_s(\boldsymbol{x};0) \equiv E_s(0)$. With all of these comments in mind, the coupled system is written as (7.1.2).

$$\left.\begin{aligned}
&\frac{\partial E_p}{\partial z} = -i\tfrac{1}{k_p}\Delta_\perp E_p - \tfrac{k_p}{2k_s} g_r |E_s|^2 E_p, \quad \boldsymbol{x} \in \Omega \\
&\frac{\partial E_s}{\partial z} = -i\tfrac{1}{k_s}\Delta_\perp E_s + \tfrac{1}{2} g_r |E_p|^2 E_s, \quad \boldsymbol{x} \in \Omega \\
&E_p(\boldsymbol{x};z) = E_s(\boldsymbol{x};z) = 0 \text{ for } \boldsymbol{x} \in \partial\Omega \\
&E_p(\boldsymbol{x};0) = E_p(0) \text{ and } E_s(\boldsymbol{x};0) = E_s(0)
\end{aligned}\right\} \qquad (7.1.2)$$

Since the electric fields E_p and E_s have compact support on Ω, then $E_p(\boldsymbol{x}; z) \equiv 0 \equiv E_s(\boldsymbol{x}; z)$ for all $\boldsymbol{x} \in \partial\Omega$. This is the *Dirichlet boundary condition* imposed on the electric fields.

Now consider the case that models a physical system. For $n_d = 2$, the change of variables Ψ in (7.1.3) will transform the unit-based form of the steady–state system (7.1.2) into the unitless system (7.1.4). More precisely, $(x_1, x_2, z, E_p, E_s) \overset{\Psi}{\mapsto} (\xi_1, \xi_2, \tau, u, v)$ is defined by

$$\left.\begin{aligned}
&(x_1, x_2, z, E_p, E_s) \overset{\Psi}{\mapsto} (\xi_1, \xi_2, \tau, u, v) \\
&\tau = \tfrac{1}{2} g_r \left|E_p(0)\right|^2 z \\
&\xi_j = \sqrt{\tfrac{1}{2} g_r \left|E_p(0)\right|^2 k_s}\; x_j, \quad j = 1,2 \\
&u = \frac{1}{\left|E_p(0)\right|} E_p, \quad v = \sqrt{\tfrac{k_p}{k_s}} \frac{1}{\left|E_p(0)\right|} E_s
\end{aligned}\right\}. \tag{7.1.3}$$

This mapping Ψ transforms (7.1.2) into the canonical Schrödinger system (7.1.4)

$$\left.\begin{aligned}
&\frac{\partial u}{\partial \tau} = i c_p \Delta u - |v|^2 u, \text{ for } \boldsymbol{\xi} \in \Psi(\Omega) \\
&\frac{\partial v}{\partial \tau} = i c_s \Delta v + |u|^2 v, \text{ for } \boldsymbol{\xi} \in \Psi(\Omega) \\
&u(\boldsymbol{\xi}, 0) = f(\boldsymbol{\xi}), v(\boldsymbol{\xi}, 0) = g(\boldsymbol{\xi}), \boldsymbol{\xi} \in \Psi(\Omega) \\
&u(\boldsymbol{\xi}, \tau) \equiv 0 \equiv v(\boldsymbol{\xi}, \tau), \text{ for } \boldsymbol{\xi} \in \partial\Psi(\Omega)
\end{aligned}\right\}. \tag{7.1.4}$$

A few definitions are in order. The constants are $c_p = \dfrac{k_s}{2k_p}$ and $c_s = ½$, the Cauchy data are $u(\boldsymbol{\xi}, 0)$ $= f(\boldsymbol{\xi}) = \dfrac{E_p(\boldsymbol{\xi};0)}{\left|E_p(\boldsymbol{\xi};0)\right|} = \dfrac{E_p(0)}{\left|E_p(0)\right|}$ and $v(\boldsymbol{\xi}, 0) = g(\boldsymbol{\xi}) = \sqrt{\dfrac{k_p}{k_s}} \cdot \dfrac{E_s(\boldsymbol{\xi};0)}{\left|E_p(\boldsymbol{\xi};0)\right|} = \sqrt{\dfrac{k_p}{k_s}} \cdot \dfrac{E_s(0)}{\left|E_p(0)\right|}$, the Laplacian Δ is defined with respect to the unitless spatial variables $\boldsymbol{\xi} = [\xi_1, \xi_2] \Rightarrow \Delta = \dfrac{\partial^2}{\partial \xi_1^2} + \dfrac{\partial^2}{\partial \xi_2^2}$, and the domain of definition of the unitless fields u and v is $\Psi(\Omega)$ = the image of the domain Ω under the mapping Ψ. The steady–state system (7.1.4) is reduced to a single (spatial) dimension by using the one–dimensional Laplacian $\dfrac{\partial^2}{\partial \xi^2}$ with respect to the sole (unitless) spatial variable ξ and is presented as equation (7.1.5). In this case, $\Omega = [a, b]$, $\Psi(\Omega) = [\Psi(a), \Psi(b)] \equiv [\psi_a, \psi_b]$, and

$$\left.\begin{aligned}
&\frac{\partial u}{\partial \tau} = i c_p \frac{\partial^2 u}{\partial \xi^2} - |v|^2 u, \text{ for } \boldsymbol{\xi} \in [\psi_a, \psi_b] \\
&\frac{\partial v}{\partial \tau} = i c_s \frac{\partial^2 v}{\partial \xi^2} + |u|^2 v, \text{ for } \boldsymbol{\xi} \in [\psi_a, \psi_b] \\
&u(\boldsymbol{\xi}, 0) = f(\boldsymbol{\xi}), v(\boldsymbol{\xi}, 0) = g(\boldsymbol{\xi}), \boldsymbol{\xi} \in [\psi_a, \psi_b] \\
&u(\psi_a, \tau) = u(\psi_b, \tau) \equiv 0 \equiv v(\psi_a, \tau) = v(\psi_b, \tau)
\end{aligned}\right\}. \tag{7.1.5}$$

Finally, it should be noted that the unitless variable τ will be referred to as a *temporal variable* even though it *does not represent physical time*. This label is used as a convenience only. It *may* be possible to apply the method of *inverse scattering* to the system (7.1.5). Rather than taking that approach, however, the remainder of this chapter will focus on developing a numerical method to approximate a solution of (7.1.4)/(7.1.5).

§7.2 A Quasi–Implicit Finite Difference Scheme

As mentioned in the previous section, this chapter will concentrate on the development of numerical methods to obtain (approximate) solutions of the steady–state Raman scattering equation (7.1.4) in the physical cases of (spatial) dimension $n_d = 1, 2$. To that end, write (7.1.4) as the vector partial differential equation (7.2.1)

$$\left.\begin{array}{l} \dfrac{\partial \boldsymbol{w}}{\partial \tau} = i\,C\,\Delta \boldsymbol{w} + N(\boldsymbol{w})\boldsymbol{w} \\ \boldsymbol{w}_o(\xi) = \begin{bmatrix} f(\xi) \\ g(\xi) \end{bmatrix} \\ \boldsymbol{w}(\xi,\tau) = 0 \quad \text{for } \xi \in \partial\Psi(\Omega) \end{array}\right\} . \tag{7.2.1}$$

Here, $\boldsymbol{w}$ is the vector $\boldsymbol{w} = \begin{bmatrix} u \\ v \end{bmatrix}$, C is the diagonal matrix $C = \begin{bmatrix} c_p & 0 \\ 0 & c_s \end{bmatrix}$, N is the nonlinear operator $N(\boldsymbol{w}) = \begin{bmatrix} -|v|^2 & 0 \\ 0 & |u|^2 \end{bmatrix}$, and $\boldsymbol{w}_o$ is the vector of Cauchy data $\boldsymbol{w}_o(\xi) = \boldsymbol{w}(\xi;0) = \begin{bmatrix} f(\xi) \\ g(\xi) \end{bmatrix}$. Establish two sets of discrete grid points $\Xi = \{\xi_1, \xi_2, \ldots, \xi_K\}$ with respect to the spatial variables $\boldsymbol{\xi} = [\xi_1, \xi_2]$ and $\mathrm{T} = \{\tau_1, \tau_2, \ldots, \tau_N\}$ with respect to the temporal variable τ. It is noted that, in the case of one spatial variable, the vector $\boldsymbol{\xi}$ becomes the single variable ξ. Following Thomas [78], define difference operators on the discretization of the unitless field vector $\boldsymbol{w}$ over the spatial Ξ and temporal T grids: $\boldsymbol{w}(\xi_k,\tau_n) = \boldsymbol{w}_k^n$. Specifically, if $\boxed{u(\xi_k,\tau_n) = u_k^n}$, then the *first order temporal difference operator* is $\boxed{\delta_\tau^+\left(u_k^n\right) \equiv \dfrac{1}{\Delta\tau}\left(u_k^{n+1} - u_k^n\right)}$.

Similarly, the *first order spatial difference operator* is $\boxed{\delta_\xi^+\left(u_k^n\right) = \dfrac{1}{\Delta\xi}\left(u_{k+1}^n - u_k^n\right)}$, while the *second order spatial difference operator* is $\boxed{\delta_\xi^2\left(u_k^n\right) = \dfrac{1}{\Delta\xi^2}\left(u_{k+1}^n - 2u_k^n + u_{k-1}^n\right)}$. The underlying assumption is that the temporal and spatial mesh spacings $\Delta\tau = \tau_{n+1} - \tau_n$ and $\Delta\xi = \xi_{k+1} - \xi_k$, respectively, are uniform across their grids. Specifically, $\tau_{n+1} - \tau_n = \tau_{m+1} - \tau_m$, for any n and m. The same holds for $\Delta\xi$. The temporal derivative is approximated as the first order difference

$\frac{\partial u}{\partial \tau}(\xi_k, \tau_n) \approx \delta_\tau^+\left(u_k^n\right)$ and the second spatial derivative by the second order difference $\frac{\partial^2 u}{\partial \xi^2}(\xi_k, \tau_n) \approx \delta_\xi^2\left(u_k^n\right)$. These difference operators can be applied directly to the field vector $\boldsymbol{w}$ elementwise.

There are at least three (temporal, spatial)–grid choices for equation (7.2.1). Namely, *explicit* in time, *implicit* in time, and the renowned *Crank–Nicolson* method. The explicit method uses the idea that the spatial grid is computed at time τ_n while the temporal grid spans time τ_{n+1}: $\delta_\tau^+\left(\boldsymbol{w}_k^n\right) = C\,\delta_\xi^2\left(\boldsymbol{w}_k^n\right) + N\left(\boldsymbol{w}_k^n\right)\boldsymbol{w}_k^n$. This is illustrated in the left-hand portion of Figure 7.2. The implicit method approximates the second spatial derivative at time τ_{n+1}: $\delta_\tau^+\left(\boldsymbol{w}_k^n\right) = C\,\delta_\xi^2\left(\boldsymbol{w}_k^{n+1}\right) + N\left(\boldsymbol{w}_k^{n+1}\right)\boldsymbol{w}_k^{n+1}$; see the middle portion of Figure 7.2. Finally, the Crank–Nicolson approach is to average the explicit and implicit methods with respect to the spatial derivative so that $\delta_\tau^+\left(\boldsymbol{w}_k^n\right) = \frac{1}{2}\left[C\left(\delta_\xi^2\left(\boldsymbol{w}_k^n\right) + \delta_\xi^2\left(\boldsymbol{w}_k^{n+1}\right)\right) + N\left(\boldsymbol{w}_k^n\right)\boldsymbol{w}_k^n + N\left(\boldsymbol{w}_k^{n+1}\right)\boldsymbol{w}_k^{n+1}\right]$ as per the right-hand portion of Figure 7.2.

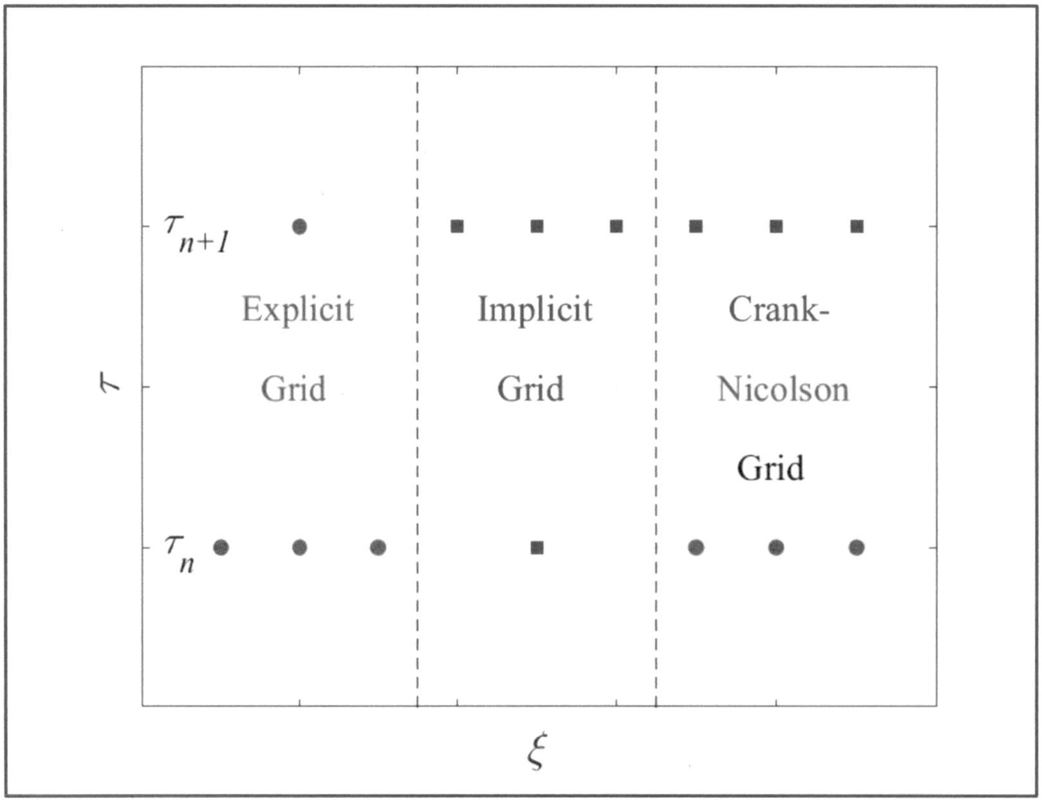

Figure 7.2. Grid for various difference methods.

With respect to purely linear systems (i.e., $N(\boldsymbol{w})$ = constant), the explicit method can become unstable depending upon the choice of the grid mesh $\Delta\tau$ and $\Delta\xi$. The implicit method, conversely, is unconditionally stable (for the linear case). The presence of the nonlinear term $N\left(\boldsymbol{w}_k^n\right)\boldsymbol{w}_k^n$ significantly complicates both stability and convergence of finite difference methods. As will be seen in this chapter, using a quasi–implicit method can produce a stable and convergent system provided the temporal mesh size $\Delta\tau$ is bounded *a priori.*

The Quasi–Implicit Method (one spatial dimension)

The method utilized is implicit in the (linear) spatial derivative and explicit with respect to the nonlinear portion. First, the one–spatial dimensional case $\boldsymbol{\xi} = \xi \in [a_o, b_o] \subset \mathbb{R}$ is developed.

$$\delta_\tau^+\left(\boldsymbol{w}_k^n\right) = C\,\delta_\xi^2\left(\boldsymbol{w}_k^{n+1}\right) + N\left(\boldsymbol{w}_k^n\right)\boldsymbol{w}_k^n \tag{7.2.2}$$

Equation (7.2.2) is referred to as the *quasi–implicit method* for the *PDE* (7.2.1). Throughout this chapter, use the notation $\lambda_p = \dfrac{(\Delta\xi)^2}{c_p\,\Delta\tau}$, $\lambda_s = \dfrac{(\Delta\xi)^2}{c_s\,\Delta\tau}$, $\zeta_p = \dfrac{(\Delta\xi)^2}{c_p}$, $\zeta_s = \dfrac{(\Delta\xi)^2}{c_s}$, $u_k^n = u(\xi_k, \tau_n)$, $v_k^n = v(\xi_k, \tau_n)$, $\boldsymbol{u}^n = \left[u_1^n, u_2^n, \cdots, u_K^n\right]^T$, $\boldsymbol{v}^n = \left[v_1^n, v_2^n, \cdots, v_K^n\right]^T$, $I_{K\times K}$ = the $K \times K$ identity matrix, $U^n = diag\left\{\left|u_1^n\right|^2, \left|u_2^n\right|^2, \cdots, \left|u_K^n\right|^2\right\}$ and $V^n = diag\left\{\left|v_1^n\right|^2, \left|v_2^n\right|^2, \cdots, \left|v_K^n\right|^2\right\}$ are $K \times K$ diagonal matrices, $A = \begin{bmatrix} 2 & -1 & 0 & \cdots & 0 \\ -1 & 2 & -1 & \cdots & 0 \\ 0 & -1 & 2 & \cdots & 0 \\ \vdots & \vdots & \vdots & \ddots & \vdots \\ 0 & 0 & 0 & \cdots & 2 \end{bmatrix}$ is the $K \times K$ tri–diagonal matrix with 2 across the main diagonal and -1 across the sub– and super–diagonals, and $A(\lambda) = A - i\cdot\lambda\, I_{K\times K}$. With this notation established, the quasi–implicit system (7.2.2) can be written as the pair of linear matrix equations

$$\left.\begin{aligned} A(\lambda_p)\boldsymbol{u}^{n+1} &= -i\left(\lambda_p I_{K\times K} - \zeta_p V^n\right)\boldsymbol{u}^n \\ A(\lambda_s)\boldsymbol{v}^{n+1} &= -i\left(\lambda_s I_{K\times K} + \zeta_s U^n\right)\boldsymbol{v}^n \end{aligned}\right\}. \tag{7.2.3}$$

Before addressing the stability and convergence of (7.2.2), a few comments are required. First, the notion of stability is formalized.

Definition 7.1: For M a continuous operator, the finite difference scheme $\boldsymbol{u}^{n+1} = M(\boldsymbol{u}^n)$ is *stable* with respect to the norm $\|\bullet\|$ provided there exist non–negative constants X_o, T_o, K, and α so that $\|\boldsymbol{u}^{n+1}\| \leq K e^{\alpha \cdot \tau} \|\boldsymbol{u}^0\|$ for all $\tau \in [0, T_o]$ and $\xi \in [0, X_o]$.

The idea of *convergence* is informally presented as requiring the solution $\boldsymbol{u}^n = \left[u_1^n, u_2^n, \cdots, u_K^n\right]^T$ of the finite difference method $\boldsymbol{u}^{n+1} = M(\boldsymbol{u}^n)$ to converge to the solution $u(\xi, \tau)$ of the corresponding *PDE*. Suppose that $\mathcal{L}$ is a differential, $\mathcal{G}$ a continuous operator, and that $\boldsymbol{u}^{n+1} = M(\boldsymbol{u}^n)$ is a finite difference scheme which approximates a solution to the partial differential equation $\mathcal{L}u = \mathcal{G}(u)$. For example, $\mathcal{L} = \Delta$ and $\mathcal{G}(u) = u^2$. If the difference between the solution of the *PDE* $\mathcal{L}u = \mathcal{G}(u)$ and the finite difference scheme $\boldsymbol{u}^{n+1} = M(\boldsymbol{u}^n)$ tends to 0 with a respect to a norm, then the finite difference scheme is *consistent* with respect to that norm. Following Thomas [78] the notions of convergence ad consistency can be formalized as below.

Definition 7.2: The finite difference scheme $\boldsymbol{u}^{n+1} = M(\boldsymbol{u}^n)$ which approximates the solution of the partial differential equation $\mathcal{L}u = \mathcal{G}(u)$ is *convergent* at time τ provided $\lim_{n\to\infty}|\tau_n - \tau| = 0$ implies $\lim_{n\to\infty}\left\|u_n^k - u(\xi,\tau)\right\| = 0$ whenever $|\xi_k - \xi| \to 0$.

Let $\mathcal{F}(u_k^n) = u_k^n - M(u_k^{n-1})$ be the finite difference operator whose solution $\boldsymbol{u}^n$ approximates the solution of the partial difference equation $\mathcal{A}(u) = \mathcal{L}u = \mathcal{G}(u)$.

Definition 7.3: The finite difference scheme $\boldsymbol{u}^{n+1} = M(\boldsymbol{u}^n)$ is *consistent* with the partial differential equation $\mathcal{L}u = \mathcal{G}(u)$ in the norm $\|\bullet\|$ if the difference between the operators $\mathcal{F}(u_k^n)$ and $\mathcal{A}(u)$ converges to 0 as the underlying variable grid spacings converge to 0. That is, the finite difference scheme is consistent provided $T(u, u_k^n) = \mathcal{A}(u) - \mathcal{F}(u_k^n)$ and $\lim_{\substack{\Delta\xi\to 0\\ \Delta\tau\to 0}}\left\|T\left(u(\xi_k,\tau_n), u_k^n\right)\right\| = 0$.

With these definitions in mind, a result from linear algebra is introduced which is crucial to proving that (7.2.3) is stable and convergent. The matrices whose diagonals are one row up and down from the main diagonal are

$$E_1 = \begin{bmatrix} 0 & 1 & 0 & \cdots & 0 \\ 0 & 0 & 1 & \cdots & 0 \\ & & \ddots & \ddots & \\ & & 0 & 0 & 1 \\ & & & 0 & 0 \end{bmatrix} \text{ and } E_{-1} = \begin{bmatrix} 0 & 0 & 0 & \cdots & 0 \\ 1 & 0 & 0 & \cdots & 0 \\ & \ddots & \ddots & \ddots & \\ & & 1 & 0 & 0 \\ & & & 1 & 0 \end{bmatrix}.$$

The tridiagonal matrix $A = \begin{bmatrix} a & b & 0 & \cdots & 0 \\ c & a & b & \cdots & 0 \\ 0 & c & a & \ddots & \vdots \\ & \ddots & \ddots & \ddots & b \\ & & & c & a \end{bmatrix}$ can now be written as $A = a \cdot I_{K \times K} + b \cdot E_1 + c \cdot E_{-1}$.

Theorem 7.1: *The eigenvalues of the K x K tridiagonal matrix* $A = a \cdot I_{K \times K} + b \cdot E_1 + c \cdot E_{-1}$ *are* $\lambda_k = a + 2\sqrt{bc}\cos\left(\frac{k\pi}{K+1}\right)$ *with corresponding eigenvectors*

$$\boldsymbol{v}_k = \frac{1}{|c_K|}\left[\rho \cdot \sin\left(\frac{k\pi}{K+1}\right), \rho^2 \cdot \sin\left(\frac{2k\pi}{K+1}\right), \cdots, \rho^K \cdot \sin\left(\frac{K \cdot k\pi}{K+1}\right)\right]^T, \ \rho = \sqrt{c/b}, \ and$$

$$c_K^2 = \sum_{\ell=1}^{K} \rho^{2\ell} \cdot \sin^2\left(\frac{\ell\pi}{K+1}\right).$$

■

The statement and proof of this theorem can be found in Smith [70] (pages 59 and 154–156). Recall that the left-hand side of (7.2.3) features the tridiagonal matrices $A(\lambda_p) = A - i\,\lambda_p I$ and $A(\lambda_p) = A - i\,\lambda_p I$. It follows immediately from Theorem 7.1 that the eigenvalues of the tridiagonal matrix $A = 2\,I_{K \times K} - E_1 - E_{-1}$ are $\mu_k = 2 + 2\cos\left(\frac{k\pi}{K+1}\right)$ with associated eigenvectors $\boldsymbol{v}_k = \frac{1}{\sqrt{\frac{1}{2}(K+1)}} \cdot \left[\sin\left(\frac{k\pi}{K+1}\right), \sin\left(\frac{2k\pi}{K+1}\right), \cdots, \sin\left(\frac{K \cdot k\pi}{K+1}\right)\right]^T$ for $k = 1, 2, \ldots, K$. Thus, $\lambda_k = \mu_k - i \cdot \lambda$ is an eigenvalue of $A(\lambda)$ since $A(\lambda) \cdot \boldsymbol{v}_k = (A - i\,\lambda\,I) \cdot \boldsymbol{v}_k = A \cdot \boldsymbol{v}_k - i\,\lambda\,I \cdot \boldsymbol{v}_k = A \cdot \boldsymbol{v}_k - i\,\lambda \cdot \boldsymbol{v}_k = (\mu_k - i \cdot \lambda) \cdot \boldsymbol{v}_k = \lambda_k \boldsymbol{v}_k$. Therefore, by the Spectral Theorem of linear algebra, the matrix $A(\lambda)$ can be decomposed as $A(\lambda) = \mathbb{U}^T\,\Lambda\,\mathbb{U}$ where $\Lambda = diag\{\mu_1 - i \cdot \lambda, \mu_2 - i \cdot \lambda, \ldots, \mu_K - i \cdot \lambda\}$ and $\mathbb{U}$ is an orthogonal matrix whose columns are the eigenvectors $\boldsymbol{v}_k$: $\mathbb{U} = \begin{bmatrix} | & | & & | \\ \boldsymbol{v}_1 & \boldsymbol{v}_2 & \cdots & \boldsymbol{v}_K \\ | & | & & | \end{bmatrix}$. Hence, the matrices $A(\lambda_p)$ and $A(\lambda_s)$ are invertible with

$$A^{-1}(\lambda_p) = \mathbb{U}^T \Lambda_p^{-1} \mathbb{U}, \ A^{-1}(\lambda_s) = \mathbb{U}^T \Lambda_s^{-1} \mathbb{U}, \ \Lambda_p^{-1} = diag\left\{\frac{1}{\mu_1 - i\lambda_p}, \frac{1}{\mu_2 - i\lambda_p}, \cdots, \frac{1}{\mu_K - i\lambda_p}\right\}$$

and a similar form for Λ_s^{-1}.

The quasi–implicit method (7.2.3) can now be written as

$$\left.\begin{aligned} \boldsymbol{u}^{n+1} &= B_{n,p}\,\boldsymbol{u}^n \\ \boldsymbol{v}^{n+1} &= B_{n,s}\,\boldsymbol{v}^n \end{aligned}\right\} \tag{7.2.4}$$

where $\boxed{B_{n,p} = -i\mathbb{U}^T\Lambda_p^{-1}\mathbb{U}\left[\lambda_p I - \zeta_p V^n\right]}$ and $\boxed{B_{n,s} = -i\mathbb{U}^T\Lambda_s^{-1}\mathbb{U}\left[\lambda_s I + \zeta_s U^n\right]}$. Using this concise form of the quasi–implicit method, the process of proving stability and convergence will be achieved via the *Lax Equivalence Theorem*, see, e.g., Ortega and Poole [58], Richtmyer and Morton [66], Thomas [78]. The theorem states that a finite difference scheme for a well-posed initial value problem is convergent if and only if it is stable and consistent. To accomplish this, additional definitions are required. Whenever possible, proofs will be extended to spatial dimension $n_d \geq 2$ so that the development of a finite difference scheme for $n_d = 2$ will not require a repeat of the theoretical foundations. Equation (7.2.1) will be the model that the analysis follows.

The remainder of this section applies the power of functional analysis to demonstrate the stability and convergence of the quasi–implicit finite difference scheme (7.2.3)/(7.2.4). To clarify the meaning of any unfamiliar symbols, the reader is referred to the *Introduction*.

In order to prove the Raman scattering problem is well-posed, the following remark is posited.

Remark 7.1: Throughout this remark, let X and Y be the separable Hilbert spaces with $X = H_2(\Omega) = \left\{f : \|f\|_{2,2}^2 = \int_\Omega \left[|f(\mathbf{x})|^2 + |\nabla f(\mathbf{x})|^2 + |\Delta f(\mathbf{x})|^2\right] d\mathbf{x} < \infty\right\}$ a Sobolev space and $Y = L_2(\Omega) = \left\{f : \|f\|_2^2 = \int_\Omega |f(\mathbf{x})|^2 d\mathbf{x} < \infty\right\}$ a Lebesgue space, respectively. It is evident that $X \subset Y$ and for any $f \in X$, $\|f\|_Y = \|f\|_2 \leq \|f\|_{2,2} = \|f\|_X$. Also, the special Sobolev space $H^1([0, T]; L_2(\Omega)) = \left\{f(\mathbf{x}, t) : \int_0^\infty \left[\|\phi(\mathbf{x},t)\|_2^2 + \left\|\frac{\partial}{\partial t}\phi(\mathbf{x},t)\right\|_2^2\right] dt < \infty\right\}$ is utilized to demonstrate the existence and uniqueness of solutions to (7.1.4)–(7.1.5). Assume that the spatial dimension n_d does not exceed 2 (i.e., $n_d = 1$ or 2).

(i) Let $\mathcal{L}$ be the Laplacian operator Δ as utilized in (7.2.3) operating from the Hilbert space X into Y. Then $\mathcal{L} \in \mathcal{Q}(X,Y)$ is the infinitesimal generator of a C_o–semigroup[19] $S(\tau)$. This is a standard result from functional analysis as per Rauch [64] or Pazy [59]. In particular, $S(\tau)$ is represented by the operator $e^{\mathcal{L}\tau}$ and there exist constants c_o and λ_o so that $\| S(\tau)\phi(\xi,\tau) \|_Y = \| e^{\mathcal{L}\tau}\phi(\xi,\tau) \|_Y \leq c_o e^{\lambda_o \tau} \| \phi(\xi,\tau) \|_Y$.

(ii) Let $\mathbf{w}(\xi,\tau) = [u(\xi,\tau),\ v(\xi,\tau)]^T$ be the vector of unitless electric fields and $\boldsymbol{\omega}(\xi,\tau) = [\omega_1(\xi,\tau),\ \omega_1(\xi,\tau)]^T$. As per (7.2.1), $N(\mathbf{w}) = \begin{bmatrix} -|v|^2 & 0 \\ 0 & |u|^2 \end{bmatrix}$ and $N(\boldsymbol{\omega}) = \begin{bmatrix} -|\omega_2|^2 & 0 \\ 0 & |\omega_1|^2 \end{bmatrix}$. Define

---

[19] See page 11 of the *Introduction*.

$\mathcal{N}(\boldsymbol{w}) = N(\boldsymbol{w})\boldsymbol{w} = \begin{bmatrix} -|v|^2 u \\ |u|^2 v \end{bmatrix}$ and $\mathcal{N}(\boldsymbol{\omega}) = N(\boldsymbol{\omega})\boldsymbol{\omega} = \begin{bmatrix} -|\omega_2|^2 \omega_1 \\ |\omega_1|^2 \omega_2 \end{bmatrix}$. To obtain an upper bound on $\| \mathcal{N}(\boldsymbol{w}) - \mathcal{N}(\boldsymbol{\omega}) \|_Y$, it is sufficient to first bound $\| -|v|^2u + |\omega_2|^2\omega_1 \|_Y = \| -|v|^2u + |v|^2v - |v|^2v + |\omega_2|^2\omega_1 - |\omega_2|^2\omega_2 + |\omega_2|^2\omega_2 \|_Y \le \| -|v|^2v + |\omega_2|^2\omega_2 \|_Y + \| -|v|^2(v-u) + |\omega_2|^2 (\omega_1 - \omega_2) \|_Y$. In a parallel fashion, $\| |u|^2v - |\omega_1|^2\omega_2 \|_Y \le \| |u|^2u - |\omega_1|^2\omega_1 \|_Y + \| |u|^2(v-u) + |\omega_1|^2 (\omega_1 - \omega_2) \|_Y$. Haraux [32] and Pazy [59] report that for any $u, v \in L_2(\Omega)$, the bounds

$$\| -|v|^2v + |u|^2u \|_Y = \| -|v|^2v + |u|^2u \|_2 \le c_o\cdot\left(\|v\|_{2,2}^2 + \|u\|_{2,2}^2\right)\|u - v\|_{2,2}$$

so that the following estimates hold.

$$\begin{aligned} \| -|v|^2u + |\omega_2|^2\, \omega_1 \|_Y &= \| -|v|^2u + |\omega_2|^2\, \omega_1 \|_2 \\ &\le c_{1,1}\cdot\left(\|v\|_{2,2}^2 + \|\omega_2\|_{2,2}^2\right)\|\omega_2 - v\|_{2,2} + c_{1,2}\cdot\left(\|v\|_{2,2}^2 + \|\omega_2\|_{2,2}^2\right)\|(\omega_1 - \omega_2) - (v-u)\|_{2,2} \\ &\le c_1\cdot\left(\|v\|_{2,2}^2 + \|\omega_2\|_{2,2}^2\right)\|\boldsymbol{\omega} - \boldsymbol{w}\|_{2,2}. \end{aligned}$$

Similarly, $\| |u|^2v - |\omega_1|^2\, \omega_2 \|_Y = \| |u|^2v - |\omega_1|^2\, \omega_2 \|_2 \le c_2\cdot\left(\|u\|_{2,2}^2 + \|\omega_1\|_{2,2}^2\right)\|\boldsymbol{\omega} - \boldsymbol{w}\|_{2,2}$.

Consequently,

$$\begin{aligned} \| \mathcal{N}(\boldsymbol{w}) - \mathcal{N}(\boldsymbol{\omega}) \|_Y &= \| \mathcal{N}(\boldsymbol{w}) - \mathcal{N}(\boldsymbol{\omega}) \|_2 \\ &\le c_1\cdot\left(\|v\|_{2,2}^2 + \|\omega_2\|_{2,2}^2\right)\|u - \omega_1\|_{2,2} + c_2\cdot\left(\|u\|_{2,2}^2 + \|\omega_1\|_{2,2}^2\right)\|v - \omega_2\|_{2,2} \\ &\le c_\omega\cdot\left(\|\boldsymbol{w}\|_{2,2}^2 + \|\boldsymbol{\omega}\|_{2,2}^2\right)\|\boldsymbol{\omega} - \boldsymbol{w}\|_{2,2}. \end{aligned}$$

That is, the operator $\mathcal{N}$ is bound by a monotonically increasing function ψ: $\mathbb{R}^+ \otimes \mathbb{R}^+ \to \mathbb{R}^+$ with $\psi(a, b) = c_\omega\cdot(a^2 + b^2)$.

(iii) Let $\boldsymbol{w} = [w_1, w_2]^T$ be a solution of (7.2.1). Then the nonlinear operator $\mathcal{N}(\boldsymbol{w}) = N(\boldsymbol{w})\cdot\boldsymbol{w} = \begin{bmatrix} -|w_2|^2 & 0 \\ 0 & |w_1|^2 \end{bmatrix}\cdot\begin{bmatrix} w_1 \\ w_2 \end{bmatrix} = \begin{bmatrix} -|w_2|^2 w_1 \\ |w_1|^2 w_2 \end{bmatrix}$ is quadratically nonlinear. That is, $\mathcal{N}(\boldsymbol{0}) = \boldsymbol{0} = [0, 0]^T$ and $\mathcal{N}'(\boldsymbol{0})(\boldsymbol{z}) = \boldsymbol{0}$, $\boldsymbol{z} = [z_1, z_2]^T$. The first assertion follows directly from the definition of $\mathcal{N}(\boldsymbol{w})$ while the second can determined from the Fréchet derivative for $\mathcal{N}(\boldsymbol{w})$: $\mathcal{N}'(\boldsymbol{w})(\boldsymbol{z}) = \begin{bmatrix} -z_1(\bar{z}_2 w_2 + \bar{w}_2 z_2) - w_1|z_2|^2 \\ z_2(\bar{z}_1 w_1 + \bar{w}_1 z_1) + w_2|z_1|^2 \end{bmatrix}$. Here, $\bar{z}_j$ is the complex conjugate of z_j. Since $\mathcal{N}(\boldsymbol{w})$ is a third order polynomial in the components of $\boldsymbol{w} = [w_1, w_2]^T$, then $\mathcal{N}(\boldsymbol{w})$ is smoothly Fréchet differentiable and $\mathcal{N}'(\boldsymbol{w}) \in \mathcal{C}(X,Y)$ = the set of all continuous maps from X into Y. Observe

that $\mathcal{N}(z) - \mathcal{N}(w) - \mathcal{N}'(w)(z-w) = \begin{bmatrix} z_1\left(|w_2|^2 - |z_2|^2\right) + w_1\left(\overline{w}_2(z_2 - w_2) + w_2(\overline{z}_2 - \overline{w}_2)\right) \\ z_2\left(|z_1|^2 - |w_1|^2\right) + w_2\left(\overline{w}_1(z_1 - w_1) + w_1(\overline{z}_1 - \overline{w}_1)\right) \end{bmatrix}$. For any distinct functions $z, w \in B \subset X$, B a bound set with bound M, then $\|z\|_X \leq \|z_1\|_X^2 + \|z_2\|_X^2 \leq \|z_1\|_B^2 + \|z_2\|_B^2 \leq M^2 + M^2 = 2M^2$ and $\|w\|_X \leq 2\ M^2$. Using these estimates and repeated applications of the triangle inequality and the Cauchy–Schwarz inequality produces

$$\begin{aligned} &\left\| -z_1\left(|z_2| - |w_2|\right)\left(|z_2| + |w_2|\right) + w_1\overline{w}_2(z_2 - w_2) + w_1 w_2(\overline{z}_2 - \overline{w}_2)\right\|_Y \\ &\leq \left(\|z_1\|_Y \cdot \left[\|w_2\|_Y + \|z_2\|_Y\right]\right)^2 \|z_2 - w_2\|_Y^2 + 2\left(\|w_1\|_Y \cdot \|w_2\|_Y\right)^2 \|z_2 - w_2\|_Y^2 \\ &\leq \left\{\left(\|z_1\|_X \cdot \left[\|w_2\|_X + \|z_2\|_X\right]\right)^2 + 2\left(\|w_1\|_X \cdot \|w_2\|_X\right)^2\right\} \|z_2 - w_2\|_X^2 \\ &\leq \left(4M^3 + 2M^2\right)\|z_2 - w_2\|_X^2 . \end{aligned}$$

Similarly, $\left\| z_2\left(|z_1|^2 - |w_1|^2\right) + w_2\left(\overline{w}_1(z_1 - w_1) + w_1(\overline{z}_1 - \overline{w}_1)\right)\right\|_Y \leq \left(4M^3 + 2M^2\right)\|z_1 - w_1\|_X^2$.

As $X = H_2(\Omega) \subset Y = L_2(\Omega)$, then $\|z\|_Y \leq \|z\|_X$. Combining these bounds yields

$$\| \mathcal{N}(z) - \mathcal{N}(w) - \mathcal{N}'(w)(z-w) \|_Y \leq c_o \|z - w\|_X^2 \text{ for } c_o = (4M^3 + 2M^2).$$

(iv) The estimates above indicate that the operators $\mathcal{L} = \Delta$ (the Laplacian in spatial dimensions 1 and 2) and $\mathcal{N}$ satisfy for formalism established in Chapter 8. As a consequence, the following theorems can be stated for the Raman scattering model established in (7.2.1).

(v) A *generalized solution* of (7.2.1) with respect to the C_o–semigroup $S(\tau) = e^{i\tau C\cdot\Delta}$, is given by the formula $w(\xi, \tau) = [S(\tau)w_o](\xi) + \int_0^\tau [S(\tau - s)N(w)](\xi, s)\, ds$. The concept of a generalized solution is detailed in §8.2 of Chapter 8; especially Definition 8.1.

Theorem 7.2 (Existence and Uniqueness): *For every* $f, g \in X = H_2(\Omega)$, *there exists* $T \in [0, \infty)$ *so that* (7.2.1) *has a unique generalized solution* $w(\xi, \tau) \in H^1([0, T]; L_2(\Omega))$ *and either* $T = \infty$ *or* $\| w(\bullet, \tau) \|_2 \to \infty$ *as* $\tau \to T$.

Proof: Since the operator $\mathcal{L} = \Delta$ satisfies *Key Condition (A)*, then so does a constant matrix multiple $iC\cdot\Delta$. Moreover, $\mathcal{N}(w) = \begin{bmatrix} -|v|^2 u \\ |u|^2 v \end{bmatrix}$ satisfies *Key Conditions (C) – (D)* of Chapter 8. Thus, by Lemmas 8.2 and 8.4, the existence and uniqueness of solutions to (7.2.1) is assured. ■

The stability of solutions theorem below follows immediately from Theorem 8.5 of Chapter 8 with $X = H_2(\Omega)$ and $Z = L_2(\Omega)$.

Theorem 7.3 (Stability of Solutions): *If* $\boldsymbol{w}(\xi,\tau) \in H^1([0, T]; L_2(\Omega))$ *is a generalized solution of* (7.2.1) *with* $\boldsymbol{w}(\xi, 0) = \boldsymbol{w}_o(\xi) = [f(\xi), g(\xi)]^T$, *then there is a neighborhood* Θ_o *of* $\boldsymbol{w}_o \in H_2(\Omega)$ *so that for any* $\boldsymbol{\omega}_o(\xi) = [\phi(\xi), \gamma(\xi)]^T \in \Theta_o$ *and every interval* $[0, T] \subset [0, \infty)$ *there is a generalized solution* $\boldsymbol{\omega}(\xi,\tau) \in H^1([0, T]; L_2(\Omega))$ *with* $\boldsymbol{\omega}(\xi, 0) = \boldsymbol{\omega}_o(\xi)$. *Moreover, there is a constant* C_o *so that*

$$\| \boldsymbol{w}(\xi,\tau) - \boldsymbol{\omega}(\xi,\tau) \|_2 \le C_o \| \boldsymbol{w}_o(\xi) - \boldsymbol{\omega}_o(\xi) \|_{2,2}$$

for all $\boldsymbol{\omega}_o \in \Theta_o$ *and* $t \in [0, T]$. ■

By applying the process of integration by parts for the one spatial–dimensional case ($n_d = 1$) and *Green's Theorem* for the two spatial–dimensional case ($n_d = 2$), then Theorem 7.4 can be achieved. This result is important since it implies that the unitless electric field solutions $u(\xi,\tau)$ and $v(\xi,\tau)$ of (7.1.4)–(7.1.5) *exist for all time* $\tau > 0$.

Theorem 7.4 (Conservation of Norms): *Let* $u(\xi,\tau) \equiv u(\tau)$ *and* $v(\xi,\tau) \equiv v(\tau)$ *be the solutions of* (7.1.4)–(7.1.5). *Then for all* $\tau \ge 0$, *the* L_2*–norms of* $u(\tau)$ *and* $v(\tau)$ *are conserved by* $u(0) \equiv f$ *and* $v(0) \equiv g$ *via* (7.2.5)–(7.2.6).

$$\| u(\tau) \|_2 \le \| f \|_2 \tag{7.2.5}$$

$$\| u(\tau) \|_2^2 + \| v(\tau) \|_2^2 = \| f \|_2^2 + \| g \|_2^2 \tag{7.2.6}$$

Proof: For the two spatial–dimensional case ($n_d = 2$), observe that $\frac{\partial}{\partial \tau} \| u(\tau) \|_2^2 = \frac{\partial}{\partial \tau} \int_{\Psi(\Omega)} |u(\tau)|^2 \, d\boldsymbol{\omega}$

where $\boldsymbol{\omega}$ is the element of area of the domain $\Psi(\Omega)$. Since $|u(\tau)|^2 = u(\tau)\bar{u}(\tau)$ then

$$\frac{\partial}{\partial \tau} \int_{\Psi(\Omega)} |u(\tau)|^2 \, d\boldsymbol{\omega} = \int_{\Psi(\Omega)} \left(u(\tau) \frac{\partial \bar{u}}{\partial \tau}(\tau) + \bar{u}(\tau) \frac{\partial u}{\partial \tau}(\tau) \right) d\boldsymbol{\omega}$$

$$= \int_{\Psi(\Omega)} \left(u \left[-ic_p \Delta \bar{u} - |v|^2 \bar{u} \right] + \bar{u} \left[ic_s \Delta u - |v|^2 u \right] \right) d\boldsymbol{\omega} \text{ which by Green's Theorem}$$

$$= ic_p \int_{\partial\Psi(\Omega)} \left(- \left[u \frac{\partial \bar{u}}{\partial \boldsymbol{n}} - \bar{u} \frac{\partial u}{\partial \boldsymbol{n}} \right] + \left[\bar{u} \frac{\partial u}{\partial \boldsymbol{n}} - u \frac{\partial \bar{u}}{\partial \boldsymbol{n}} \right] \right) d\sigma + ic_p \int_{\Psi(\Omega)} \left(\nabla u \bullet \nabla \bar{u} - \nabla u \bullet \nabla \bar{u} \right) d\boldsymbol{\omega} - 2 \int_{\Psi(\Omega)} |v|^2 |u|^2 \, d\boldsymbol{\omega} .$$

Now the first integral over the boundary of the domain $\partial\Psi(\Omega)$ is 0 by the Dirichlet conditions imposed in (7.1.4) while the second integral over the domain $\Psi(\Omega)$ is 0 since the integrand is 0. Therefore,

$$\frac{\partial}{\partial\tau}\| u(\tau) \|_2^2 = -2 \int_{\Psi(\Omega)} |v|^2 |u|^2 \, d\omega \le 0 \text{ for all } \tau \ge 0.$$

In a parallel manner, it can be shown that

$$\frac{\partial}{\partial\tau}\| v(\tau) \|_2^2 = 2 \int_{\Psi(\Omega)} |v|^2 |u|^2 \, d\omega \ge 0.$$

The first inequality indicates that, as a function of τ, $\| u(\tau) \|_2^2$ (and hence $\| u(\tau) \|_2$) is a positive, *decreasing* function so that $\| u(\tau) \|_2 \le \| u(0) \|_2 = \| f \|_2$. This establishes (7.2.5). Since $\boldsymbol{w}(\xi,\tau) = [u(\xi,\tau), v(\xi,\tau)]^T$ is the solution of (7.2.1), then $\|\boldsymbol{w}(\xi,\tau)\|_2^2 \equiv \|\boldsymbol{w}(\tau)\|_2^2 = \|u(\tau)\|_2^2 + \|v(\tau)\|_2^2$ and $\frac{\partial}{\partial\tau}\|\boldsymbol{w}(\tau)\|_2^2 = \frac{\partial}{\partial\tau}\|u(\tau)\|_2^2 + \frac{\partial}{\partial\tau}\|v(\tau)\|_2^2 = -2\int_{\Psi(\Omega)} |v|^2|u|^2\, d\omega + 2\int_{\Psi(\Omega)} |v|^2|u|^2\, d\omega = 0$ which means that $\|\boldsymbol{w}(\tau)\|_2^2 = \|\boldsymbol{w}(0)\|_2^2 = \|u(0)\|_2^2 + \|v(0)\|_2^2 = \|f\|_2^2 + \|g\|_2^2$ which proves (7.2.6).

To prove the one spatial–dimensional case, repeat the integral calculations above with integration by parts replacing Green's Theorem. The results are the same. ■

Remark 7.2: Define the function $H(\tau) = \|u(\tau)\|_2^2 + \|v(\tau)\|_2^2$ for $\tau \ge 0$. Then by Theorem 7.4, $H(\tau) = \|f\|_2^2 + \|g\|_2^2$ so that $\frac{dH}{d\tau}(\tau) = H'(\tau) = 0$ for all $\tau \ge 0$. The function H is called the *Hamiltonian* for (7.1.4)–(7.1.5). As $H'(\tau) = 0$, the Hamiltonian is *conserved*. As a direct result of this observation, the upper limit for existence of the solution of (7.1.4)–(7.1.5) in time is seen to be $T = \infty$. Therefore, the solution is global in time and exists for all $\tau \in [0, \infty)$.

§7.3 Stability, Consistency, and Convergence

In the proof of the Conservation of Norms Theorem 7.4, the inequalities (7.3.1)–(7.3.2) are established.

$$\frac{\partial}{\partial\tau}\| u(\tau) \|_2^2 = -2 \int_{\Psi(\Omega)} |v|^2 |u|^2 \, d\omega \le 0 \text{ for all } \tau \ge 0 \qquad (7.3.1)$$

$$\frac{\partial}{\partial\tau}\| v(\tau) \|_2^2 = 2 \int_{\Psi(\Omega)} |v|^2 |u|^2 \, d\omega \ge 0 \text{ for all } \tau \ge 0 \qquad (7.3.2)$$

Thus, as functions of τ, $u(\bullet,\tau)$ is decreasing while $v(\bullet,\tau)$ is increasing. Since the quasi–implicit method (7.2.2) is constructed for the one–spatial dimensional case only, this section is restricted to $n_d = 1$. Therefore, to prove that (7.2.4) and hence (7.2.3) is stable, it is sufficient to show that the matrix norms on $B_{n,p}$ and $B_{n,s}$ are bounded via (7.3.3)–(7.3.4), see, e.g., Smith [70] for details.

$$\| B_{n,p} \| \leq 1 \text{ for all } n \geq 1 \tag{7.3.3}$$
$$\| B_{n,s} \| \leq 1 + c_o \Delta\tau \text{ for all } n \geq 1,\ c_o \text{ a positive constant} \tag{7.3.4}$$

Recall the definitions of $B_{n,p}$, $B_{n,s}$ (page 152), and $\mu_k = 2 + 2\cos\left(\frac{k\pi}{K+1}\right)$ to make the following observations. Both $B_{n,p}$ and $B_{n,s}$ are composed of unitary matrices $\mathbb{U}$ which have norm 1. Also, the μ_k are positive and *decreasing* so that $\mu_1^2 + \lambda_p^2 \geq \mu_2^2 + \lambda_p^2 \geq \cdots \geq \mu_K^2 + \lambda_p^2$ and $\frac{1}{\sqrt{\mu_K^2 + \lambda_p^2}} \geq \cdots \geq \frac{1}{\sqrt{\mu_2^2 + \lambda_p^2}} \geq \frac{1}{\sqrt{\mu_1^2 + \lambda_p^2}}$. Figure 7.3 illustrates this remark.

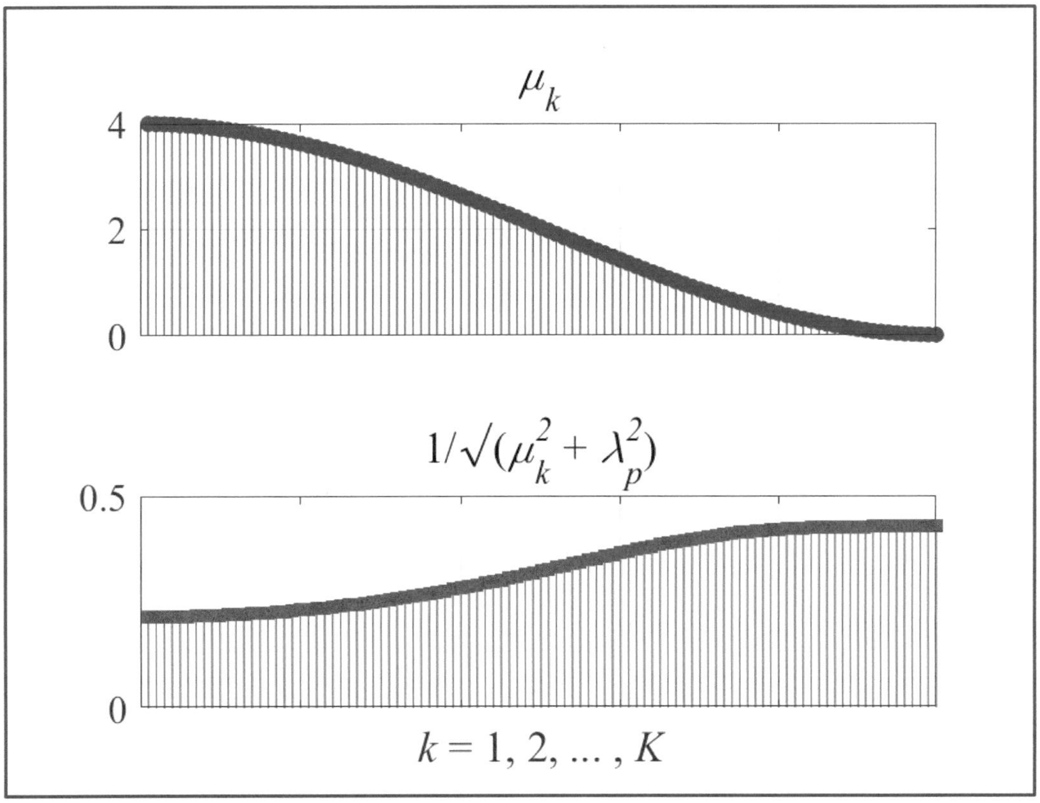

Figure 7.3. Decreasing μ_k and increasing $1/\sqrt{\mu_k^2 + \lambda_p^2}$.

Therefore,

$$\| B_{n,p} \| \leq 1 \cdot \max_{1\leq k\leq K} \left| \frac{1}{\mu_k - i\lambda_p} \right| \cdot 1 \cdot \max_{1\leq k\leq K} \left| \lambda_p - \zeta_p \left| v_k^{n-1} \right|^2 \right| \leq \frac{|\lambda_p|}{\sqrt{\mu_K^2 + \lambda_p^2}} \max_{1\leq k\leq K} \left| 1 - \Delta\tau \left| v_k^{n-1} \right|^2 \right|$$

$$\leq \max_{1\leq k\leq K}\left|1-\Delta\tau\left|v_k^{n-1}\right|^2\right|. \tag{7.3.5}$$

In a similar manner, it is shown that

$$\| B_{n,s} \| \leq \max_{1\leq k\leq K}\left|1+\Delta\tau\left|u_k^{n-1}\right|^2\right|. \tag{7.3.6}$$

To prove that the quasi–implicit difference scheme (7.2.4) is stable, it is sufficient to demonstrate that the right-hand sides of (7.3.5) and (7.3.6) can be bounded by 1 and $1 + c_o\cdot\Delta\tau$, respectively. This is achieved, for the one–spatial dimensional case, via the L_∞ norms on the Cauchy data $f(\xi)$ and $g(\xi)$ and Theorem 7.5. Specifically, if the Cauchy data of (7.2.1) are bounded on $I_\psi \equiv [\psi_a, \psi_b]$ and the time step is controlled, then the quasi–implicit scheme is stable. The notation I_ψ means the image of the interval $[a, b]$ under the change of variables Ψ and is used throughout the remainder of this section.

Theorem 7.5 (Stability): *If* $f, g \in L_\infty(I_\psi)$ *and* $\Delta\tau \leq \frac{2}{\|g\|_\infty^2}\exp\left(-2T\|f\|_\infty^2\right)$, *then the finite difference scheme* (7.2.4) *is stable.*

Proof: By (7.3.5), $\| B_{n,p} \| \leq 1$ provided $\left\|\boldsymbol{v}^{n-1}\right\|_{\ell_\infty}^2 \leq \frac{2}{\Delta\tau}$ where $\boldsymbol{v}^{n-1} = \left[v_1^{n-1}, v_2^{n-1}, \cdots, v_K^{n-1}\right]^T$ is a vector. Indeed, $\left\|\boldsymbol{v}^{n-1}\right\|_{\ell_\infty}^2 \leq \frac{2}{\Delta\tau}$ implies $\left|v_k^{n-1}\right|^2 \leq \frac{2}{\Delta\tau}$ for any k, and therefore $\Delta\tau\left|v_k^{n-1}\right|^2 \leq \frac{2}{\Delta\tau}\cdot\Delta\tau \leq 2$ for all k. Hence, $\| B_{n,p} \| \leq \max_{1\leq k\leq K}\left|1-\Delta\tau\left|v_k^{n-1}\right|^2\right| \leq |1 - 2| = 1$. Similarly, $\left\|\boldsymbol{u}^{n-1}\right\|_{\ell_\infty}^2 \leq c_o$ implies $\| B_{n,s} \| \leq 1 + c_o\cdot\Delta\tau$. The ℓ_∞–norm on a vector $\boldsymbol{w} = [w_1, w_2, \ldots, w_K]$ is the maximum element: $\|\boldsymbol{w}\|_{\ell_\infty} = \max_{1\leq k\leq K}|w_k|$. To find the required *a priori* bound on $\Delta\tau$, proceed sequentially.

For $n = 1$, $\left\|\boldsymbol{u}^0\right\|_{\ell_\infty}^2 \leq \|f\|_\infty^2$ and $\left\|\boldsymbol{v}^0\right\|_{\ell_\infty}^2 \leq \|g\|_\infty^2$ by *Conservation of Norms* Theorem 7.4 and (7.2.5)–(7.2.6). Thus, it is required that $c_o = \|f\|_\infty^2$ and $\frac{2}{\Delta\tau} \geq \|g\|_\infty^2$.

For $n = 2$, $\left\|\boldsymbol{u}^1\right\|_{\ell_\infty}^2 \leq \left\|B_{0,p}\right\|_{\ell_\infty}^2 \cdot \left\|\boldsymbol{u}^0\right\|_{\ell_\infty}^2 \leq \left(\max_{1\leq k\leq K} \leq \left|1-\Delta\tau\left|v_k^0\right|^2\right|\right)^2 \|f\|_{\ell_\infty}^2 \leq \left(\max_{1\leq k\leq K} \leq \left|1-\Delta\tau\left|g_k\right|^2\right|\right)^2 \|f\|_{\ell_\infty}^2$. If $\frac{2}{\Delta\tau} \geq \|g\|_\infty^2$, then $\max_{1\leq k\leq K}\left|1-\Delta\tau\left|g_k\right|^2\right| \leq 1$, and consequently

$$\left\|\boldsymbol{v}^1\right\|_{\ell_\infty}^2 \leq \left\|B_{0,s}\right\|_{\ell_\infty}^2 \cdot \left\|\boldsymbol{v}^0\right\|_{\ell_\infty}^2 \leq \left(\max_{1\leq k\leq K}\left|1+\Delta\tau\left|u_k^0\right|^2\right|\right)^2 \|g\|_{\ell_\infty}^2 \leq \left(\max_{1\leq k\leq K}\left|1+\Delta\tau\left|f_k\right|^2\right|\right)^2 \|g\|_{\ell_\infty}^2 \leq \|g\|_\infty^2\left(1+\Delta\tau\|f\|_\infty^2\right)^2.$$

Continuing in this manner, an induction argument for $n = N$ leads to

$$\left\|\boldsymbol{u}^N\right\|_{\ell_\infty}^2 \le \left\|B_{N,p}\right\|_{\ell_\infty}^2 \cdot \left\|\boldsymbol{u}^{N-1}\right\|_{\ell_\infty}^2 \le \left(\max_{1\le k\le K}\left|1-\Delta\tau\left|v_k^{N-1}\right|^2\right|\right)^2 \left\|\boldsymbol{u}^{N-1}\right\|_{\ell_\infty}^2$$

and

$$\left\|\boldsymbol{v}^N\right\|_{\ell_\infty}^2 \le \left\|B_{N,s}\right\|_{\ell_\infty}^2 \cdot \left\|\boldsymbol{v}^{N-1}\right\|_{\ell_\infty}^2 \le \left(\max_{1\le k\le K}\left|1+\Delta\tau\left|u_k^{N-1}\right|^2\right|\right)^2 \left\|\boldsymbol{v}^{N-1}\right\|_{\ell_\infty}^2 .$$

But $\dfrac{2}{\Delta\tau} \ge \left\|\boldsymbol{v}^{N-1}\right\|_\infty^2$ implies that

$$\left\|\boldsymbol{u}^N\right\|_{\ell_\infty}^2 \le \left\|\boldsymbol{u}^{N-1}\right\|_{\ell_\infty}^2 \le \cdots \le \left\|\boldsymbol{u}^1\right\|_{\ell_\infty}^2 \le \left\|\boldsymbol{u}^0\right\|_{\ell_\infty}^2 \le \left\|f\right\|_\infty^2 \tag{7.3.7}$$

and

$$\left\|\boldsymbol{v}^N\right\|_{\ell_\infty}^2 \le \left\|g\right\|_\infty^2\left(1+\Delta\tau\left\|f\right\|_\infty^2\right)^{2N}. \tag{7.3.8}$$

Recall that $\Delta\tau = \dfrac{T}{N}$ so that $\dfrac{2}{\Delta\tau} \ge \left\|g\right\|_\infty^2\left(1+\Delta\tau\left\|f\right\|_\infty^2\right)^{2N} = \left\|g\right\|_\infty^2\left(1+\dfrac{2T}{2N}\left\|f\right\|_\infty^2\right)^{2N}$. Observe that, as $N \to \infty$, the right-hand side of the inequality above becomes $\left\|g\right\|_\infty^2 \exp\left(2T\left\|f\right\|_\infty^2\right)$. Thus, the requirements $\dfrac{1}{\Delta\tau} \ge \tfrac{1}{2}\left\|g\right\|_\infty^2 \exp\left(2T\left\|f\right\|_\infty^2\right) \Leftrightarrow \Delta\tau \le \dfrac{2}{\left\|g\right\|_\infty^2}\exp\left(-2T\left\|f\right\|_\infty^2\right)$ and $c_o = \left\|f\right\|_\infty^2$ insure that (7.2.4) is stable. ■

To establish the consistency of the finite difference scheme (7.2.4), as per Definition 7.3, recall the notation of (7.2.1): $\boldsymbol{w} = [u,\ v]^T$, $\boldsymbol{w}_o = [f,\ g]^T$, $C = \begin{bmatrix} c_p & 0 \\ 0 & c_s \end{bmatrix}$, and $\Gamma = \begin{bmatrix} \zeta_p^{-1} & 0 \\ 0 & \zeta_s^{-1} \end{bmatrix}$. Let the differential operator $\mathscr{L}(\boldsymbol{w}) = \dfrac{\partial \boldsymbol{w}}{\partial\tau} - iC\dfrac{\partial^2 \boldsymbol{w}}{\partial\xi^2} - N(\boldsymbol{w})\boldsymbol{w}$ have solution $\boldsymbol{w}(\xi,\ \tau)$ and difference operator $\mathscr{K}(\boldsymbol{w}_k^n) = \dfrac{1}{\Delta\tau}\left(\boldsymbol{w}_k^n - \boldsymbol{w}_k^{n-1}\right) - i\Gamma\left(\boldsymbol{w}_{k+1}^n - 2\boldsymbol{w}_k^n + \boldsymbol{w}_{k-1}^n\right) - N(\boldsymbol{w}_k^{n-1})\boldsymbol{w}_k^{n-1}$ have solution $\boldsymbol{w}_k^n = \left[u_k^n, v_k^n\right]^T$. Finally, set $\mathcal{T}(\boldsymbol{w}, \boldsymbol{w}_k^n) = \mathscr{L}(\boldsymbol{w}) - \mathscr{K}(\boldsymbol{w}_k^n)$. The finite difference scheme (7.2.4) is consistent provided $\lim_{\substack{\Delta\tau\to 0\\ \Delta\xi\to 0}}\left\|\mathcal{T}\left(\boldsymbol{w}(\xi_k,\tau_n), \boldsymbol{w}_k^n\right)\right\| = 0$ which is demonstrated in the next theorem.

Theorem 7.6 (Consistency): *The finite difference scheme* (7.2.4) *is consistent whenever f, g* $\in L_2(I_\psi)$.

Proof: Expanding $\boldsymbol{w}_{k+1}^n \equiv \boldsymbol{w}(\xi_k + \Delta\xi, \tau_n)$ and $\boldsymbol{w}_k^{n-1} \equiv \boldsymbol{w}(\xi_k, \tau_n - \Delta\tau)$ via Taylor's Theorem component wise produces

$$\boldsymbol{w}_{k+1}^n = \boldsymbol{w}_k^n + \Delta\xi \frac{\partial \boldsymbol{w}}{\partial \xi}(\xi_k, \tau_n) + \tfrac{1}{2}(\Delta\xi)^2 \frac{\partial^2 \boldsymbol{w}}{\partial \xi^2}(\xi_k, \tau_n) + \tfrac{1}{6}(\Delta\xi)^3 \frac{\partial^3 \boldsymbol{w}}{\partial \xi^3}(\xi_k, \tau_n) + \cdots$$

and

$$\boldsymbol{w}_k^{n-1} = \boldsymbol{w}_k^n - \Delta\tau \frac{\partial \boldsymbol{w}}{\partial \tau}(\xi_k, \tau_n) + \tfrac{1}{2}(\Delta\tau)^2 \frac{\partial^2 \boldsymbol{w}}{\partial \tau^2}(\xi_k, \tau_n) + \cdots.$$

Therefore,

$$\begin{aligned}
\mathcal{T}\left(\boldsymbol{w}(\xi_k,\tau_n), \boldsymbol{w}_k^n\right) &= \frac{\partial \boldsymbol{w}}{\partial \tau}(\xi_k,\tau_n) - \frac{1}{\Delta\tau}\left(\Delta\tau \frac{\partial \boldsymbol{w}}{\partial \tau}(\xi_k,\tau_n) + O(\Delta\tau^2)\right) - iC\frac{\partial^2 \boldsymbol{w}}{\partial \xi^2}(\xi_k,\tau_n) \\
&\quad + \frac{i}{\Delta\xi^2} C\left(\Delta\xi^2 \frac{\partial^2 \boldsymbol{w}}{\partial \xi^2}(\xi_k,\tau_n) + O(\Delta\xi^4)\right) - N(\boldsymbol{w}_k^n)\boldsymbol{w}(\xi_k,\tau_n) \\
&\quad + N(\boldsymbol{w}_k^{n-1})\left(\boldsymbol{w}(\xi_k,\tau_n) - O(\Delta\tau)\right) \\
&= O(\Delta\tau) + O(\Delta\xi^2) + \left(N(\boldsymbol{w}_k^{n-1}) - N(\boldsymbol{w}_k^n)\right)\boldsymbol{w}_k^n.
\end{aligned}$$

Since

$$\begin{aligned}
N(\boldsymbol{w}_k^{n-1}) - N(\boldsymbol{w}_k^n) &= \begin{bmatrix} |v_k^n|^2 - |v_k^{n-1}|^2 & 0 \\ 0 & |u_k^{n-1}|^2 - |u_k^n|^2 \end{bmatrix} \\
&= \begin{bmatrix} |v(\xi_k,\tau_n)|^2 - |v(\xi_k,\tau_n - \Delta\tau)|^2 & 0 \\ 0 & |u(\xi_k,\tau_n - \Delta\tau)|^2 - |u(\xi_k,\tau_n)|^2 \end{bmatrix}
\end{aligned}$$

and, as the absolute value operator is continuous, then $\lim_{\Delta\tau\to 0}\left(N(\boldsymbol{w}_k^{n-1}) - N(\boldsymbol{w}_k^n)\right) = \boldsymbol{0}$. But the L_2–norm is a continuous operator, so that

$$\lim_{\substack{\Delta\tau\to 0\\ \Delta\xi\to 0}} \left\|\mathcal{T}\left(\boldsymbol{w}(\xi_k,\tau_n), \boldsymbol{w}_k^n\right)\right\| \le \lim_{\substack{\Delta\tau\to 0\\ \Delta\xi\to 0}} \|O(\Delta\tau)\|_2 + \lim_{\substack{\Delta\tau\to 0\\ \Delta\xi\to 0}} \|O(\Delta\xi^2)\|_2 + \lim_{\substack{\Delta\tau\to 0\\ \Delta\xi\to 0}} \left\|\left(N(\boldsymbol{w}_k^{n-1}) - N(\boldsymbol{w}_k^n)\right)\boldsymbol{w}_k^n\right\|_2 = 0. \quad \blacksquare$$

§7.4 Some Results

The previous sections of this chapter have focused on the development of a stable and convergent quasi–implicit scheme for the steady–state stimulated Raman scattering model (7.1.5) \ (7.2.1). Next, the method is implemented and some observations on model behavior can be stated. First note that, for technical reasons, the difference equation (7.2.3) is used to compute solutions to (7.1.5)/(7.2.1) rather than the more direct method (7.2.4). As an initial value of the pump field, select the *pulse* or *taperless beam* $f(\xi) = 1$ for $\xi \in [-\xi_o, \xi_o]$ and $f(\xi) = 0$ otherwise with a scaled version of f as the initial Stokes beam $g(\xi)$. A reasonable physical scaling is $g(\xi) = 0.1638 \times f(\xi)$. Therefore, the $\| f \|_\infty = 1$ and $\| g \|_\infty = 0.1638$. Figure 7.4 illustrates the pulse beams.

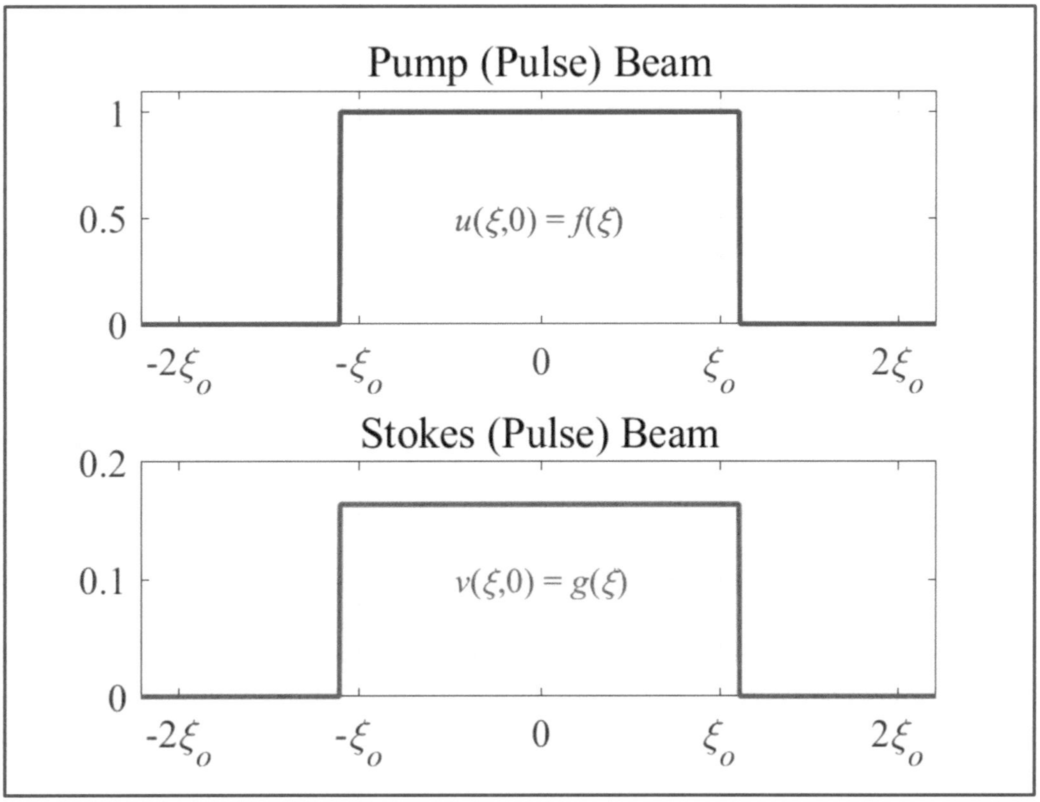

Figure 7.4. Initial pulse beams for steady–state stimulated Raman scattering.

Set the *Stokes gain constant* g_r to 0.004 cm^2/Mw, an initial pump intensity $|E_p(0)|^2 = 10$ Mw/cm^2, and wave numbers $k_p = 2\pi/\lambda_p = (2\pi/3.5) \times 10^5$ $rad/cm = 1.7951 \times 10^5$ rad/cm, $k_s = 2\pi/\lambda_s = (2\pi/4.1) \times 10^5$ $rad/cm = 1.5324 \times 10^5$ rad/cm. With these quantities defined, the unitless spatial and temporal extents are $\xi \in [-2\xi_o, 2\xi_o]$, $\xi_o = 55.36$ and $\tau \in [0, T]$, $T = 8.0$ as per the change of variables (7.1.3). Since the quasi–implicit method is $O(\Delta\tau) + O(\Delta\xi^2)$ accurate, then N_ξ should be selected so that $\Delta\xi^2$ is the same order of magnitude as $\Delta\tau$. This is summarized in Table 7.1.

Physical Constants	Unitless Constants
Beam radius 1 *cm* Propagation length $x = 400$ *cm* Stokes gain $g_r = 0.0004$ cm^2/Mw Initial pump beam intensity $\lvert E_p(0)\rvert^2 = 10$ Mw/cm^2 Pump and Stokes wavelengths: $\lambda_p = 3.5 \times 10^{-5}$ *rad/cm*, $\lambda_s = 4.1 \times 10^{-5}$ *rad/cm* Pump and Stokes wavenumbers $k_p = 1.7951 \times 10^5$ *rad/cm*, $k_s = 1.7951 \times 10^5$ *rad/cm*	$\xi_o = 55.36$ $\lVert f \rVert_\infty = 1$, $\lVert g \rVert_\infty = 0.1638$ $T = 8$ $\Delta\tau \leq \dfrac{2e^{-T\cdot\lVert f\rVert_\infty^2}}{\lVert g\rVert_\infty^2} = 0.025 \Rightarrow N_\tau = 320$ $N_\xi = 1200 \Rightarrow \Delta\xi = 0.1845$ and $\Delta\xi^2 = 0.0341$
Accuracy in τ: $O(0.025)$	Accuracy in ξ: $O(0.1845^2) = O(0.0341)$

Table 7.1. Constants to be used in the quasi–implicit method.

It is worth noting that the Dirichlet boundary condition is implicitly included in the quasi–implicit scheme (since the boundary values are zero at the endpoints $-2\xi_o$ and $2\xi_o$). The construction of the Cauchy data $f(\xi)$ and $g(\xi)$ are such that $f(\xi) \equiv 0 \equiv g(\xi)$ for $\xi \in [-2\xi_o, -\xi_o) \cup (\xi_o, 2\xi_o]$. This will ensure that sufficient distance is inserted between the non-zero portion of the initial conditions to mitigate any "aliasing" due to the difference scheme.

With these constants and comments in mind, solutions to the steady–state Raman system (7.1.5) are approximated via the quasi–implicit method. The intensities (that is, the square of the amplitudes) of the pump $u(\xi, \tau)$ and Stokes $v(\xi, \tau)$ fields are depicted in Figures 7.5 and 7.6.

Observe the overall behavior of the fields. As displayed in Figure 7.5, the pump beam has maximal energy at $\tau = 0$ and depletes its intensity as τ approaches $T = 8$. This is amplified in Figure 7.6. Unlike the initial pulse condition, the beam edges build up a set of oscillatory lobes (again, see Figure 7.6) which decrease rapidly on the interior interval $(-\xi_o, \xi_o)$. The opposite occurs for the Stokes beam $v(\xi, \tau)$ as the intensity increases with τ. This is precisely how Raman scattering is supposed to function: The pump beam injects energy into the Stokes beam. The conserved Hamiltonian means that the dissipated pump energy flows into the Stokes beam. Further, the oscillations noted on the pump beam are present for the Stokes beam (Figures 7.5–7.6) with greater amplitude near $-\xi_o$ and ξ_o. These oscillations are known as *Fresnel diffraction* or *Fresnel ringing*. This ringing is due to the nonlinear interaction of the pump and Stokes beam and is *not* an artifact of the calculations. Indeed, the bound on the time increment $\Delta\tau$ is enforced as per Theorem 7.5 so that, by the Lax Equivalence Theorem, the quasi–implicit method (7.2.3) is convergent. Moreover, increasing the number of points for the temporal and spatial computations *does not* alter the results presented in Figures 7.5–7.6. This observation is left as an exercise. Therefore, the observed phenomenon of Fresnel ringing is genuine.

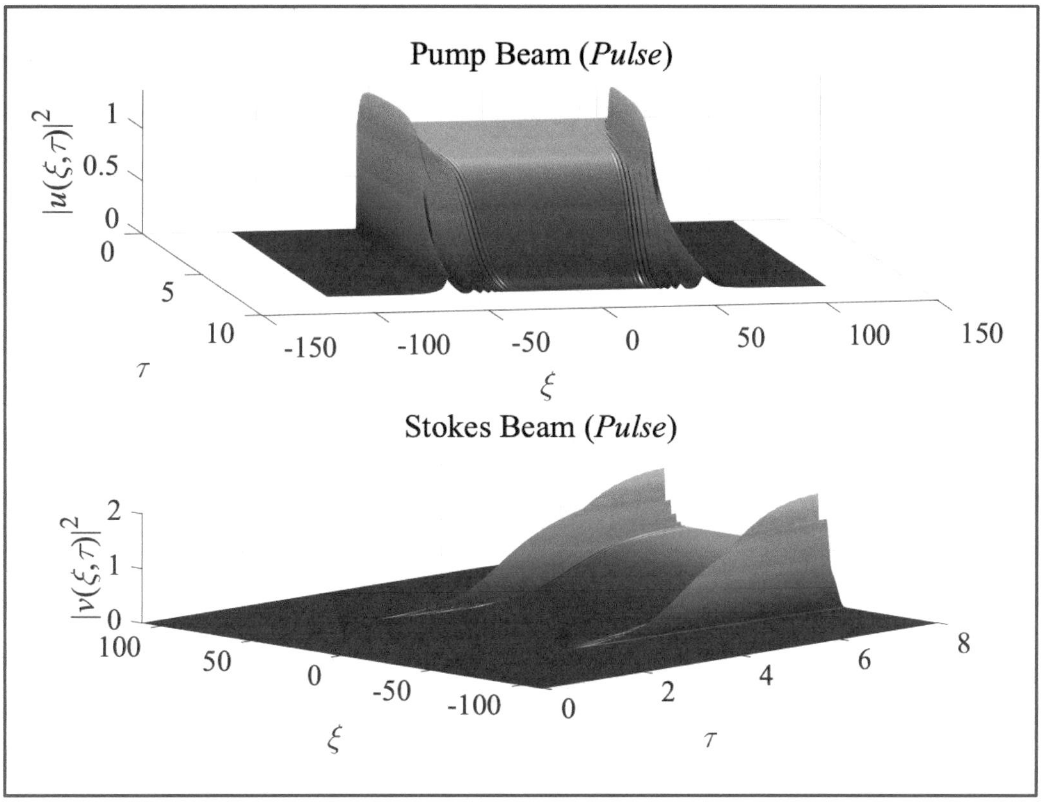

Figure 7.5. Pump and Stokes amplitudes for initial pulse beams.

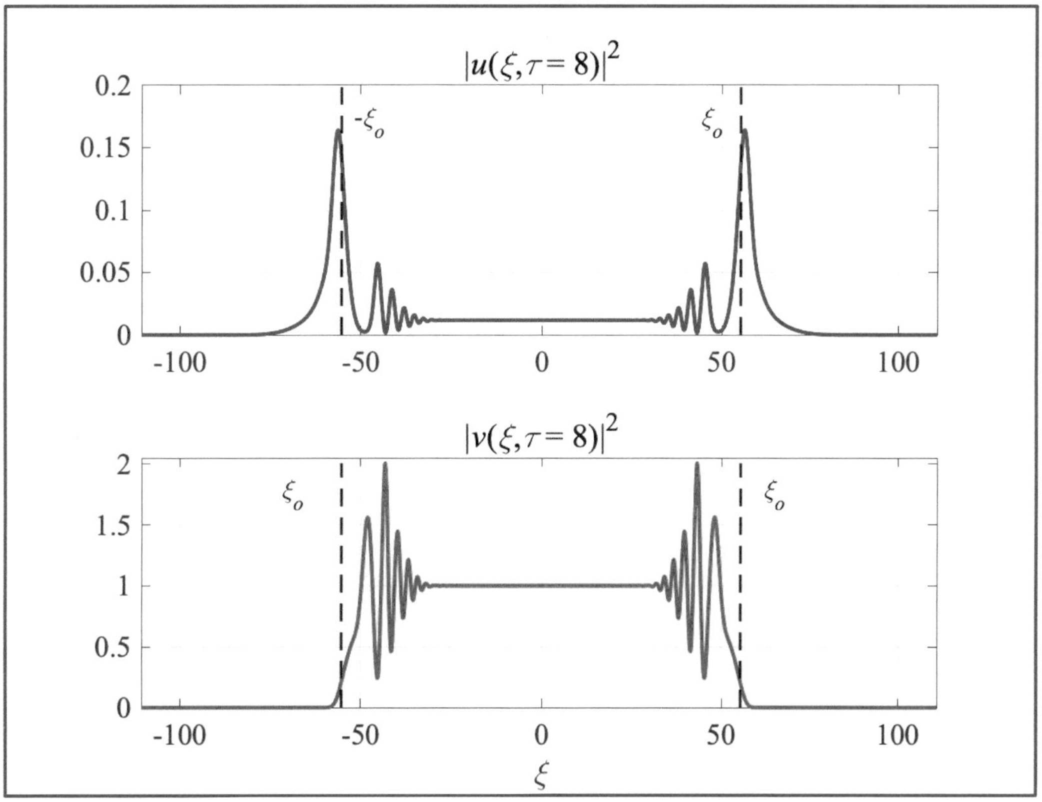

Figure 7.6. Fresnel ringing for initial pulse beams.

This begs the question "Can Fresnel ringing be mitigated?" The answer is "yes" by either *tapering* or smoothing the initial pulse beams. A *tapered pulse beam* is defined by (7.4.1) and graphed in upper portion of Figure 7.7 for $s = 5.623$.

$$f_{taper}(\xi) = \begin{cases} 1 & \text{for } -\xi_o \leq \xi \leq \xi_o \\ \exp\left(-\dfrac{(\xi+\xi_o)^2}{s^2}\right) & \text{for } -\tfrac{3}{2}\xi_o \leq \xi < -\xi_o \\ \exp\left(-\dfrac{(\xi-\xi_o)^2}{s^2}\right) & \text{for } \xi_o \leq \xi < \tfrac{3}{2}\xi_o \\ 0 & \text{otherwise} \end{cases} \tag{7.4.1}$$

A *smoothed pulse* is a continuous function with compact support defined via (7.4.2) and displayed in the lower portion of Figure 7.7 with $s = 25$. The smoothed pulse is also referred to as the *exponential pulse*.

$$f_{smooth}(\xi) = \begin{cases} 1 - \exp\left(-\dfrac{(|\xi|-\xi_o)^2}{s^2}\right) & \text{for } -\xi_o \leq \xi \leq \xi_o \\ 0 & \text{otherwise} \end{cases} \tag{7.4.2}$$

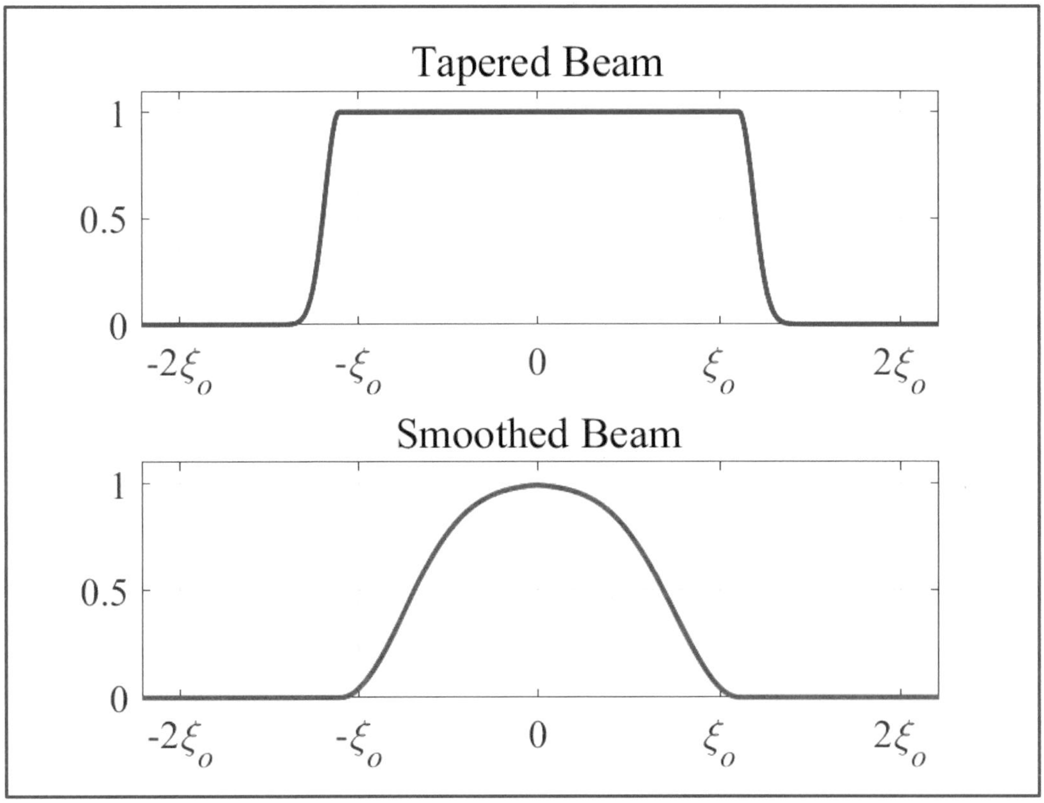

Figure 7.7. Initial beam edge smoothing to mitigate Fresnel ringing.

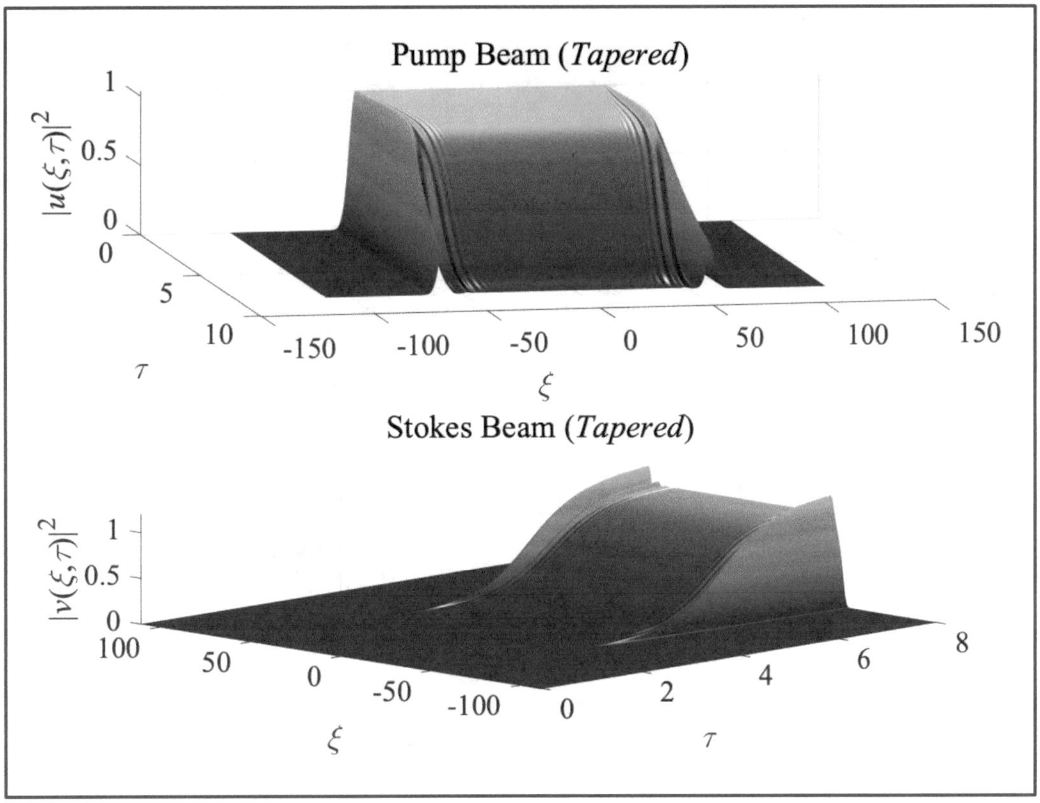

Figure 7.8. Pump and Stokes amplitudes for tapered beams.

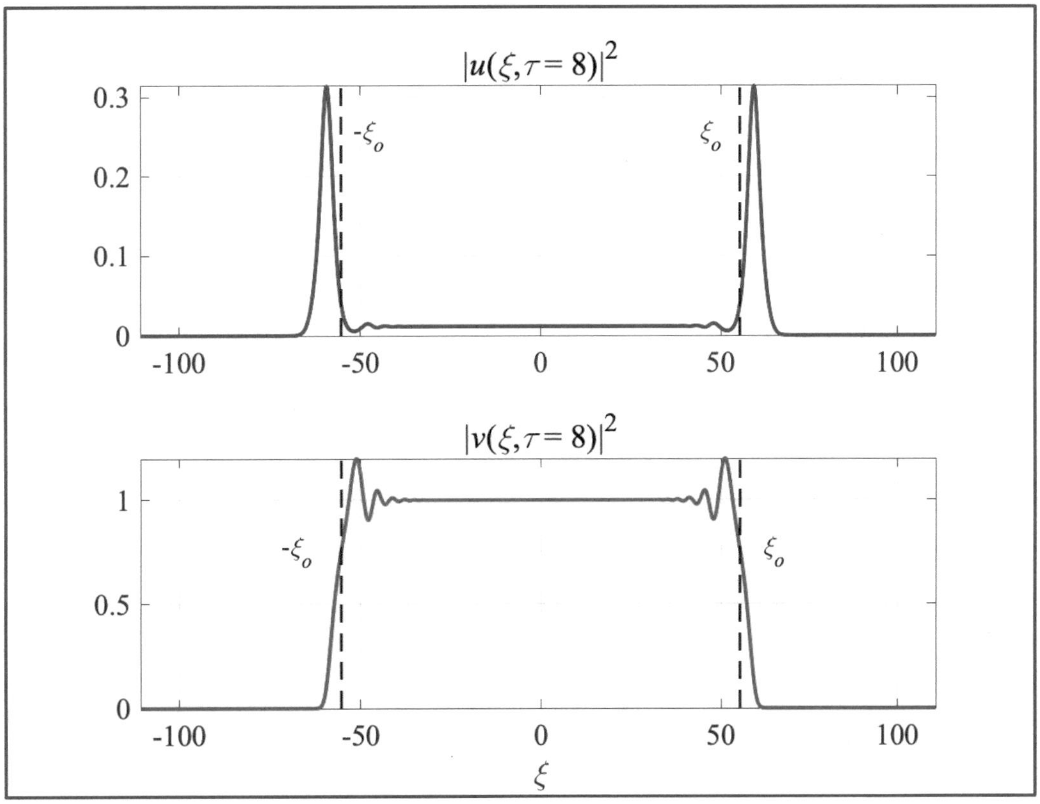

Figure 7.9. Fresnel ringing for tapered beams.

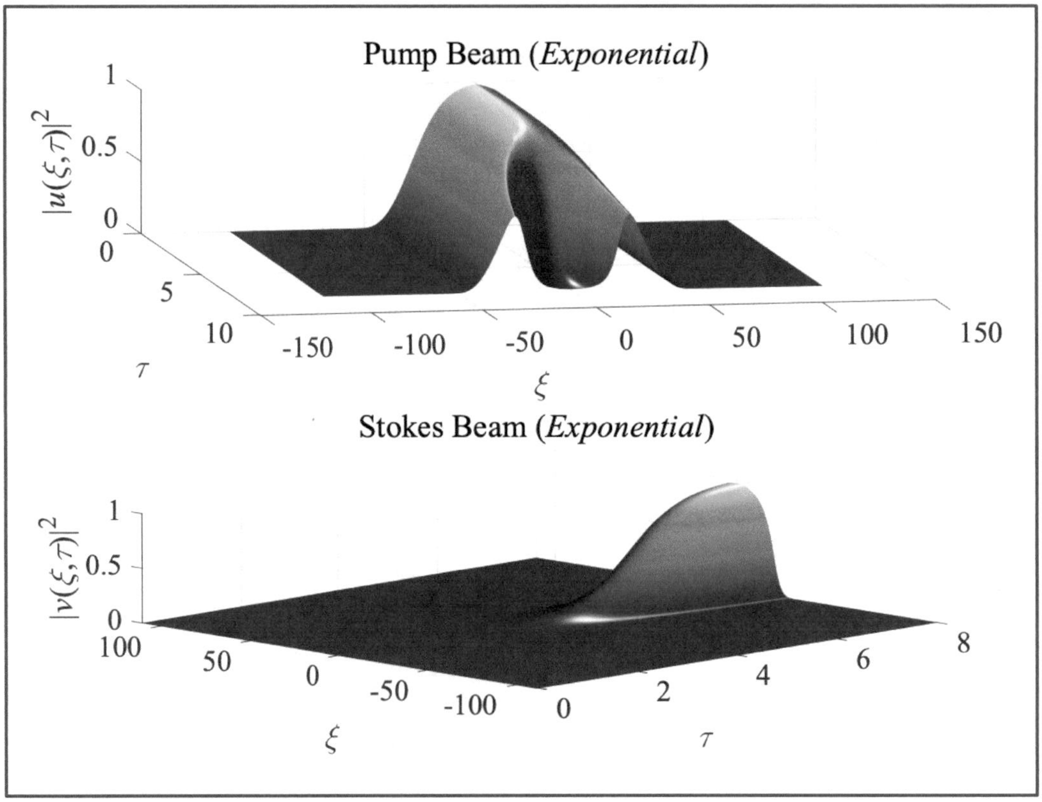

Figure 7.10. Pump and Stokes amplitudes for exponentially smoothed beams.

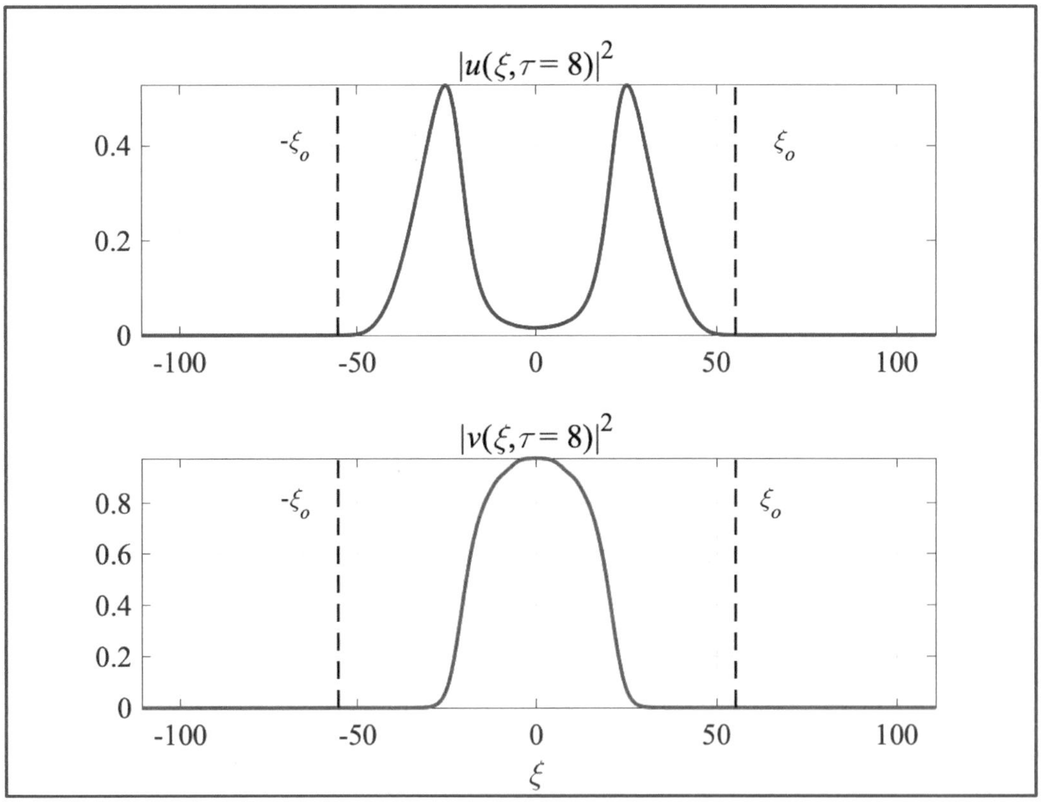

Figure 7.11. Fresnel ringing for exponentially smoothed beams.

Notice that, for the tapered beam of (7.4.1), the number of Fresnel oscillations are reduced and the ringing drops off more rapidly within $(-\xi_o, \xi_o)$. The intensity of the primary oscillation for the *pump field* at $u(-\xi_o, \tau = 8)$ and $u(\xi_o, \tau = 8)$ are considerably higher for the tapered beam than for the pulse beam. Compare Figures 7.6 and 7.9. The opposite is true for the Stokes field. In this case, the Stokes field for the tapered beam has a smaller oscillation peak than does the pulse beam.

Further, the Fresnel ringing is dampened and flattened more rapidly for the tapered beam. More of the intensity attains its level peak at 1 over the interval $[-\xi_o, \xi_o]$ for the tapered beam than for the pulse beam. Again, Figures 7.6 and 7.9 illustrate this observation.

The exponential beam of (7.4.2) results in pump $u(\xi, \tau)$ and Stokes $v(\xi, \tau)$ fields whose intensities lie entirely within the interval $[-\xi_o, \xi_o]$. The pump retains its energy for a longer time frame and then rapidly and smoothly decays to near zero intensity. The opposite occurs for the Stokes field which has a more focused but less broad intensity across $[-\xi_o, \xi_o]$. These comments are reflected in Figures 7.10–7.11.

The tapered beam appears to give the best laser performance in delivering the widest, flattest Stokes beam over the desired spatial extent $[-\xi_o, \xi_o]$.

§7.5 Alternating Direction Implicit Methods

In the case of two spatial variables $\boldsymbol{\xi} = [\xi_1, \xi_2]$, the quasi–implicit method cannot be utilized. Instead, a variation of this approach is implemented. Specifically, a quasi–implicit method *for each spatial variable* is applied while alternately "freezing" the time value of the variable not being updated. More precisely, adapt the notation from §7.2 and consider the two-spatial dimensional version of the Raman scattering model established in equation (7.2.1). In this section, assume that the spatial vector $\boldsymbol{\xi} = [\xi_1, \xi_2]$ is defined in the domain $\Psi(\Omega)$. In turn, take the domain to be the Cartesian product of two compact intervals: $\Psi(\Omega) = [a_1, b_1] \times [a_2, b_2]$. As above, take $\boldsymbol{w}$ to be the vector of the dimensionless fields $\boldsymbol{w}(\boldsymbol{\xi}, \tau) = [u(\boldsymbol{\xi}, \tau), v(\boldsymbol{\xi}, \tau)]^T$. Define the computational grids $\Xi_1 = \{\xi_{11}, \xi_{12}, \ldots, \xi_{1N_1}\}$, $\Xi_2 = \{\xi_{21}, \xi_{22}, \ldots, \xi_{2N_2}\}$, and $\vartheta = \{\tau_1, \tau_2, \ldots, \tau_{N_\tau}\}$, $\xi_{1,k} = \xi_{1,1} + (k-1)\cdot\Delta\xi_1$, $\xi_{2,k} = \xi_{2,1} + (k-1)\cdot\Delta\xi_2$, and $\tau_\ell = \tau_1 + (\ell-1)\cdot\Delta\tau$ where $\Delta\xi_j = \dfrac{b_j - a_j}{N_j}$ for $j = 1, 2$ and $\Delta\tau = \dfrac{T}{N_\tau}$. With the usual notation $\boldsymbol{w}(\xi_{1j}, \xi_{2k}, \tau_n) = \boldsymbol{w}^n_{j,k}$, the difference operators for each spatial variable are

$$\left.\begin{aligned} \delta_1^2\left(\boldsymbol{w}^{n+1}_{j,k}\right) &= \frac{1}{\Delta\xi_1^2}\left(\boldsymbol{w}^{n+1}_{j+1,k} - 2\boldsymbol{w}^{n+1}_{j,k} + \boldsymbol{w}^{n+1}_{j-1,k}\right) \\ \delta_2^2\left(\boldsymbol{w}^{n+2}_{j,k}\right) &= \frac{1}{\Delta\xi_2^2}\left(\boldsymbol{w}^{n+2}_{j,k+1} - 2\boldsymbol{w}^{n+2}_{j,k} + \boldsymbol{w}^{n+2}_{j,k-1}\right) \end{aligned}\right\}. \tag{7.5.1}$$

The strategy is to advance the time step first in the ξ_1–direction while holding the grid about the ξ_2 variable fixed. Then switch or *alternate* directions by advancing the time step in ξ_2 while holding ξ_1 fixed. This is the long–established *alternating direction method.* By making the linear portion of the partial differential equation (that is the Laplacian) implicit in time and the nonlinear portion explicit in time, the *alternating direction quasi–implicit method* (*ADQI*) can be described by (7.5.2)

$$\left.\begin{aligned} \delta_\tau^+\left(\mathbf{w}_{j,k}^n\right) &= C\,\delta_1^2(\mathbf{w}_{j,k}^{n+1}) + N\left(\mathbf{w}_{j,k}^n\right)\mathbf{w}_{j,k}^n \\ \delta_\tau^+\left(\mathbf{w}_{j,k}^{n+1}\right) &= C\delta_2^2(\mathbf{w}_{j,k}^{n+2}) + N\left(\mathbf{w}_{j,k}^{n+1}\right)\mathbf{w}_{j,k}^{n+1} \end{aligned}\right\}. \tag{7.5.2}$$

It is left to the (brave–hearted) reader to extend the proofs of Theorems 7.5 (Stability) and 7.6 (Consistency) to the two–dimensional case. These results along with the Lax Equivalence Theorem, will lead to the bound from Theorem 7.6 on the temporal space $\Delta\tau$ which ensures that the *ADQI* method (7.5.2) is convergent.

Mickens [51] suggests making the finite difference schemes implicit in the linear portion but explicit only with respect to the nonlinear operator *N*. Consequently, the Mickens' scheme for the dimensionless Raman scattering equation (7.2.1) is

$$\left.\begin{aligned} \delta_\tau^+\left(\mathbf{w}_{j,k}^n\right) &= C\,\delta_1^2(\mathbf{w}_{j,k}^{n+1}) + N\left(\mathbf{w}_{j,k}^n\right)\mathbf{w}_{j,k}^{n+1} \\ \delta_\tau^+\left(\mathbf{w}_{j,k}^{n+1}\right) &= C\delta_2^2(\mathbf{w}_{j,k}^{n+2}) + N\left(\mathbf{w}_{j,k}^{n+1}\right)\mathbf{w}_{j,k}^{n+2} \end{aligned}\right\}. \tag{7.5.3}$$

Other researchers such as Hoff [39], Nishida and Smoller [56], Reynolds [65], and Tadmor, Sackler, and Sackler [77] have developed convergent finite different schemes for a wider class of nonlinear problems.

Exercise

7.1 Run the quasi–implicit scheme (7.2.3) via `ramaqi.m` for different grid spacings. For example, $N_\xi = 500$ and $N_\xi = 1500 \Rightarrow \Delta\tau = 0.016$ and $\Delta\xi = 0.1476$. Is there a qualitative difference for the pulse beam?

Chapter 8

The Hartman–Grobman Theorem

By generalizing the Hartman–Grobman Theorem for the flows of vector fields, it is seen that a class of nonlinear evolution equations admit a local linearization. Under the restriction that the nonlinear part of the solution operator is compact, the linearization turns out to be global. This local linearization is applied to the periodic Korteweg de Vries, Burgers', and cubic Schrödinger equations.

§8.1 Introduction

The problem of linearizing nonlinear (ordinary and partial) differential equations has been examined by research scientists over the last 60 years. In particular, Hartman [33–35] and Grobman [30], independently in 1959, gave criteria for the local linearization of ordinary differential equations. In 1950, Hopf [40] discovered a change of variables that transformed the nonlinear partial differential equation of Burgers into the (linear) heat equation. Indeed, as seen in §6.2, the change of variables $u(x, t) = -2\sigma^2 \frac{\partial}{\partial x} \ln(v(x, t))$ provides the solution of Burgers' equation $u_t = \sigma^2 u_{xx} - u\, u_x$, $u(x, 0) = f(x)$ by solving the (linear) heat equation $v_t = \sigma^2 v_{xx}$, $v(x, 0) = \exp\left(-\frac{1}{2\sigma^2}\int_{-\infty}^{x} f(\xi)\,d\xi\right)$.

Gardner, Greene, Kruskal, and Miura [27, 28] announced the inverse scattering theory for the Korteweg de Vries (*KdV*) equation in 1967. The *KdV* equation and *inverse scattering transform* are detailed in §6.3. These are examples of *integrable PDE*s. Integrable equations can be solved by transformations that reduce the nonlinear equations to linear equations. Berger, Church, and Timourian [10] proved that for $q \in L_2(\mathbb{R})$, the Ricatti equation $u'(t) + u^2(t) = q(t)$ is integrable via topological transformation. Nikolenko [55] announced a theorem concerning the local linearization of certain partial differential equations. There are several other nonlinear *PDE*s which admit a linearizing change of variables/transformation. Is there a general theory for such transformations? The answer is "yes." It is the focus of this chapter to define the *a priori* conditions necessary to achieve both local and global linearization of semi–linear equations. Furthermore, this local linearization is applied to the nonlinear (cubic) Schrödinger equation, the periodic *KdV* equation, and Burgers' equation. This chapter will follow the work of Costa [19].

More precisely, let L: $X \to Y$ be a linear operator between the separable Hilbert spaces X and Y, and let N: $X \to Y$ be a nonlinear operator. The focus of this research is the semi–linear Cauchy problem (8.1.1).

$$\left.\begin{aligned} &\frac{\partial u}{\partial t} = Lu + N(u), \quad t \in [0,T] \\ &u(\boldsymbol{x},0) = f(\boldsymbol{x}), \quad \boldsymbol{x} \in \Omega \subset \mathbb{R}^n \end{aligned}\right\} \tag{8.1.1}$$

The question examined is: "Does there exist a local or global change of variables $u = H(v)$ that reduces the semi–linear equation (8.1.1) to the linear equation (8.1.2) below?"

$$\left.\begin{aligned} &\frac{\partial v}{\partial t} = Lv, \quad t \in [0,T] \\ &v(\boldsymbol{x},0) = H^{-1}\left[f(\boldsymbol{x})\right], \quad \boldsymbol{x} \in \Omega \subset \mathbb{R}^n \end{aligned}\right\}. \tag{8.1.2}$$

This is equivalent to asking whether (8.1.1) is integrable. Naturally, the answer is "no" for arbitrary L and N. If certain smoothness and spectral restrictions are placed on L and N, however, then the answer becomes "yes." In what follows, a list of requirements on L and N are given in order to achieve local linearization. This is accomplished by generalizing the *Hartman–Grobman Theorem* for maps and flows. From the body of the proof of the Mapping Theorem, the additional requirements on L and N to insure a global linearization are obtained. The *Flow Theorem* will then be applied to the *periodic generalized KdV* equation (8.1.3), Burgers' equation (8.1.4), and the *cubic nonlinear Schrödinger* equation (*NLS*) (8.1.5) as stated below.

Generalized KdV

$$\left.\begin{aligned} &\frac{\partial u}{\partial t} = u - \frac{\partial^3 u}{\partial x^3} + \Psi(u) - u\frac{\partial u}{\partial x} \\ &u(x,0) = f(x) \\ &u(x+2\pi,t) = u(x,t) \end{aligned}\right\} \tag{8.1.3}$$

Here $\Psi(u)$ is a nonlinear term that satisfies the smoothness conditions (C) – (D) given in the next section.

Burgers' Equation

$$\left.\begin{aligned} &\frac{\partial u}{\partial t} = \lambda\frac{\partial^2 u}{\partial x^2} - u\frac{\partial u}{\partial x} \\ &u(x,0) = f(x) \end{aligned}\right\} \tag{8.1.4}$$

NLS

$$\left.\begin{aligned} &\frac{\partial u}{\partial t} = \Delta u - ku\,|\,u\,|^2, && t > 0 \\ &u(\boldsymbol{x},0) = f(\boldsymbol{x}), && \boldsymbol{x} \in \Omega \\ &u(\boldsymbol{x},t) = 0, && \boldsymbol{x} \in \partial\Omega \end{aligned}\right\} \tag{8.1.5}$$

§8.2 Definitions, Notation, and Key Conditions

The following symbols and notation will be used throughout this chapter. The symbols X, Y, and Z represent separable Hilbert spaces with X continuously embedded in Z. The symbol $\sigma(A)$ is the set of all eigenvalues of the linear operator A, also called the *spectrum* of A. The symbol $\sigma_n(\mathbb{R})$ is defined to be the set of all $n \times n$ matrices with real coefficients whose eigenvalues are *not* purely imaginary: $\sigma_n(\mathbb{R}) = \{A \in Mat_{n\times n}(\mathbb{R}) : \lambda \in \sigma(A) \Rightarrow \mathrm{Re}(\lambda) \neq 0\}$. Some of the other notations used in this section are as follows.

$\mathscr{C}(X,Y)$ is the set of all continuous maps from X into Y

$\mathscr{C}^m(X,Y)$ is the set of all m–times continuously Fréchet–differentiable maps from X into Y

$\mathscr{L}(X,Y)$ is the set of all linear maps from X into Y

$\mathscr{C}^0_*(X,Y) = \{f \in \mathscr{C}(X,Y)$: f is uniformly bounded and uniformly continuous$\}$

$f'(u)v$ is the Fréchet derivative of the operator $f(u)$ evaluated at v

$\mathscr{M}(X,Y) = \{f \in \mathscr{C}^1(X,Y)$: f is bounded and $f(0) = 0 = f'(0)v$ for all $v \in Dom[f'(0)]\}$

$\mathscr{B}_X(0,r) = \{\boldsymbol{x} \in X$: $\|\boldsymbol{x}\|_X < r\} \equiv$ the sphere of radius r in X–space

$$H^1([0,T];Z) = \left\{ f : \int_0^T \left(\|f(t)\|_Z^2 + \left\| \frac{\partial f}{\partial t}(t) \right\|_Z^2 \right) dt < \infty \right\}$$

Additional definitions of function spaces are found in the *Introduction*. The Hartman–Grobman Theorem for ordinary differential equations is now stated with respect to the Banach space E.

The Hartman–Grobman Theorem: *Let $A \in \sigma_n(\mathbb{R})$ and $F \in C^2(E)$ define the differential vector field $\boldsymbol{x}'(t) = A\boldsymbol{x} + F(\boldsymbol{x})$, $\boldsymbol{x}(0) = \boldsymbol{x}_o$, with generalized solution $X^t \equiv e^{At}\boldsymbol{x}_o + \int_0^t e^{A(t-\tau)} F(\boldsymbol{x}(\tau))\, d\tau$. Let $\boldsymbol{u}(t) = e^{At}\boldsymbol{u}_o \equiv L^t\boldsymbol{u}_o$ be the solution of the ordinary differential equation $\boldsymbol{u}'(t) = A\boldsymbol{u}$, $\boldsymbol{u}(0) = \boldsymbol{u}_o$. Then there exists a homeomorphism H mapping a neighborhood of $\boldsymbol{x} = \boldsymbol{0}$ onto a neighborhood of $\boldsymbol{u} = \boldsymbol{0}$ so that $H \circ X^t \circ H^{-1} = L^t$.*

In order to generalize the Hartman–Grobman Theorem to partial differential equations, critical restrictions must be placed on L and N. These *key conditions* are listed below. Some of these ideas, such as *semigroups* and *contraction* or *expansion mappings*, are discussed in the *Introduction.*

Key Conditions

(A) L is the infinitesimal generator of a C_o–semigroup $S(t)$. Informally, this means that $S(t) = e^{tL}$ is the solution operator of the linear *PDE* $\frac{\partial u}{\partial t} = Lu$ so that $u(t) = e^{tL}u(0) \equiv S(t)\, u(0)$. See the *Introduction*, page 11, for more details on semigroups.

(B) The spectrum $\sigma(L) = \{\lambda\colon Lu = \lambda u\}$ of L has non-zero real parts.

(C) The map $N\colon X \to Z$ is locally Lipschitz. More specifically, there is a continuous, monotonically increasing (in each variable) function $\psi\colon \mathbb{R}^+ \oplus \mathbb{R}^+ \to \mathbb{R}$ so that (C.1) holds.

$$\|N(u) - N(v)\|_Z \leq \psi(\|u\|_X, \|v\|_X)\, \|u - v\|_X \tag{C.1}$$

(D) N is quadratically nonlinear and smoothly Fréchet differentiable. That is, $N(0) = 0 = N'(0)v$ for all $v \in \mathrm{Dom}[\,N'(0)\,]$ and $N'(u) \in \mathcal{C}(X, \mathcal{L}(X,Z))$. Further, N is quadratically bounded on bounded sets in X. This means that, for any bounded set $B \subset X$, there is an absolute constant c_o so that for any $u, v \in B$

$$\| N(u) - N(v) - N'(v)(u-v) \|_Z \leq c_o \|u - v\|_X^2\,. \tag{D.1}$$

Remark 8.1: *(i)* By the Hille–Yosida Theorem, L can be an *unbounded* operator and still satisfy condition (A) provided it generates a C_o–semigroup of contractions. The operator L is a contraction provided the spectrum $\sigma(L)$ is always less than 0. Equivalently, $\lambda \in \sigma(L)$ implies $\lambda < 0$.
(ii) Condition (B) means that $\mathrm{Re}(\lambda) \neq 0$, whenever $\lambda \in \sigma(L)$.

Definition 8.1: A function $u(\boldsymbol{x}, t) \in H^1([0, T]; Z)$ is a *generalized solution* of (8.1.1) with $S(t) = e^{tL}$ provided (8.2.1) holds for every $f \in X$ and $t \in [0, T]$.

$$u(\boldsymbol{x},t) = [S(t)f](\boldsymbol{x}) + \int_0^t [S(t-s)N(u)](\boldsymbol{x},s)\,ds \tag{8.2.1}$$

For continuous, differential operators (such as $L = \frac{\partial}{\partial x}$, Δ, or $\frac{\partial^3}{\partial x^3}$), the solution operator S is the semigroup $S(t) = e^{tL}$. In the formal language of *Key Condition* (A), L is the infinitesimal generator of a C_o–semigroup $S(t)$.

Definition 8.2: An invertible isomorphism $L \in \mathcal{L}(X,Y)$ is called an *expansive–contractive map* provided $X = X_1 \oplus X_2$, $Y = Y_1 \oplus Y_2$, $X_j = L^{-1}(Y_j)$, and $L_1 = L|_{X_1}$ is an expansion while $L_2 = L|_{X_2}$ is a contraction. A map $\Lambda_e: X \to Y$ is an *expansion* provided $\|\Lambda_e(x)\|_Y > \|x\|_X$ while a map $\Lambda_c: X \to Y$ is a *contraction* provided $\|\Lambda_c(x)\|_Y < \|x\|_X$.

Remark 8.2: (*i*) These definitions above are crucial to the generalization of the Hartman–Grobman Theorem (*HGT*) to infinite dimensions between distinct Hilbert spaces. Irwin [42], Pugh and Shub [38, 62] have extended the *HGT* to maps of a Banach space into itself. The novelty of this chapter is the generalization of the *HGT* to distinct Hilbert spaces and consequent applications to nonlinear partial differential equations.

(ii) Other authors refer to generalized solutions as weak or mild solutions. Similarly, expansive–contractive maps have also been called hyperbolic linear isomorphisms, see, e.g., Irwin [42], Pugh, and Shub [38, 62].

(iii) If the spectrum $\sigma(L)$ of the operator L avoids the unit circle, then for an appropriate norm, L is an expansive–contractive map. Moreover, the splitting of the spaces X and Y into expansive and contractive subspaces is invariant with respect to the map L.

§8.3 The Mapping Theorem

In this section, a generalization of Hartman's Theorem for Maps is given. While the proof is a generalization of the one given by Pugh [62] and Hartman [35], there are subtle changes and sufficient novelty to merit its exposition. More precisely, Pugh and Hartman prove the Mapping Theorem for maps operating on the same Banach space E: $L \in \mathcal{L}(E,E)$, $N \in \mathcal{M}(E,E)$. The theorem below is presented for maps operating between *distinct* Hilbert spaces. Furthermore, due to the Hilbert space setting, a global version of the *Mapping Theorem* is achieved. The theorem for flows on vector fields is discussed in the next section.

The Hartman Mapping Theorem: *Let $L \in \mathcal{L}(X,Y)$ be an expansive–contractive map and $N \in \mathcal{M}(X,Y)$. Then there exists a unique homeomorphism H so that $H \circ (L+N) = L \circ H$ on sufficiently small neighborhoods U of $0 \in X$ and V of $0 \in Y$. That is, Diagram 8.1 below commutes.*

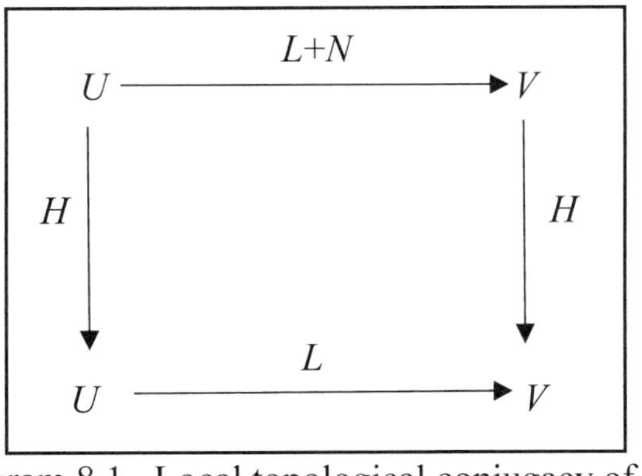

Diagram 8.1. Local topological conjugacy of $L+N$ to L.

Proof: For technical reasons, it is easier to prove the local topological conjugacy of $T \equiv L + N$ to $R \equiv L + M$. To that end, let $U = \mathcal{B}_X(0,r)$ and $V = \mathcal{B}_Y(0,s)$. Take r and s sufficiently small and N, $M \in \mathcal{M}(X,Y)$, so that the operators T and R map U homeomorphically onto V. That is, $(L + N)U = (L + M)U = V$. This result is a direct application of the Inverse Function (see, e.g., Schwartz [68]).

Let $H = I + h$, where I is the identity operator. Choose r and s sufficient small so that T and R map U onto V. Then, as T is a homeomorphism, the estimate $\|T(\boldsymbol{x}) - T(\boldsymbol{x}')\|_V \le 2s$ holds for any $\boldsymbol{x}, \boldsymbol{x}' \in U$. Similar estimates are obtained for the homeomorphism R: $\|R(\boldsymbol{x}) - R(\boldsymbol{x}')\|_V \le 2s$. Moreover, for any $\boldsymbol{y}, \boldsymbol{y}' \in V$, $\|T^{-1}(\boldsymbol{y}) - T^{-1}(\boldsymbol{y}')\|_U \le 2r$, with identical results for R^{-1}. These estimates will be utilized later on in the proof.

The conjugacy equation

$$HT = RH \tag{8.3.1}$$

and identification $H = I + h$ are equivalent to both (8.3.2) and (8.3.3) below.

$$h = (MH + Lh - N)T^{-1} \tag{8.3.2}$$
$$h = L^{-1}(hT + N - MH) \tag{8.3.3}$$

Indeed, (8.3.1) implies $(I + h)T = R(I + h)$ which means $T + hT = R + Rh$. Now $R - T = M - N$ so that $hT = M - N + Rh = M - N + (L + M)h = M(I + h) - N + Lh$ or $h = (MH - N + Lh)T^{-1}$. This verifies (8.3.2). By proceeding in a similar manner, (8.3.1) implies $T + hT = HT = RH = R + Rh$. Subtracting L from both sides of this relation produces $N + hT = M + (L + M)h = M(I + h) + Lh$ so that $N + hT - MH = Lh \Rightarrow L^{-1}(N + hT - MH) = h$. This is (8.3.3).

Since X and Y are separable spaces, they are isometrically isomorphic. Therefore, as U and V are homeomorphic and as L is an expansive–contractive map from X to Y, then U and V inherit the expansive–contractive splitting. Thus, Diagram 8.2 commutes.

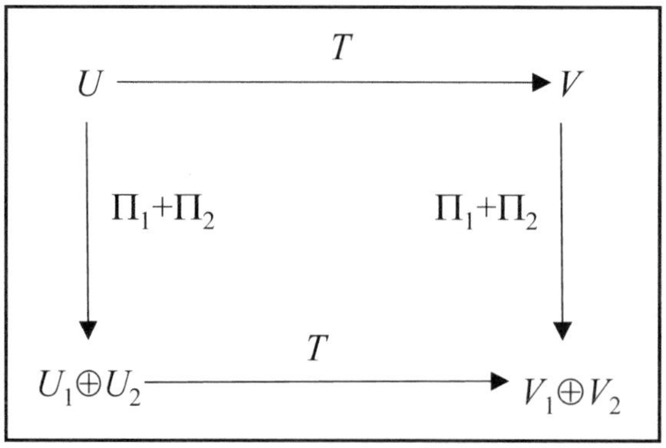

Diagram 8.2. Projection into expansive–contractive coordinates.

More precisely, let Π_1 and Π_2 be the projections of the expansive–contractive map L. That is, $L_1 = \Pi_1 L$ is the expansive portion of the mapping and $L_2 = \Pi_2 L$ is the contractive element. Applying Π_1 and Π_2 as left–hand operators to (8.3.1) produces $\Pi_1 HT = \Pi_1 RH$ and $\Pi_2 HT = \Pi_2 RH$. The first of these equations results in

$$\Pi_1 HT = \Pi_1(I + h)(L + N) = \Pi_1(L + N + h[L + N]) = \Pi_1(L + M)H = (L_1 + M_1)H.$$

Equivalently, $(L_1 + N_1 + h_1[L + N]) = (L_1 + M_1)H$. Canceling common terms yields $(N_1 + h_1 T) = M_1 + (L_1 + M_1)h \Rightarrow h_1 T = M_1 H + L_1 h - N_1$, and finally

$$h_1 T = (M_1 H + L_1 h - N_1)T^{-1}. \tag{8.3.4}$$

The same procedure applied to $\Pi_2 HT = \Pi_2 RH$ results in

$$h_2 T = (M_2 H + L_2 h - N_2)T^{-1}. \tag{8.3.5}$$

By applying the projection Π_j on the left in (8.3.3) and using the methodology above, it is seen that $h_j = \Pi_j h = L_j^{-1}\left(hT + N_j - M_j h\right)$ for $j = 1, 2$. This results in the pair of equations

$$h_1 = L_1^{-1}\left(hT + N_1 - M_1 H\right) \tag{8.3.6}$$
$$h_2 = L_2^{-1}\left(hT + N_2 - M_2 H\right). \tag{8.3.7}$$

To prove that the mapping h (and hence H) exists, utilize just equations (8.3.5) and (8.3.6). Define $V^* = \mathscr{C}_*^0(U, U_1 \oplus U_2) \cap \mathscr{C}_*^0(V, V_1 \oplus V_2) \equiv V_1^* \oplus V_2^*$. Apply the uniform C_o–topology to V^* so that it becomes a Banach space with norm $\|v\|_{V^*} = \|v_1\|_{V_1^*} + \|v_2\|_{V_2^*}$ for any $v = v_1 \oplus v_2 \in V^*$. Equations (8.3.5)–(8.3.6) define a map τ from V^* into V^* via equation (8.3.8).

$$(h_1, h_2) \overset{\tau}{\mapsto} \left(L_1^{-1}(hT + N_1 - M_1(I + h)), (M_2(I + h) + L_2 h - N_2)T^{-1}\right) \tag{8.3.8}$$

For any two elements $h = (h_1, h_2)$, $h' = (h_1', h_2') \in V^*$, then

$$\begin{aligned}\|\tau(h) - \tau(h')\|_{V^*} &= \left\|L_1^{-1}\left(hT + N_1 - M_1(I + h)\right) - L_1^{-1}\left(h'T + N_1 - M_1(I + h')\right)\right\|_{V_1^*} \\ &\quad + \left\|\left(M_2(I + h) + L_2 h - N_2\right)T^{-1} - \left(M_2(I + h') + L_2 h' - N_2\right)T^{-1}\right\|_{V_2^*} \\ &= \left\|L_1^{-1}(h - h')T + L_1^{-1}M_1(h - h')\right\|_{V_1^*} + \left\|M_2(h - h')T^{-1} + L_2(h - h')T^{-1}\right\|_{V_2^*}.\end{aligned}$$

Since L_1 is an expansive map, then L_1^{-1} is contractive and consequently, $\|L_1^{-1}(\xi-\xi')\|_{V_1^*} \le \lambda_1 \|\xi-\xi'\|_{V_1^*}$ for some $\lambda_1 < 1$. As noted in the first portion of the proof, the mappings T^{-1} and R^{-1} are bounded by $2r$ while T and R are bounded by $2s$. Hence,

$$\begin{aligned}\|\tau(h)-\tau(h')\|_{V*} &\le \lambda_1 \cdot 2s \cdot \|h-h'\|_{V*} + \lambda_1 \cdot 2s \cdot \|h-h'\|_{V*} + 2r \cdot \|M_2(h-h') + L_2(h-h')\|_{V*} \\ &\le 4s\lambda_1 \|h-h'\|_{V*} + (2r \cdot 2s + 2r \cdot \lambda_2)\|h-h'\|_{V*} \\ &\le (4s\lambda_1 + 4rs + 2r\lambda_2)\|h-h'\|_{V*}. \end{aligned} \tag{8.3.9}$$

The factor $M_2(h-h') + L_2(h-h')$ is bounded via

$$\begin{aligned}\|M_2(h-h') + L_2(h-h')\|_{V*} &\le \|M_2(h-h')\|_{V*} + \|L_2(h-h')\|_{V*} \le \|R(h-h')\|_{V*} + \lambda_2 \|(h-h')\|_{V*} \\ &\le (2s + \lambda_2)\ \|(h-h')\|_{V*}\end{aligned}$$

which completes the estimate obtained in (8.3.9). As the second coordinate in the projection L_2 is contractive, then $\lambda_2 < 1$. Therefore, $\|\tau(h)-\tau(h')\|_{V*} \le (4s + 4rs + 2r)\|h-h'\|_{V*}$. By selecting $s < 1/8$ and $r < 1/6$, the estimate now becomes $\|\tau(h)-\tau(h')\|_{V*} \le \omega\|h-h'\|_{V*}$ and $\omega < 1$. In this case, τ is a contraction. By the Contraction Mapping Theorem (see, e.g., Berger [8], page 112), τ has a unique fixed point $h = (h_1, h_2)$ which satisfies (8.3.5)–(8.3.6). Moreover, by a corollary to the Contraction Mapping Theorem (see, again Berger [8]), $h = h(T, R)$ depends continuously on the pair of mappings (T, R). So then does $H = (\mathrm{I} + h) = (I + h(T, R)) \equiv H(T, R)$. Having established a unique solution to (8.3.5)–(8.3.6), there is subsequently a unique solution to (8.3.2)–(8.3.3) and therefore to (8.3.1). That is, $H = H(T, R)$.

It remains to show that $H(T, R)$ is a homeomorphism. Observe that by applying $H(R, T)$ to (8.3.1) on the left yields $H(R, T)\circ H(T, R)\circ T = H(R, T)\circ R\circ H(T, R) = T\circ H(R, T)\circ H(T, R)$. As the solution of (8.3.1) is unique, then it must be the case that $H(R, T)\circ H(T, R) = I$ = the identity operator. Therefore, the left–inverse of $H(T, R)$ is $H(R, T)$. By symmetry, $H(T, R)\circ H(R, T) = I$ and $H(R, T)$ is also the right inverse of $H(T, R)$. Therefore, $H = H(T, R)$ is a homeomorphism. ■

Remark 8.3: Notice from the proof above, h is a local homeomorphism precisely because $T = L + N$ and $R = L + M$ are local homeomorphisms on the neighborhood U of $0 \in X$. This statement is a consequence of the *Inverse Function Theorem*. If the additional restriction that T and R are *proper* maps is imposed, then the *Global Inverse Mapping Theorem* (see Berger [8], page 221) insures that T and R are global homeomorphisms. Therefore, under this additional requirement, h and thereby H are global homeomorphisms. An invertible continuous map from $X \to Y$ is *proper* provided the inverse image of a compact set in Y is compact in X. See Remark 8.5 for greater detail. This comment is summarized as the *Global Mapping Theorem*.

Global Mapping Theorem: *Let $L \in \mathcal{L}(X,Y)$ be an expansive–contractive map, $N \in \mathcal{M}(X,Y)$, and $T = L + N$ be a proper map. Then there exists a unique global homeomorphism H so that $HT = LH$ and Diagram 8.3 commutes.*

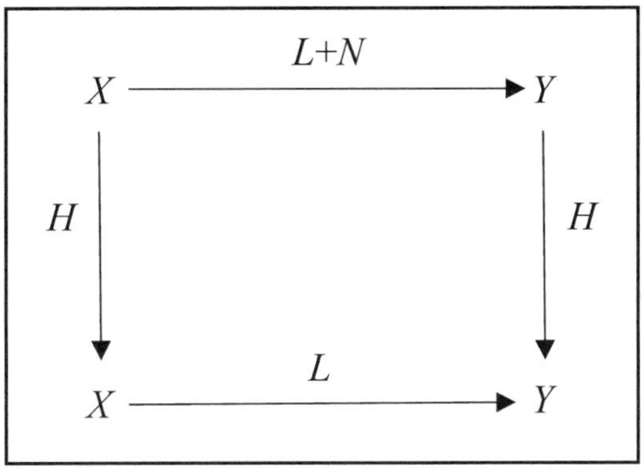

Diagram 8.3. Global topological conjugacy of L to $T = L + N$.

In the next section, the *Mapping Theorem* is applied to nonlinear evolution equations. This helps to establish a mathematical theory of the existence of a linearizing change of variables as exemplified by Burgers' equation.

§8.4 The Flow Theorem

Consider the evolution equation (8.1.1) as a vector field defined by the maps L and N on the separable Hilbert space X. Let F^t_{L+N} and F^t_L be the flows on the vector fields defined by equations (8.1.1) and (8.1.2), respectively. That is, $F^t_{L+N}\left[u(\boldsymbol{x},0)\right] = e^{tL}\left[u(\boldsymbol{x},0)\right] + \int_0^t e^{(t-s)L}\left[N\left(u(\boldsymbol{x},s)\right)\right]ds$ and $F^t_L\left[v(\boldsymbol{x},0)\right] = e^{tL}\left[v(\boldsymbol{x},0)\right]$. The *Hartman–Grobman Flow Theorem* states that the two flows F^t_{L+N} and F^t_L are locally topologically conjugate. More colloquially, the theorem states that locally there is a change of variables in which a solution of the linear equation (8.1.2) can be mapped into a solution of the nonlinear equation (8.1.1).

To prove the *Flow Theorem*, it must first be established that F^t_{L+N} is a *flow* on (8.1.1). This means it must be shown that there is a unique solution to (8.1.1). *Key Conditions* (*A*), (*C*), and (*D*) of §8.2 guarantee the existence of a unique solution to (8.1.1) in the separable Hilbert space $H^1([0, T]; Z)$.

Existence and Uniqueness Theorem: *Let $L \in \mathcal{L}(X,Z)$ satisfy Key Condition (A) and $N \in \mathcal{M}(X,Z)$ satisfy Key Conditions (C) – (D). For every $f \in X$, there exists a time $T_m \in \mathbb{R}^+$ so that (8.1.1) has a unique generalized solution $u \in H^1([0, T]; Z)$ for every subinterval $[0, T] \subset [0, T_m)$.*

The proof of this theorem will be carried out via a series of lemmas. These lemmas, in turn, are generalizations of the work of Rauch [64], Henry [37], and Haraux [32].

Lemma 8.1 (Uniqueness): *If $u \equiv u(\mathbf{x}, t)$ and $v = v(\mathbf{x}, t)$ are solutions of* (8.1.1) *in $H^1([0, T]; Z)$ with L satisfying Key Condition (A) and N satisfying (C) – (D), then u = v almost everywhere with respect to the Z norm.*

Proof: Since e^{tL} is a C_o–semigroup, then there are positive constants C and λ so that

$$\left\| e^{tL} \right\|_{\mathcal{Q}(X,Y)} \leq C e^{t\lambda} \text{ for } t \geq 0.$$

As N satisfies *Key Condition* (*C*), there exists a function ψ which satisfies (*C*.1). Also, since u and v are solutions of (8.1.1) in $H^1([0, T]; Z)$, then the monotonic increasing function ψ is bounded by the bounded solutions u and v. That is, $\|u\|_X, \|v\|_X < \infty$ implies $\psi\left(\|u\|_X, \|v\|_X\right) < \infty$. Take $C_o > 0$ sufficiently large so that

$$C e^{t\lambda} \cdot \psi\left(\|u\|_X, \|v\|_X\right) \leq C_o \tag{8.4.1}$$

for $0 \leq t \leq T$. By supposition, u and v are solutions of (8.1.1) so that by (8.2.1), $u(\mathbf{x}, t) - v(\mathbf{x}, t) = [S(t)f](\mathbf{x}) + \int_0^t [S(t-s)N(u)](\mathbf{x},s)\,ds - [S(t)f](\mathbf{x}) + \int_0^t [S(t-s)N(v)](\mathbf{x},s)\,ds$, which by the linearity of $S(t) = e^{tL}$ becomes $u(\mathbf{x}, t) - v(\mathbf{x}, t) = \int_0^t e^{(t-s)L}[N(u) - N(v)](\mathbf{x},s)\,ds$. Consequently, by (8.4.1) and *Key Condition* (*C*),

$$\begin{aligned} \|u(\mathbf{x},t) - v(\mathbf{x},t)\|_Z &\leq \int_0^t C e^{(t-s)\lambda}\, \psi\left(\|u\|_X, \|v\|_X\right) \|u(\mathbf{x},s) - v(\mathbf{x},s)\|_X\, ds \\ &\leq C_o \int_0^t \|u(\mathbf{x},s) - v(\mathbf{x},s)\|_X\, ds \quad . \end{aligned} \tag{8.4.2}$$

By Gronwall's inequality, $\|u(\mathbf{x},t) - v(\mathbf{x},t)\|_Z \leq 0$ for $0 \leq t \leq T$. Hence, $\|u(\mathbf{x},t) - v(\mathbf{x},t)\|_Z = 0$ for $0 \leq t \leq T$. ■

Lemma 8.2 (Local Existence of Solutions): *For any $f \in X$, there is a number $\delta\left(\|f\|_X\right) > 0$ so that* (8.1.1) *has a solution $u \in H^1([0, \delta]; Z)$.*

Proof: For any $\tau > 0$, define the mapping $\mathcal{Q}_\tau$: $H^1([0,\tau]; Z) \mapsto H^1([0,\tau]; Z)$ by $\mathcal{Q}_\tau[u] = e^{\tau L} f(\mathbf{x}) + \int_0^\tau e^{(\tau-s)L} N(u(\mathbf{x},s))\,ds$ and set $\Theta_\tau = \{u \in H^1([0,\tau]; Z)$: $\| u(t) - e^{tL}u(0) \|_Z \leq 1$ for $t \in [0,\tau]\}$. Also, let $M_\tau = 1 + \sup_{0 \leq \tau \leq 1}\left(\|e^{\tau L}u(0)\|_Z\right)$ and $\sqrt{M} = M_\tau \cdot \psi(M_\tau, M_\tau) \cdot \sup_{\|w\|_X \leq M_\tau}\left(\|N(w)\|_Z\right)$. In this setting ψ is the function from the *Key Condition* (*C*). For any $u, v \in \Theta_\tau$ and $\tau \in [0, \delta]$ with $1 \geq \delta > 0$, it is seen that $\| \mathcal{Q}_\tau[u] - \mathcal{Q}_\tau[v] \|_Z = \left\| \int_0^\tau e^{(\tau-s)L}\left(N(u(s)) - N(v(s))\right) ds \right\|_Z$. Now by the Cauchy–Schwarz inequality,

$$\left\| \int_0^\tau e^{(\tau-s)L}\left(N(u(s)) - N(v(s))\right) ds \right\|_Z \leq \int_0^\tau \left(M_\tau \cdot \psi(\| u \|_X, \| v \|_X) \cdot \| u - v \|_X\right)^2 ds \leq \tau \cdot M \leq \delta \cdot M.$$

Select $\delta \in (0,1]$ so that $\delta \cdot M < 1$. Then $\mathcal{Q}_\tau$ is a contraction of Θ_τ into Θ_τ. Consequently, by the *Contraction Mapping Theorem* (see, e.g., Berger [8], page 112), $\mathcal{Q}_\tau$ has a unique fixed point $u \in \Theta_\tau$:

$u(\boldsymbol{x}, t) = \mathcal{Q}_\tau[u(\boldsymbol{x}, t)] = e^{\tau L} f(\boldsymbol{x}) + \int_0^\tau e^{(\tau-s)L} N\left(u(\boldsymbol{x}, s)\right) ds$ which is a generalized solution of (8.1.1).■

Lemma 8.3 (Local Patching): *If* u_1 *and* u_2 *are solutions of* (8.1.1) *on the time intervals* $[0, t_1]$ *and* $[t_1, t_2]$, *respectively, so that* $u_1(t_1) = u_2(t_1)$, *then* $u(t) = \begin{cases} u_1(t) & 0 \leq t \leq t_1 \\ u_2(t) & t_1 \leq t \leq t_2 \end{cases}$ *is a solution of* (8.1.1) *on the time interval* $[0, t_2]$.

Proof: To prove the lemma, the generalized solution $u(t) = e^{tL} u(0) + \int_0^t e^{(t-s)L} N\left(u(s)\right) ds$ must be verified to hold for $0 \leq t \leq t_2$. By construction, $u(t) = u_1(t)$ for $t \in [0, t_1]$ so that on this interval, $u(t)$ solves (8.1.1). Now, for $t_1 \leq t \leq t_2$,

$$u(t) = u_2(t) = e^{(t-t_1)L} u_2(t_1) + \int_{t_1}^t e^{(t-s)L} N\left(u_2(s)\right) ds = e^{(t-t_1)L} u_1(t_1) + \int_{t_1}^t e^{(t-s)L} N\left(u_2(s)\right) ds$$

since $u_2(t_1) = u_1(t_1)$. But, as u_1 satisfies (8.1.1), $u_1(t_1) = e^{t_1 L} u_1(0) + \int_0^{t_1} e^{(t_1-s)L} N\left(u_1(s)\right) ds$.

Therefore, $u_2(t) = e^{(t-t_1)L} \left(e^{t_1 L} u_1(0) + \int_0^{t_1} e^{(t_1-s)L} N\left(u_1(s)\right) ds\right) + \int_{t_1}^t e^{(t-s)L} N\left(u_2(s)\right) ds$

$$= e^{tL} u_1(0) + \int_0^{t_1} e^{(t-s)L} N\left(u_1(s)\right) ds + \int_{t_1}^t e^{(t-s)L} N\left(u_2(s)\right) ds = e^{tL} u(0) + \int_0^t e^{(t-s)L} N\left(u(s)\right) ds. \quad ■$$

Lemma 8.4 (Maximal Orbits): *For any* $f \in X$, *there is a* $T \in \mathbb{R}^+ \cup \{+\infty\}$ *and a generalized solution* $u \in H^1([0, T]; Z)$ *of* (8.1.1) *for every internal* $[0, T_o] \subset [0, T]$ *so that either* $T = +\infty$ *or* $\| u(t) \|_z \to \infty$ *as* $t \to T$.

Proof: Suppose T is finite and that $\| u(t) \|_z \to \infty$ as $t \to T$. Let $B, U \in \mathbb{R}^+ \otimes Z$ be a closed, bounded set and an open set, respectively, with $B \subset U$. Furthermore, suppose $(t, u(t)) \in B$ for $t_o \leq t \leq T$. By the Heine–Borel Theorem, B can be assumed to have the form $B = [a, b] \otimes B_Z$ where B_Z is a bound set in the Hilbert space Z. The intention is to prove that there exists a function $u_1 \in B_Z$ so that $\lim_{t \to T^-} u(t) = u_1$. The expression $t \to T^-$ means that t approaches T from the left so that t is always less than T. Once the existence of u_1 is established, then the value of *Lemma* 8.3 comes into play. That is, the solution $u(t)$ can be extended to another function $u_2(t)$ over $[T, t_2]$. This contradicts the maximality of T. Hence, $T = +\infty$.

To show that u_1 exists, set $M_1 = \sup\{\| N(t, u) \|_X : (t, u) \in B\}$. The aim is to show $\| u(t) \|_Z$ is bounded as $t \to T^-$. Following Henry [37] (page 56), it is seen that for $t_o \le t \le T$ and $\beta \in [0, 1)$, the following estimates hold.

$$\begin{aligned}
\| u(t) \|_Z &\le \| e^{tL} f(\mathbf{x}) \|_X + \int_0^t \left\| (I+L)^\beta e^{(t-s)L} N(s, u(s)) \right\|_X ds \\
&\le M e^{t\lambda} \| f(\mathbf{x}) \|_X + \int_0^t M e^{(t-s)\lambda} (t-s)^{-\beta} M_1 \, ds \\
&= M e^{t\lambda} \| f(\mathbf{x}) \|_X + M \cdot M_1 \left| \int_0^t \frac{e^{(t-s)\lambda}}{(t-s)^\beta} ds \right| \\
&\le M e^{t\lambda} \| f(\mathbf{x}) \|_X + \frac{M \cdot M_1}{|\lambda|^{1-\beta}} \left| \gamma(1-\beta, -\lambda t) - \gamma(1-\beta, 0) \right| < \infty.
\end{aligned}$$

Here $\gamma(\alpha, \zeta) = \int_0^\zeta t^{\alpha-1} e^{-t} \, dt$ is the *lower incomplete gamma function*. Subsequently, $\| u(t) \|_Z$ is bounded as $t \to T^-$.

Next, suppose $t_o \le t \le \tau \le T$. As is seen in the proof of *Lemma* 8.3, $u(t) - u(\tau) = e^{\tau L} \left(e^{(t-\tau)L} - I \right) u(0) + \int_\tau^t e^{(t-s)L} N(s, u(s)) ds$. Consequently,

$$\| u(t) - u(\tau) \|_Z \le \| e^{\tau L} \|_{\mathcal{L}(X,Z)} \cdot \| e^{(t-\tau)L} - I \|_{\mathcal{L}(X,Z)} \cdot \| u(0) \|_Z + M \cdot M_1 \left| \int_\tau^t e^{(t-s)\lambda} ds \right|$$

$$\le \| e^{\tau L} \|_{\mathcal{L}(X,Z)} \cdot \| e^{(t-s)L} - I \|_{\mathcal{L}(X,Z)} \cdot \left(M e^{\tau\lambda} \| f(\mathbf{x}) \|_X + M \cdot M_1 \left| \int_0^\tau \frac{e^{(\tau-s)\lambda}}{(\tau-s)^\beta} ds \right| \right) + M \cdot M_1 \left| \int_\tau^t e^{(t-s)\lambda} ds \right|.$$

Since e^{tL} is a C_o–semigroup, there exist constants M_o and λ_o so that $\| e^{\tau L} \|_{\mathcal{L}(X,Z)} \le M_o \, e^{\tau \lambda_o}$, and $\lim_{\tau \to T^-} \left\| e^{(t-\tau)L} - I \right\|_{\mathcal{L}(X,Y)} = 0$ for $t_o \le t \le \tau \le T$. This is so because, as $t \to T^-$, τ must also approach t in value. Thus, the continuous nature of the semigroup compels $e^{(t-\tau)L}$ to approach the identity operator I as $t \to T^-$. Hence, $\lim_{t \to T^-} u(t)$ exists in the Hilbert space Z. Therefore, there exists $u_1 \in Z$ so that $\lim_{t \to T^-} u(t) = u_1$. ■

The combined result of *Lemmas* 8.1–8.4 is to establish the *local* existence and uniqueness of solutions to (8.1.1) on $H^1([0, T]; Z)$. Furthermore, as stated in *Lemma* 8.4, a solution either "blows up" (i.e., $\| u(t) \| \to \infty$) in finite time *or* exists globally in time. The final lemma of §8.4 describes the behavior of solutions to (8.1.1) in *space Z*.

Lemma 8.5 (Stability of Solutions): *Let $u(\boldsymbol{x}, t) \in H^1([0, T]; Z)$ be the generalized solution of (8.1.1) with $u(\boldsymbol{x}, 0) = f(\boldsymbol{x})$. Then there is a neighborhood $\mathcal{O}_f$ of $f \in X$ so that for any $g \in \mathcal{O}_f$ and every interval $[0, T] \subset [0, T_m)$ there is a generalized solution $v(\boldsymbol{x}, t) \in H^1([0, T]; Z)$ with $v(\boldsymbol{x}, 0) = g(\boldsymbol{x})$. Moreover, there is a constant C_o such that*

$$\|u(\boldsymbol{x}, t) - v(\boldsymbol{x}, t)\|_Z \le C_o \|f(\boldsymbol{x}) - g(\boldsymbol{x})\|_X \tag{8.4.3}$$

for all $g \in \mathcal{O}_f$ and $t \in [0, T]$.

Proof: Let $u(\boldsymbol{x}, t)$ be a generalized solution of (8.1.1) with Cauchy data $f \in X$ defined by (8.4.4). Next, let $v(\boldsymbol{x}, t)$ be defined by (8.4.5).

$$u(\boldsymbol{x}, t) = e^{tL} f(\boldsymbol{x}) + \int_0^t e^{(t-s)L} N\big(u(\boldsymbol{x}, s)\big)\,ds \tag{8.4.4}$$

$$v(\boldsymbol{x}, t) = e^{tL} g(\boldsymbol{x}) + \int_0^t e^{(t-s)L} N\big(v(\boldsymbol{x}, s)\big)\,ds \tag{8.4.5}$$

Then v is a generalized solution of (8.1.1) with initial value $g(\boldsymbol{x})$. Further, let g be in a sufficiently small neighborhood $\mathcal{O}_f$ of f so that $\|f(\boldsymbol{x}) - g(\boldsymbol{x})\|_X$ can be made less than $1/C_o$ for a preselected constant C_o. To prove (8.4.3), observe that

$$\|v(\boldsymbol{x}, t) - u(\boldsymbol{x}, t)\|_Z \le \|e^{tL}(g(\boldsymbol{x}) - f(\boldsymbol{x}))\|_Z + \|\int_0^t e^{(t-s)L}\big(N\big(v(\boldsymbol{x}, s) - N\big(u(\boldsymbol{x}, s)\big)\big)ds\,\|_Z\,.$$

As noted in the proof of *Lemma* 8.1, the fact that e^{tL} is a semigroup means that there are positive constants C and λ so that $\left\|e^{tL}\right\|_{\mathcal{L}(X,Y)} \le C e^{t\lambda}$ for all $t \ge 0$. Consequently, for all $t \in [0, T]$, the constant $M_1 = C\,e^{T\lambda}$ is such that $\|e^{tL}(g(\boldsymbol{x}) - f(\boldsymbol{x}))\|_Z \le M_1 \|g(\boldsymbol{x}) - f(\boldsymbol{x})\|_X$. Further, $N \in \mathcal{M}(X,Z)$ means that N is a bounded operator. This combined with *Key Condition* (*C*) means that

$$\begin{aligned}\|\int_0^t e^{(t-s)L}\big[N\big(v(\boldsymbol{x}, s)\big) - N\big(u(\boldsymbol{x}, s)\big)\big]ds\,\|_Z &\le \|\,e^{tL}\,\|_{\mathcal{L}(X,Z)} \cdot \|\int_0^t e^{-sL}\big[N\big(v(\boldsymbol{x}, s)\big) - N\big(u(\boldsymbol{x}, s)\big)\big]ds\,\|_Z \\ &\le M_1 \cdot \int_0^t M_2 \cdot \psi(M_v, M_u) \cdot \|v(\boldsymbol{x}, s) - u(\boldsymbol{x}, s)\|_X\, ds\,.\end{aligned}$$

Here M_2 is the bound on $\|\,e^{-sL}\,\|_{\mathcal{L}(X,Z)}$ for all $s \in [0, t] \subset [0, T]$ while M_u and M_v are the bounds on $N(u)$ and $N(v)$, respectively. Therefore,

$$\|v(\boldsymbol{x}, t) - u(\boldsymbol{x}, t)\|_Z \le M_1 \|g(\boldsymbol{x}) - f(\boldsymbol{x})\|_X + M_3 \cdot \int_0^t \|v(\boldsymbol{x}, s) - u(\boldsymbol{x}, s)\|_X\, ds \tag{8.4.6}$$

where $M_3 = M_1 \cdot M_2 \cdot \psi(M_v, M_u)$. By Gronwall's inequality, (8.4.3) holds for all $t \in [0, T]$ when $C_o = M_1 \cdot \exp\left(\frac{M_3 T}{C_a}\right)$ and C_a is an absolute constant between the norms of the embedded Hilbert spaces $Z \subset X$. That is, $\|u\|_Z \leq C_a \cdot \|u\|_X$. In summary,

$$\|v(\mathbf{x}, t) - u(\mathbf{x}, t)\|_Z \leq C_o \|g(\mathbf{x}) - f(\mathbf{x})\|_X .$$ ■

Remark 8.4: The *Fréchet derivative* of an operator $N \in \mathcal{M}(X,Z)$ exists if N can be locally bound by a linear operator. Moreover, being *Fréchet differentiable* implies that the operator is *Gateaux differentiable*. Formally, $N \in \mathcal{M}(X,Z)$ means that N is *Gateaux differentiable* at $v \in X$ provided $\lim_{h\to 0}\|N(u+h\cdot v) - N(u) - h\cdot N'(u)v\| = 0$. Informally, the *Gateaux derivative* is defined as $N'(u)v = \lim_{h\to 0}\frac{N(v+h\cdot u)-N(v)}{h}$. Whenever $N'(u) \in \mathcal{L}(X,Z)$ on a sufficiently small neighborhood of v, then the *Gateaux derivative* is equal to the *Fréchet derivative.* By *Key Condition* (*D*), N is Fréchet differentiable. Examining the linearized equation $\frac{\partial v}{\partial t} = Lw + N'\big(u(t)\big)v$ leads to the last lemma of the section.

Lemma 8.6: *Let the neighborhood* $\mathcal{O}_f$ *of* $f \in X$ *be as in Lemma 8.5. Then the flow* F_{L+N}^t *is continuously Fréchet differentiable on* $\mathcal{O}_f$ *for all* $t \in [0, T]$.

Proof: Let $S_L^j(t)$ be the generalized solution operator of $\frac{\partial v}{\partial t} = Lw + N'\big(u_j(t)\big)v$ with respect to $v(0) = f \in \mathcal{O}_f$. Specifically, $S_L^j(t)(v) = e^{tL}v(0) + \int_0^t e^{(t-s)L}N'\big(u_j(s)\big)v(s)\,ds$. If $\varphi_j(t) \equiv S_L^j(t)(v)$ and $\Delta\varphi(t) = \varphi_1(t) - \varphi_2(t)$, then as $N'(u)$ is locally linear,

$$\Delta\varphi(t) = \int_0^t e^{(t-s)L}\left[N'\big(u_1(s)\big) - N'\big(u_2(s)\big)\right]v(s)\,ds . \tag{8.4.7}$$

But since the Fréchet and Gateaux derivatives are equivalent in this setting, $N'(u_1)v - N'(u_2)v =$
$$\lim_{h\to 0}\frac{[N(v+h\cdot u_1)-N(v)]-[N(v+h\cdot u_2)-N(v)]}{h} = \lim_{h\to 0}\frac{N(v+h\cdot u_1)-N(v+h\cdot u_2)}{h}.$$
Now *Key Condition* (*C*) implies

$$\|N(v+h\cdot u_1) - N(v+h\cdot u_2)\|_Z \leq \psi\left(\|v+h\cdot u_1\|_X, \|v+h\cdot u_2\|_X\right)\|h\cdot u_1 - h\cdot u_2\|_X$$

and subsequently

$$
\begin{aligned}
\|N'(u_1)v - N'(u_2)v\|_Z &\le \lim_{h\to 0}\frac{\|N(v+h\cdot u_1) - N(v+h\cdot u_2)\|}{h} \\
&\le \lim_{h\to 0}\frac{1}{h}\psi\left(\|v+h\cdot u_1\|_X, \|v+h\cdot u_2\|_X\right)\cdot h\cdot\|u_1-u_2\|_X \\
&\le \psi\left(\|v\|_X, \|v\|_X\right)\|u_1-u_2\|_X \quad .
\end{aligned}
\tag{8.4.8}
$$

Now choose $\mathcal{O}_f$ sufficiently small so that $\|v\|_X \le M$, $\psi(\|v\|_X, \|v\|_X) \le \psi(M, M) < C_o$ and $\|u_1 - u_2\|_X < \delta/C_o$. Then for all $t \in [0, T]$, $\|N'(u_1)v - N'(u_2)v\|_Z \le \delta$ and

$$
\begin{aligned}
\|\Delta\varphi(t)\| &\le M_o e^{\lambda_o t}\int_0^t e^{-\lambda_o s}\,\|N'(u_1)v - N'(u_2)v\|_Z\, ds \\
&\le M_o e^{\lambda_o T}\frac{1}{\lambda_o}\left(1-e^{-\lambda_o T}\right)\delta \qquad \text{for all } t\in[0,T].
\end{aligned}
$$

Hence, N is continuously Fréchet differentiable. ■

The combination of *Key Conditions* (*A*) – (*D*), the Hartman Mapping Theorem, and *Lemmas* 8.1–8.6 yield the generalized *Flow Theorem*. Before stating the Theorem, adopt the notation $Y = H^1([0, T]; Z)$ and take $T_m \in \mathbb{R}^+$ as in the *Existence and Uniqueness Theorem* (from the beginning of §8.4).

The Hartman–Grobman Flow Theorem: *Let* $L \in \mathcal{L}(X,Y)$ *satisfy Key Conditions* (*A*) – (*B*) *and* $N \in \mathcal{M}(X,Y)$ *satisfy Key Conditions* (*C*) – (*D*). *Then there exists a unique homeomorphism* H, *independent of time* t, *so that* $H \circ F^t_{L+N} = F^t_L \circ H$ *on sufficiently small neighborhoods* $\mathcal{O}_X$ *of* $0 \in X$ *and* $\mathcal{O}_Z$ *of* $0 \in Z$. *That is, Diagram 8.4 below commutes for all* $t \in [0, T] \subset [0, T_m)$.

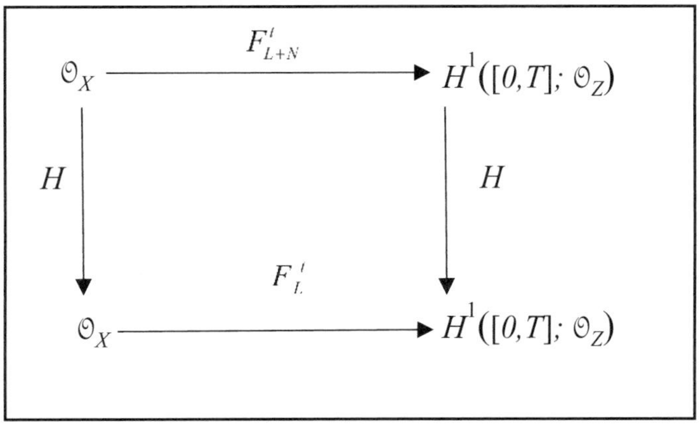

Diagram 8.4. Local topological conjugacy of F^t_{L+N} to F^t_L.

Proof: Let $\mathcal{L}^t = F_L^t$ and $\mathcal{N}^t u(\mathbf{x}, 0) = \int_0^t e^{(t-s)L} N(u(\mathbf{x},s))\,ds$. Then $F_{L+N}^t = \mathcal{L}^t + \mathcal{N}^t$. First notice that, by *Key Condition* (*A*), *L* is the generator of a C_o–semigroup. Therefore, $\mathcal{L}^t$ is a linear map from *X* into $Y = H^1([0, T]; Z)$ for each $t \in [0, T]$. To show that $\mathcal{L}^t$ is an expansive–contractive map, recall that *Key Condition* (*B*) insures that the spectrum of *L*, $\sigma(L)$, has non-zero real–parts: $\lambda \in \sigma(L) \Rightarrow \mathrm{Re}(\lambda) \neq 0$. Hence, for all $t \in [0, T]$, $|e^{t\lambda}| \neq 1$. Since $\sigma(\mathcal{L}^t) = \{e^{t\lambda}: \lambda \in \sigma(L)\}$, then by Remark 8.2 (*iii*) $\mathcal{L}^t$ is an expansive–contractive map for all $t \in [0, T]$.

Next, the goal is to show $\mathcal{N}^t \in \mathcal{M}(X,Y)$ for all $t \in [0, T]$. By hypothesis, $N \in \mathcal{C}^1(X,Z)$ and $\mathcal{L}^t$ is a C_o–semigroup acting on *Y*. Therefore, $\mathcal{L}^t \in \mathcal{C}^1(X,Z)$ for all *t*. Consequently, by *Lemma* 8.6, $F_{L+N}^t = \mathcal{L}^t + \mathcal{N}^t \in \mathcal{C}^1(\mathcal{O}_X, \mathcal{O}_Y)$ where $\mathcal{O}_Y = H^1([0, T]; \mathcal{O}_Z)$. The flow $F_{L+N}^t: u(\mathbf{x}, 0) \mapsto u(\mathbf{x}, t) \in \mathcal{O}_Y$ for all $t \in [0, T]$ is constructed as

$$u(\mathbf{x},t) \equiv F_{L+N}^t\left(u(\mathbf{x},0)\right) = e^{tL}u(\mathbf{x},0) + \int_0^t e^{(t-s)L} N\left(u(\mathbf{x},s)\right)ds = \mathcal{L}^t u(\mathbf{x}, 0) + \mathcal{N}^t u(\mathbf{x}, 0).$$

Now by the *Existence and Uniqueness Theorem*, there exists $T_m \in \mathbb{R}^+$ so that $\| u(\mathbf{x}, t) \|_Y \leq C_o < \infty$ for all $t \in [0, T] \subset [0, T_m)$. Subsequently,

$$\| \mathcal{N}^t u(\mathbf{x}, t) \|_{\mathcal{O}_Y} = \| F_{L+N}^t u(\mathbf{x}, 0) - \mathcal{L}^t u(\mathbf{x}, 0) \|_{\mathcal{O}_Y} \leq \| u(\mathbf{x}, t) \|_{\mathcal{O}_Y} + C(T)\cdot\| u(\mathbf{x}, 0) \|_{\mathcal{O}_X} < C_1 < \infty$$

for all $t \in [0, T] \subset [0, T_m)$. Here $C(T) = M e^{\lambda T}$ when $L \in \mathcal{L}(X,Y)$ bounded and $T \leq T_m < \infty$ and $C(T) = 1$ when *L* is the unbounded generator of a C_o–semigroup of contractions and $T_m = \infty$.

Observe that this bound on $\mathcal{N}^t$ is uniform with respect to *t*. Therefore, $\mathcal{N}^t$ is a uniformly bounded map from $\mathcal{O}_X$ to $\mathcal{O}_Y$ for all $t \in [0, T]$.

By *Key Condition* (*D*), $N(0) = N'(0)v = 0$. Therefore, $\mathcal{N}^t[0] = \int_0^t e^{(t-s)L} N(0)\,ds = 0$. Furthermore, $(\mathcal{N}^t)'(u)v = \int_0^t e^{(t-s)L} N'(u)v\,ds$ means that $(\mathcal{N}^t)'(0)v = \int_0^t e^{(t-s)L} N'(0)v\,ds = \int_0^t e^{(t-s)L}\, 0\,ds = 0$ for all $v \in \mathrm{Dom}[(\mathcal{N}^t)'(u)]$. Hence, $\mathcal{N}^t \in \mathcal{M}(X,Y)$ for all $t \in [0, T]$. By the *Hartman Mapping Theorem*, there exists a unique homeomorphism $H: \mathcal{O}_X \to \mathcal{O}_Y$ so that $H \circ F_{L+N}^t = H \circ (\mathcal{L}^t + \mathcal{N}^t) = H = F_L^t \circ H$.

■

Remark 8.5: The *Flow Theorem* can be extended if an additional condition is imposed upon $\mathcal{N}^t$ and hence F_{L+N}^t. An invertible continuous map *M* between the function spaces *X* and *Y* is *proper* provided the inverse image of a compact set is also compact. Specifically, if $M \in \mathcal{C}(X,Y)$, $C \subset Y$

is compact, and $M^{-1}(C) \subset X$ is also compact, then M is *proper*. If $\mathcal{N}^t$ is proper for each $t \in [0, T]$ then so is $F^t_{L+N} = \mathcal{L}^t + \mathcal{N}^t$ since $\mathcal{L}^t = e^{tL}$ generates a C_o–semigroup. In the proof of the *Hartman–Grobman Flow Theorem* (above), the estimate on $\| \mathcal{N}^t u(\mathbf{x}, 0) \|_{\mathcal{O}_Y}$ can be extended. Define the solution set of an operator F to be $\mathfrak{S}_F = \{f \in X\colon f(\mathbf{x}) = u(\mathbf{x}, 0)$ and $F(f) = u(\mathbf{x}, t)$ is a generalized solution of (8.1.1)$\}$. By Theorem 2.7.1. of Berger [8], if F is proper then F is a closed map whose solution set $\mathfrak{S}_F$ is compact. If $Z = \mathfrak{S}_F$ and $F = F^t_{L+N}$ is proper, then the inverse image $X = F^{-1}(Y)$ is compact. Here, $Y = H^1\left([0,T]; \mathfrak{S}_{F^t_{L+N}}\right)$. Hence, the estimate

$$\| \mathcal{N}^t u(\mathbf{x}, 0) \|_Y = \| F^t_{L+N}\, u(\mathbf{x}, 0) - \mathcal{L}^t u(\mathbf{x}, 0) \|_Y \leq \| u(\mathbf{x}, t) \|_Y + C(T) \cdot \| u(\mathbf{x}, 0) \|_X < C_1 < \infty$$

holds uniformly for all $t \in [0, T]$. As long as all of the other hypotheses of the *Hartman–Grobman Flow Theorem* hold, then the unique homeomorphism H exists mapping X into Y so that $H \circ F^t_{L+N} = F^t_L \circ H$. As above, $Y = H^1\left([0,T]; \mathfrak{S}_{F^t_{L+N}}\right)$. This remark is now summarized as a theorem.

A Global Hartman–Grobman Flow Theorem: *Let* $L \in \mathcal{Q}(X,Y)$ *satisfy Key Conditions* $(A) - (B)$ *and* $N \in \mathcal{M}(X,Y)$ *satisfy Key Conditions* $(C) - (D)$. *If, in addition, the semi–linear* flow F^t_{L+N} *is proper for all* $t \in [0, T]$, *then there exists a unique homeomorphism H, independent of time t, so that* $H \circ F^t_{L+N} = F^t_L \circ H$ *on X and* $H^1\left([0,T]; \mathfrak{S}_{F^t_{L+N}}\right)$. *That is, Diagram 8.5 below commutes for all* $t \in [0, T] \subset [0, \infty)$.

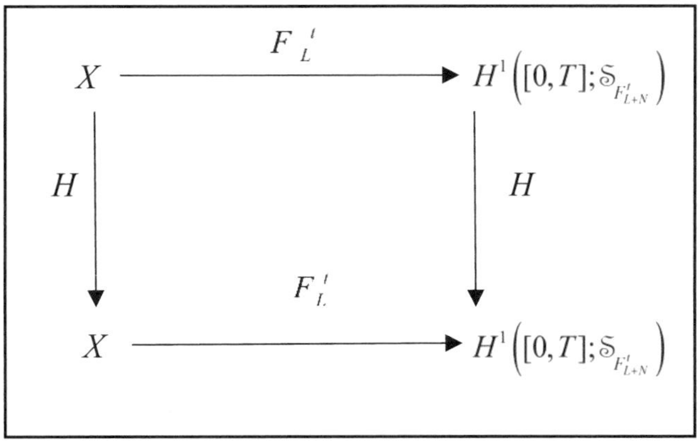

Diagram 8.5. Global topological conjugacy of F^t_{L+N} to F^t_L.

In the next section, the *Flow Theorem* will be applied to several well–known nonlinear partial differential equations.

§8.5 Applications

In this final section, the *Flow Theorem* will be applied to three well–known nonlinear evolution equations: The periodic *Korteweg de Vries* (*KdV*), *Burgers'*, and the *cubic nonlinear Schrödinger* (*NLS*) equations.

§8.5.1 A Generalized Korteweg de Vries Equation

The *Korteweg de Vries* equation, first noted by Joseph Boussinesq in 1877 [11] later reintroduced by Diederik Korteweg and Gustav de Vries in 1895 [47], models the action of waves in shallow water. The canonical form of the *KdV*, as expressed in (8.5.1), is discussed in Chapter 6 along with an analytic approach to its solution via the *inverse scattering transform*.

$$\left.\begin{aligned} &\frac{\partial u}{\partial t} - 6u\frac{\partial u}{\partial x} + \frac{\partial^3 u}{\partial x^3} = 0 \\ &u(x,0) = f(x) \end{aligned}\right\} \tag{8.5.1}$$

In this section, a generalized version of the *KdV* is explored within the context of the *Hartman–Grobman Theorem*. To that end, recall equation (8.1.3) from §8.1

$$\left.\begin{aligned} &\frac{\partial u}{\partial t} = u - \frac{\partial^3 u}{\partial x^3} + \Psi(u) - u\frac{\partial u}{\partial x} \\ &u(x,0) = f(x) \\ &u(x+2\pi,t) = u(x,t) \end{aligned}\right\}. \tag{8.1.3}$$

In this context, the mapping Ψ satisfies the *Key Conditions* (*C*) – (*D*). If $L = I - \dfrac{\partial^3}{\partial x^3}$ and $N(u) = \Psi(u) - u\dfrac{\partial u}{\partial x}$, then (8.1.3) can be re–written in the form of the semi–linear evolution equation

$$\left.\begin{aligned} &\frac{\partial u}{\partial t} = Lu + N(u) \\ &u(x,0) = f(x) \\ &u(x+2\pi,t) = u(x,t) \end{aligned}\right\}. \tag{8.5.2}$$

The only restriction on f is that it needs to be a smooth 2π–periodic real–valued function. In this setting, the following results are established.

Theorem 8.1: *For every $f \in H_3[0, 2\pi]$ there is a $T > 0$ so that* (8.1.3)/(8.5.2) *has a unique generalized solution $u \in H^1([0, T]; H_3[0, 2\pi])$.*

Proof: By the *Existence and Uniqueness Theorem*, it need only be shown that L satisfies *Key Condition* (A) and that $N(u)$ satisfies (C) – (D). As per Pazy [59], the operator $\frac{\partial^3}{\partial x^3} : H_3[0, 2\pi] \to L_2[0, 2\pi]$ is the generator of a C_o–semigroup of isometries on the Lebesgue space $L_2[0, 2\pi]$. Moreover, the identity operator I is a bounded linear operator so that $L = I - \frac{\partial^3}{\partial x^3}$ is the generator of a C_o–semigroup. Therefore, L satisfies *Key Condition* (A).

Next, let Ψ satisfy *Key Condition* (C) with the monotone operator ψ, and let $\| \bullet \|_q$ and $\| \bullet \|_{p,2}$ be the norms on the spaces $L_q[0, 2\pi]$ and $H_p[0, 2\pi]$, respectively. Using the notation $u_x = \partial u/\partial x$, then the following estimates hold due to the *Cauchy–Schwarz–Buniakowski* inequality (see, e.g., Kantorovich and Akilov [45]), along with the facts that $H_p[0, 2\pi]$ is embedded in $L_{p,}[0, 2\pi]$ and $H_q[0, 2\pi]$ is embedded in $H_p[0, 2\pi]$ for $q > p$: $H_p[0, 2\pi] \subset\!\!\to L_p[0, 2\pi]$ and $H_q[0, 2\pi] \subset\!\!\to H_p[0, 2\pi]$.

$$\begin{aligned} \| u\, u_x - v\, v_x \|_2 &= \| u(u_x - v_x) + (u - v)v_x \|_2 \le \| u \|_2 \cdot \| \partial(u - v)/\partial x \|_2 + \| u - v \|_2 \cdot \| \partial v/\partial x \|_2 \\ &\le C_1 \| u \|_2 \cdot \| u - v \|_{1,2} + C_1 \| u - v \|_2 \cdot \| v \|_{1,2} \\ &\le C_2 \| u \|_{1,2} \cdot \| u - v \|_{3,2} + C_2 \| u - v \|_{1,2} \cdot \| v \|_{3,2} \\ &\le C_3 \left(\| u \|_{3,2} + \| v \|_{3,2}\right) \cdot \| u - v \|_{3,2}. \end{aligned}$$

The C_j are all positive constants. Combining this estimate with *Key Condition* (C) on $\Psi(u)$ results in

$$\begin{aligned} \| N(u) - N(v) \|_2 &= \| \Psi(u) - u\, u_x - \Psi(v) - v\, v_x \|_2 \le \| \Psi(u) - \Psi(v) \|_2 + \| u\, u_x - v\, v_x \|_2 \\ &\le C_o\, \psi(\| u \|_{3,2}, \| v \|_{3,2}) \cdot \| u - v \|_{3,2} + C_3 \left(\| u \|_{3,2} + \| v \|_{3,2}\right) \cdot \| u - v \|_{3,2} \\ &\le \Theta(\| u \|_{3,2}, \| v \|_{3,2}) \cdot \| u - v \|_{3,2} \end{aligned}$$

where $\Theta(a, b) = C_3 (a + b) + \psi(a, b)$ is a monotone increasing function. Therefore, N satisfies *Key Condition* (C) with $X = H_{3,2}[0, 2\pi]$ and $Z = L_2[0, 2\pi]$.

Set $\Xi(u) = u\frac{\partial u}{\partial x} = \frac{\partial}{\partial x}\left(u^2\right)$. Then $N(u) = \Psi(u) - \Xi(u)$ and $N(0) = \Psi(0) - \Xi(0) = 0$ since $\Psi(u)$ is assumed to satisfy *Key Condition* (D). Observe that the Fréchet derivative of $\Xi(u)$ is calculated as follows.

$$\Xi'(u)v = \lim_{h\to 0}\frac{\Xi(u+h\cdot v)-\Xi(u)}{h} = \lim_{h\to 0}\frac{\frac{\partial}{\partial x}\left([u+h\cdot v]^2\right)-\frac{\partial}{\partial x}\left(u^2\right)}{h} = \lim_{h\to 0}\frac{h\left(v\cdot u_x + u\cdot v_x\right)-h^2(v\cdot v_x)}{h}$$

$= v\cdot u_x + u\cdot v_x = \frac{\partial}{\partial x}(u\cdot v)$. Therefore, $N'(u)v = \Psi'(u)v - \Xi'(u)v = \Psi'(u)v - \frac{\partial}{\partial x}(u\cdot v)$ and $\Xi'(0)v = 0$. Thus, $N'(0)v = \Psi'(0)v = \Xi'(0)v = 0$. By the Sobolev Imbedding Theorem (see, e.g.,

Berger [8]), $u \in X = H_3[0, 2\pi]$ implies that $u \in C^2([0, 2\pi])$ so that $u_x \in C^1([0, 2\pi])$ which is embedded in $L_2[0, 2\pi]$. Subsequently, the linear mapping $u \mapsto \Xi'(u) \equiv \frac{\partial}{\partial x} + u_x I$ is continuous from X to $\mathscr{L}(X,Z)$ with $Z = L_2[0, 2\pi]$. Hence, $N'(u) \in \mathscr{C}(X, \mathscr{L}(X,Z))$ and, by the estimates made above, it is seen for all $u, v \in B \subset X$, B a bounded set,

$$\begin{aligned} \| \Xi(u) - \Xi(v) - \Xi'(v)(u-v) \|_2 &= \| (u-v)(u_x - v_x) \|_2 \le C_o \| u - v \|_2 \cdot \| u_x - v_x \|_2 \\ &\le C_2 \| u - v \|_{1,2} \| u - v \|_{1,2} \le C_2 \|u-v\|_{1,2}^2 \le C_3 \|u-v\|_{3,2}^2 . \end{aligned}$$

This finally leads to

$$\begin{aligned} \|N(u) - N(v) - N'(u)(u-v)\|_2 &\le \|\Psi(u) - \Psi(v) - \Psi(u)(u-v)\|_2 + \|\Xi(u) - \Xi(v) - \Xi(u)(u-v)\|_2 \\ &\le C_4 \|u-v\|_{3,2}^2 + C_3 \|u-v\|_{3,2}^2 . \end{aligned}$$

That is, N satisfies *Key Condition* (*D*). ■

Remark 8.6: Pazy [59] has shown that (8.5.2) has a unique generalized solution $u \in \mathscr{C}(X,Y)$ whenever $u(x, 0) \in H_s[0, 2\pi]$ with $X = H^1([0, T]; H_s[0, 2\pi])$, $Y = L_2[0, 2\pi]$, and $s \ge 3$. The next stage in the program is to show that the flow F^t_{L+N} on (8.5.2) is locally conjugate to the linear flow F^t_L on the periodic Airy's equation (8.5.3). Here, as above, $L = I - \frac{\partial^3}{\partial x^3}$.

$$\left.\begin{aligned} &\frac{\partial v}{\partial t} = Lv \\ &v(x,0) = g(x) \\ &v(x + 2\pi, t) = v(x,t) \end{aligned}\right\} \qquad (8.5.3)$$

Theorem 8.2: *There exists a unique homeomorphism H, independent of t* $\in [0, T]$, *so that the flow* F^t_{L+N} *on* (8.5.2) *is topologically conjugate to the flow* F^t_L *on* (8.5.3) *on sufficiently small neighborhoods* $\mathcal{O}_{3,2}$ *of* $0 \in H_3[0, 2\pi]$ *into* $\mathcal{O}_2$ *of* $0 \in L_2[0, 2\pi]$.

Proof: Since $L = I - \frac{\partial^3}{\partial x^3}$ satisfies *Key Condition* (*A*) and $N(u) = \Psi(u) - u \cdot \frac{\partial u}{\partial x}$ satisfies (*C*) – (*D*) by the proof of Theorem 8.1 above, then it remains only to show that $\mathcal{L}^t = e^{tL} = F^t_L$ is expansive–contractive. Since the set $\mathcal{E} = \{e^{inx} : n \in \mathbb{Z}\}$ is composed of 2π–periodic functions and is dense in $H_3[0, 2\pi]$, then it can be used as an eigenbasis for L. Now the characteristic equation $L\, e^{inx} = \lambda\, e^{inx}$

means that $(1 - (i{\cdot}n)^3)\, e^{inx} = \lambda\, e^{inx}$ or $\lambda = (1 + i{\cdot}n^3)$. Therefore, the spectrum of the operator L is $\sigma(L) = \{1 + i{\cdot}n^3 : n \in \mathbb{Z}\}$ and consequently, for any $\lambda \in \sigma(L)$, $\mathrm{Re}(\lambda) = 1 \neq 0$. Therefore, L is an expansive–contractive linear map. ■

Next it is shown that Burgers' equation of §6.2 is a fundamental example of the *Hartman–Grobman* Theorem.

§8.5.2 Burgers' Equation

As noted in Chapter 6, J. M. Burgers [13, 14] modeled turbulence by examining the eponymous equation (8.5.4).

$$\left.\begin{array}{l} \dfrac{\partial u}{\partial t} = \sigma^2 \dfrac{\partial^2 u}{\partial x^2} - u\dfrac{\partial u}{\partial x} \\ u(x,t) = f(x) \end{array}\right\} \tag{8.5.4}$$

Burgers' equation is a second order (spatial derivative) version of the *KdV* $\dfrac{\partial u}{\partial t} = 6u\dfrac{\partial u}{\partial x} - \dfrac{\partial^3 u}{\partial x^3}$. As demonstrated in §6.2, there is a change of variables $u = -2\sigma^2 \dfrac{\partial}{\partial x}\ln(v)$ so that (8.5.4) becomes the linear (one–dimensional heat) equation (8.5.5).

$$\left.\begin{array}{l} \dfrac{\partial v}{\partial t} = \sigma^2 \dfrac{\partial^2 v}{\partial x^2} \\ v(x,0) = \exp\left(-\frac{\sigma^2}{2}\int_0^x f(\xi)\,d\xi\right) \end{array}\right\} \tag{8.5.5}$$

This change of variables $u = -2\sigma^2 \dfrac{\partial}{\partial x}\ln(v)$ is called the *Hopf–Cole transformation*.

The existence and uniqueness of a global (in time) solution of (8.5.4) is established via analytic methods in §6.2. To show that the *Hartman–Grobman* Theorem can be applied to Burgers' equation, it is required to show that $L = \sigma^2 \dfrac{\partial^2}{\partial x^2}$ and $N(u) = u\cdot\dfrac{\partial u}{\partial x}$ satisfy *Key Conditions* (*A*) – (*B*) and (*C*) – (*D*), respectively. It is well known that L is the generator of a C_o–compact analytic semigroup from $H_2(\mathbb{R})$ to $L_2(\mathbb{R})$ so that *Key Condition* (*A*) is satisfied. See, for example, Pazy [59]. Moreover, as demonstrated in §8.5.1 above, the nonlinear operator $N(u)$ satisfies *Key Conditions* (*C*) – (*D*). Now observe that, for any sequence of positive numbers $<\alpha_n>$, $n \in \mathbb{Z}^+$, the set of functions $e^{-i\sqrt{\alpha_n}x}$ and $e^{i\sqrt{\alpha_n}x}$ are the fundamental solutions of the characteristic equation $Lu = \lambda u$. Indeed, $u_n(x) = e^{-i\sqrt{\alpha_n}x}$ means that $Lu_n(x) = -\sigma^2\alpha_n\, u_n(x) = \lambda\, u_n(x)$ provided $\lambda_n = -\sigma^2\alpha_n$ for each $n \in \mathbb{Z}^+$. The same eigenvalues are generated when $u_n(x) = e^{i\sqrt{\alpha_n}x}$. Hence, $\sigma(L) = \{\lambda_n\colon \lambda_n = -\sigma^2\alpha_n$ and $n \in \mathbb{Z}^+\}$ so that $\mathrm{Re}(\lambda) < 0$ for all $\lambda \in \sigma(L)$. Therefore, $\mathrm{Re}(\lambda) \neq 0$ for any $\lambda \in \sigma(L)$ and consequently L is expansive–contractive. In fact, in this case, L is strictly contractive. Therefore, *Key Condition* (B) is satisfied and the results below hold.

Theorem 8.3: *For every $f \in H_2(\mathbb{R})$ there is a unique homeomorphism H, independent of $t \in [0, T]$, so that the flow F^t_{L+N} on (8.5.4) is topologically conjugate to the flow F^t_L on (8.5.5) on sufficiently small neighborhoods $\mathcal{O}_{2,2}$ of $0 \in H_2(\mathbb{R})$ into $\mathcal{O}_2$ of $0 \in L_2(\mathbb{R})$.*

Corollary 8.1: *Since the homeomorphism H in the Hartman–Grobman Flow Theorem is unique, then it must be equal to the Hopf–Cole transformation $u = -2\sigma^2 \frac{\partial}{\partial x}\ln(v)$. Therefore, the topological conjugacy is global.*

Thus far, the examples have been restricted to a single spatial variable $x \in \mathbb{R}$. The final example of this chapter will expand these ideas to spatial variables in $\mathbb{R}^n$.

§8.5.3 Nonlinear Schrödinger Equation

One of the most widely studied and important equations of physics is the (cubic) *nonlinear Schrödinger equation* or *NLS*. This equation is used to model the propagation of light in optical fibers, the flow of electrons in superconductors, and nonlinear plasmas. For this section, the *NLS* will be defined as the partial differential equation on a complex valued function $u(\boldsymbol{x}, t) = u_1(\boldsymbol{x}, t) + i\cdot u_2(\boldsymbol{x}, t)$, $u_1(\boldsymbol{x}, t)$ and $u_2(\boldsymbol{x}, t)$ real–valued functions, and $\boldsymbol{x} \in \mathbb{R}^n$.

$$\frac{\partial u}{\partial t} = \Delta u - k|u|^2 u, \quad u(\boldsymbol{x},0) = f(\boldsymbol{x}), \quad \boldsymbol{x} \in \mathbb{R}^n$$

The constant k is called the *wavenumber*. To apply the Hartman–Grobman theory to the *NLS*, the linear operator is the Laplacian $Lu = \Delta u$ while the nonlinear functional is the third order operator $N(u) = -k|u|^2u$. The nonlinear evolution equation and its linear counterpart then are listed as (8.5.6)–(8.5.7). Observe that (8.5.7) is the n–dimensional *heat equation*.

$$\left.\begin{aligned} &\frac{\partial u}{\partial t} = Lu + N(u) \\ &u(\boldsymbol{x},0) = f(\boldsymbol{x}), \quad \boldsymbol{x} \in \Omega \subset \mathbb{R}^n \\ &u(\boldsymbol{x},t) \equiv 0, \quad \text{for all } \boldsymbol{x} \in \partial\Omega \text{ and } t \geq 0 \end{aligned}\right\} \tag{8.5.6}$$

$$\left.\begin{aligned} &\frac{\partial v}{\partial t} = Lv \\ &v(\boldsymbol{x},0) = \phi(\boldsymbol{x}), \quad \boldsymbol{x} \in \Omega \subset \mathbb{R}^n \\ &v(\boldsymbol{x},t) \equiv 0, \quad \text{for all } \boldsymbol{x} \in \partial\Omega \text{ and } t \geq 0 \end{aligned}\right\} \tag{8.5.7}$$

Note that the additional Dirichlet boundary condition of the solution vanishing on the boundary $\partial\Omega$ of the domain of definition Ω is imposed. Also, the supposition $u \in X$, a real Hilbert space, *means* that $u_1, u_2 \in X$. Rather than expressing u as the complex sum of the two real–valued functions u_1 and u_2 via $u = u_1 + i \cdot u_2$, instead define u as the pair $u = (u_1, u_2)$. To show that the flow F^t_{L+N} on (8.5.6) is locally topologically conjugate to the flow F^t_L on (8.5.7), let $\overset{\circ}{H}_2(\Omega)$ be the Hilbert space comprised of the set of all functions in $H_2(\Omega)$ which vanish on the boundary of Ω. That is, $\overset{\circ}{H}_2(\Omega) = \{f = (f_1 . f_2) \in H_2(\Omega): f(\boldsymbol{x}) \equiv 0 \text{ for all } \boldsymbol{x} \in \partial\Omega\}$. With this notation in mind, the existence and uniqueness of solutions to (8.5.6) for dimensions less than or equal to three ($n \leq 3$) has been well established. See, for example, Haraux [32], Henry [37], or Pazy [59]. For $n > 3$, the existence and uniqueness proof follows in Corollary 8.2.

Theorem 8.4 (Existence and Uniqueness of the NLS): *For every* $f = (f_1, f_2) \in \overset{\circ}{H}_2(\Omega)$ and $k \neq 0$, *the NLS equation* (8.5.6) *has a unique generalized solution* $u = (u, u_2) \in H^1([0, \infty); L_2(\Omega))$ *for* $n = 1, 2,$ *or* 3,

Proof: As noted, existence and uniqueness on $H^1([0, \infty); L_2(\Omega))$ has been established (e.g., Haraux [32], Pazy [59]) for each interval $[0, T] \subset [0, T_m)$. To show $T_m \to \infty$, it must be established that $\|u(\bullet, t)\|_2 < \infty$ for all $t < \infty$. Throughout this chapter, the well–known behavior of $L = \Delta$ as the generator of a C_o–semigroup has been brought to bear. In this example, L is the generator of a C_o–contractive semigroup from $\overset{\circ}{H}_2(\Omega)$ to $L_2(\Omega)$. Consequently, the following estimate holds for any generalized solution $u(\boldsymbol{x}, t)$ of the *NLS* equation (8.5.6).

$$\|u(\bullet,t)\|_2^2 \leq \|f\|_2^2 + \int_0^t \|N(u(\bullet,s))\|_2^2 \, ds \tag{8.5.8}$$

To show that $\|u(\bullet,t)\|_2 < \infty$, it suffices to show that the Lyapunov function $\|N(u(\bullet,t))\|_2^2$ is decreasing in time. Equivalently, it is sufficient to show $\frac{\partial}{\partial t}\|N(u(\bullet,t))\|_2^2 \le 0$ for all t.

Notice that $\frac{\partial}{\partial t}\|N(u(\bullet,t))\|_2^2 = \frac{\partial}{\partial t}\int_\Omega \left|-k|u|^2 u\right|^2 d\mathbf{x} = k^2 \int_\Omega \frac{\partial}{\partial t}|u|^6 d\mathbf{x}$ by Leibniz's Rule.

But $|u|^6 = (u\cdot\bar{u})^3$ so that $\frac{\partial}{\partial t}|u|^6 = \frac{\partial}{\partial t}(u\cdot\bar{u})^3 = 3(u\cdot\bar{u})^2\left(u\frac{\partial \bar{u}}{\partial t} + \bar{u}\frac{\partial u}{\partial t}\right) =$

$$3|u|^4\left[(u_1+i\cdot u_2)\left(\frac{\partial u_1}{\partial t} - i\cdot\frac{\partial u_2}{\partial t}\right) + (u_1 - i\cdot u_2)\left(\frac{\partial u_1}{\partial t} + i\cdot\frac{\partial u_2}{\partial t}\right)\right] = 6|u|^4\left(u_1\frac{\partial u_1}{\partial t} + u_2\frac{\partial u_2}{\partial t}\right).$$

Now, as $u = u_1 + i\cdot u_2$ is a solution of (8.5.6), then $\frac{\partial u}{\partial t} = \Delta u - k|u|^2 u$, and subsequently $\frac{\partial u_1}{\partial t} = \Delta u_1 - k|u|^2 u_1$ and $\frac{\partial u_2}{\partial t} = \Delta u_2 - k|u|^2 u_2$. Therefore, $u_1\frac{\partial u_1}{\partial t} + u_2\frac{\partial u_2}{\partial t} = u_1\Delta u_1 - k|u|^2 u_1^2 + u_2\Delta u_2 - k|u|^2 u_2^2 = u_1\Delta u_1 + u_2\Delta u_2 - k|u|^2(u_1^2+u_2^2) = u_1\Delta u_1 + u_2\Delta u_2 - k|u|^4$. These computations lead to

$$\frac{\partial}{\partial t}\|N(u(\bullet,t))\|_2^2 = k^2\int_\Omega \frac{\partial}{\partial t}|u|^6 d\mathbf{x} = 6k^2\int_\Omega |u|^4\left(u_1\Delta u_1 + u_2\Delta u_2 - k|u|^4\right)d\mathbf{x}. \tag{8.5.9}$$

As $u(\bullet,t)\in \overset{\circ}{H}_2(\Omega)$, then $u(\bullet,t)\in C^1(\bar{\Omega})$ so that $u(\bullet,t)\in L_\infty(\bar{\Omega})$ since Ω is a bounded domain. Let $M = \|u\|_\infty$. Then by (8.5.9) and *Green's Theorem*,

$$\frac{\partial}{\partial t}\|N(u(\bullet,t))\|_2^2 = k^2\int_\Omega \frac{\partial}{\partial t}|u|^6 d\mathbf{x} \le 6M^4k^2\int_\Omega (u_1\Delta u_1 + u_2\Delta u_2)\, d\mathbf{x} - 6M^8k^2 vol(\Omega)$$
$$= -6M^4k^2\int_\Omega\left(|\nabla u_1|^2 + |\nabla u_2|^2\right)d\mathbf{x} - 6M^8k^2 vol(\Omega) < 0$$

whenever $k \neq 0$. Hence, $\|N(u(\bullet,t))\|_2^2 \le \|N(f)\|_2^2 \le \|f\|_2^6$ for all $t > 0$, and the right–hand side of (8.5.8) is bounded for all $t > 0$. ■

For $n \ge 4$, the *Sobolev Imbedding Theorem* insures that $u(\bullet,t)\in C^1(\bar{\Omega})$ and hence in $L_\infty(\bar{\Omega})$ whenever $u(\bullet,t)\in \overset{\circ}{H}_2(\Omega)$ for $q = [\frac{n}{2}]$, $[\bullet]$ = the nearest integer function. This is summarized as the result below.

Corollary 8.2: *For every* $f = (f_1, f_2) \in \overset{\circ}{H}_q(\Omega)$, $q = [\frac{n}{2}]$, $k \neq 0$, *and* $n \geq 4$, *the NLS equation* (8.5.6) *has a unique generalized solution* $u = (u_1, u_2) \in H^1([0, \infty); L_2(\Omega))$.

Finally, the *Hartman–Grobman Theorem* is applied to the *NLS*.

Theorem 8.5: *For every* $f = (f_1, f_2) \in \overset{\circ}{H}_q(\Omega)$, $q = \begin{cases} [\frac{n}{2}] & \text{for } n \geq 4 \\ 2 & n = 1,2,3 \end{cases}$ and $k \neq 0$, *there exists a unique homeomorphism H, independent of* $t \in [0, \infty)$, *so that the flow* F^t_{L+N} *on the NLS equation* (8.5.6) *is topologically conjugate to the linear* (*heat*) *flow* F^t_L *on sufficiently small neighborhoods* $\mathcal{O}_{q,2}$ *of* $0 \in \overset{\circ}{H}_2(\Omega)$ *into* $\mathcal{O}_2$ *of* $0 \in L_2(\Omega)$.

Proof: As discussed in the proof of *Theorem 8.4*, $L = \Delta$ is the generator of a contractive C_o–semigroup from $X = \overset{\circ}{H}_q(\Omega)$ into $Z = L_2(\Omega)$ for $q \geq 2$. The Dirichlet boundary condition $u(\boldsymbol{x}, t) \equiv 0$ for $\boldsymbol{x} \in \partial\Omega$ means that L is a negative operator. That is, the spectrum of L is comprised of negative numbers: $\sigma(L) = (-\infty, 0)$. Therefore, $F^t_L = e^{tL}$ is a contractive map so that L satisfies *Key Conditions* (*A*) – (*B*). For $N(u) = -k\,|u|^2 u$, the following estimates hold as per Haraux [32] and Pazy [59].

$$\| N(u) - N(v) \|_2 \leq \| N(u) - N(v) \|_{q,2} \leq C\left(\|u\|^2_{q,2} + \|v\|^2_{q,2}\right)\|u - v\|_{q,2}$$

Here, C is an absolute constant. This satisfies *Key Condition* (*C*) with $\psi\left(\|u\|_{q,2}, \|v\|_{q,2}\right) = C\left(\|u\|_{q,2} + \|v\|_{q,2}\right)$. Furthermore,

$$\begin{aligned} N'(u)v &= \lim_{h\to 0}\frac{N(u+h\cdot v)-N(u)}{h} = \lim_{h\to 0}\frac{-k\left|u+h\cdot v\right|^2(u+h\cdot v)+k\left|u\right|^2 u}{h} \\ &= \lim_{h\to 0}\frac{-k\left(h\left[2\left|u\right|^2 v+\bar{v}u^2\right]+h^2\left[2\left|v\right|^2 u+\bar{u}v^2\right]+h^3\left|v\right|^2 v\right)}{h} \\ &= \lim_{h\to 0} -k\left(\left[2\left|u\right|^2 v+\bar{v}u^2\right]+h\left[2\left|v\right|^2 u+\bar{u}v^2\right]+h^2\left|v\right|^2 v\right) \\ &= -k\left(2\left|u\right|^2 v+\bar{v}u^2\right). \end{aligned}$$

Consequently, $N'(v)(u-v) = 2\left|v\right|^2 u - 3\left|v\right|^2 v + \bar{u}v^2$ and, after some effort the calculation of $N(u) - N(v) - N'(v)(u-v)$ leads to $N(u) - N(v) - N'(v)(u-v) = \left(\bar{u}(u+v) - 2\left|v\right|^2\right)(u-v)$.

Thus, for all u, v in the bounded set $B = \left\{ \phi \in \overset{\circ}{H}_q(\Omega) : \|\phi\|_{q,2} \leq M < \infty \right\}$, repeated applications of the triangle and Cauchy–Schwarz inequalities yield

$$\begin{aligned}
\|N(u) - N(v) - N'(v)(u-v)\|_2 &\leq |k| \cdot \left(\|\bar{u}(u+v)\|_2 + 2\|v\|_2^2 \right)^2 \cdot \|u-v\|_2^2 \\
&\leq |k| \cdot \left(\|\bar{u}\|_{q,2} \|u+v\|_{q,2} + 2\|v\|_{q,2}^2 \right) \cdot \|u-v\|_{q,2}^2 \\
&\leq |k| \cdot \left(2M^2 + 2M^2 \right)^2 \cdot \|u-v\|_{q,2}^2 \leq C \cdot \|u-v\|_{q,2}^2 \quad \text{where } C = |k| \cdot (4M^2)^2 = 16|k| \cdot M^4.
\end{aligned}$$

Finally, $N(0) = 0 = N'(0)v$ for any $v \in Dom(N'(v))$. The mapping $u \mapsto N'(u) \equiv 2|u|^2 I + u^2 \overline{(\bullet)}$ where $\overline{(\bullet)}$ is the complex conjugate operator is smooth from $X = \overset{\circ}{H}_2(\Omega)$ into $\mathcal{L}(X, L_2(\Omega))$. Hence, $N(u) = -k\,|u|^2 u$ satisfies *Key Condition (D)*. By the *Hartman-Grobman Theorem*, the flow on the *NLS* (8.5.6) is locally topologically conjugate to the linear heat flow on (8.5.7). ∎

In summary, the results in this chapter extend the *Hartman–Grobman Theorem* from finite–dimensional flows (namely, *ordinary differential equations*) on a Banach space E into E to infinite dimensional flows (e.g., *partial differential equations*) between distinct Hilbert spaces. This theory is successfully applied to a generalized period *KdV*, Burgers', and the *NLS* partial differential equations. The *Hartman–Grobman Theorem* gives a general criterion for the existence and uniqueness of a change of variables $u = H(v)$ so that the nonlinear evolution equation $\frac{\partial u}{\partial t} = Lu + N(u)$ is transformed into the linear equation $\frac{\partial v}{\partial t} = Lv$. Provided the solution operator F_{L+N}^t is a proper mapping, then the change of variables is global. These results establish a general theory for the existence of a linearizing change of variables with respect to nonlinear evolution equations.

Appendix: MATLAB® Commands and Functions

Descriptions of the MATLAB files (M–files) used within this text are listed below. Complete M–files along with the MATLAB code blocks used to generate the figures and make the computations can be downloaded from the website associated with this book.

http://www.morganclaypoolpublishers.com/Costa/

To reproduce the analytic computations, the Symbolic Math Toolbox is required.

M–files

CIRCDRUM.m

```
circdrum  Circular Drum Solution

  [u,OR] = circdrum(ro,Nb,to) returns the solution u(r,theta,t) of the
  circular drum problem with respect to the initial conditions
  u(r,theta,0) = -exp(-2*(r^2 + theta^2)) and du(r,theta,0)/dt = 1
  over the domain [0,ro] x [-pi,pi] x [0,to].  In particular, the structure
  OR has the fields

        R: [1×100 double] = vector of r-values over [0,ro]
    Theta: [1×200 double] = vector of theta-values [-pi, pi]
        T: [1×50 double]  = vector of t-values over [0,to]

  The second output is the structure of radial, angular, and temporal
  grid points.
  The radius of the drum is ro while Nb is the number of Bessel
  functions of the first kind (besselj) to be used in the series
  solution (a per equations 4.6.6a-4.6.6b).

  [u,OR] = circdrum(ro,Nb,to,sigma) returns the solution u and grid
  structure OR as above with respect to the wave speed sigma

  Example:  [u,OR] = circdrum(30,[0:4],4);

See also besselj.m, integral2.m, example_4_6_3.m, chebfun.m
```

Note: In order to use this M–file, the Chebfun suite of functions must be downloaded as per the following instructions.

```
Reference to Nick Trefethan's "Chebfun" M-file
https://www.mathworks.com/matlabcentral/mlc-
downloads/downloads/submissions/23972/versions/22/previews/chebfun/examples/r
oots/html/BesselRoots.html?access_key=

Instructions for downloading Chebfun.m
```

```
>> unzip('https://github.com/chebfun/chebfun/archive/master.zip')
>> movefile('chebfun-master', 'chebfun'); addpath(fullfile(cd,'chebfun'));
   savepath
```

DALEMBERT.m

```
dalembert  d'Alembert's solution to the one-dimensional wave equation

  u = dalembert(f,g,sigma,xRange,tRange) returns d'Alembert's traveling
  wave solution to the one-dimensional wave equation

  u(x,t) = 1/2[f(x+sigma*t)+f(x-sigma*t) + F(x+sigma*t)-F(x*sigma*t)]

  F(x) = integral(g,0,x)/sigma

  The input xRange = [ax, bx] and tRange = [at,bt] are the intervals over
  which a solution is sought.  Therefore, xRange and tRange MUST be vectors
  of length 2.

  [u,x,t] = dalembert(f,g,sigma,xRange,tRange) returns d'Alembert's solution
  as above along with X and T grids (for use in plotting).

  [u,x,t] = dalembert(f,g,sigma,xRange,tRange,Nx,Nt) returns d'Alembert's
  solution as above using Nx and Nt points for the X- and T-grids,
  respectively.

  Example:  xRange = [-2*pi,2*pi]; tRange = [0,3];
            f = @(x)(exp(-x.^2)); g = @(x)(-tanh(x).*sech(x)); sigma = 2;

            [U,X,T] = dalembert(f,g,sigma,xRange,tRange);

See also waveplot.m
```

EXAMPLE_4_6_3.m

```
example_4_6_3  Nodes of the Circular Drum

  U = example_4_6_3(ro,Nbessel,T) returns the nodes of the vibrating
  circular drum of radius ro (in centimeters) with respect to the
  Nbessel nodes at time T.  Note that Nbessel can be a scalar or a
  vector.  If Nbessel = n (a scalar), then Bessel functions besselj(0,x),
  besselj(1,x), ... , bessselj(n,x) are used in the computation of the
  nodes of the circular drum.

  Example:  U = example_4_6_3(30,0:3,4);
  figure;
  for j = 1:4
    subplot(2,2,j); str = ['u_' num2str(j-1)]; Z = U.(str);
    h = surf(U.r,U.theta,Z);shading('interp'); camlight;
    set(gca,'FontSize',16,'FontName','Times New Roman');
    xlabel('\itr'); ylabel('\it\theta');
    title(['\rmNode = ' num2str(j-1) ', \itt\rm = 4']); view([10,25]);
  end

See also chebfun.m
```

Note: This M–file (**example_4_6_3**), requires that the Chebfun suite of functions must be downloaded as per the instructions contained in **circdrum**.m above.

EXAMPLE_4_6_4.m

```
example_4_6_4  Solution of the vibrating volume

 [U,S] = example_4_6_4(L,T,N) returns approximate solution of the vibrating
 volume over the parallelepiped W = [0,L(1)]x[0,L(2)]x[0,L(3)] up to
 time T and with respect to N = [n,m,k] number of terms consistent with
 the formulae at the end of the example.  The second output S is the
 structure of spatial and temporal domains and grid values which has the
 fields

        X: [1×Nx double]
    Xgrid: [Nx×Ny×Nz×Nt double]
        Y: [1×Ny double]
    Ygrid: [Nx×Ny×Nz×Nt double]
        Z: [1×Nz double]
    Zgrid: [Nx×Ny×Nz×Nt double]
        T: [1×Nt double]
    Tgrid: [Nx×Ny×Nz×Nt double]

 Nominally, Nx = 50, Ny = Nx*L(2)/L(1), Nz = Nx*L(3)/L(1), and Nt = 50.

 [U,S] = example_4_6_4(L,T,N,sigma) returns the solution U and grid
 structure S as above for the wave speed sigma.  Nominally, sigma = 1.

 Example: tic; [U,S] = example_4_6_4([2,4,6],5,[8,8,8]); toc
          Elapsed time is 369.114506 seconds.

       U % A [50×100×150×50 double] and
       S =
 struct with fields:
       X: [1×50 double]
   Xgrid: [50×100×150×50 double]
       Y: [1×100 double]
   Ygrid: [50×100×150×50 double]
       Z: [1×150 double]
   Zgrid: [50×100×150×50 double]
       T: [1×50 double]
   Tgrid: [50×100×150×50 double]%

See also example_4_6_3.m
```

GLMODE.m

```
glmode  The ODE X''(x) + mu^2*X'(x) - f(x)*X(x) = 0 used to construct the kernel
        K(x,y) of the Gel'fand-Levitan-Marchenko integral equation.

 The time-independent GLM kernel function K(x,y) satisfies the elliptic
```

```
partial differential equation Kxx - Kxy = f(x)*K(x,y) and K(x,0) = f(x).
Under a separation of variables solution K(x,y) = X(x)*Y(y), the x-valued
function X satisfies the ODE X''(x) + mu^2*X'(x) = f(x)*X(x) for arbitrary
real constant mu.

dX = glmode(x,y,mu,F,X) returns the required ODE as use for input into
MATLAB's ODE-solver suite.  In particular, glmode solves the first order
system

 X1'(x) = X2(x)
 X2'(x) = -(mu^2)*X2(x) + f(x)*X

for a function f(x) which acts as the initial condition to the Korteweg
de Vries equation dU/dt = 6*U*dU/dx - Uxxx.

Example:  f = @(x)(cos(2*pi*x));
          X = linspace(-4,4,800); F = f(X); mu = 2;
          opts = odeset('RelTol',1e-02,'AbsTol',1e-04); % Optional ODE inputs

          [x,X] = ode45(@(x,y)(glmode(x,y,mu,F,X)),[-4,4],f(0)*[1,1],opts);
```

IDFCN.m

```
idfcn  Identifier function over the interval [a, b]

  y = idfcn(x,a,b) returns the value 1 for a <= x <= b and 0 otherwise.

  Example:  x = linspace(-5,5,100);
            y = idfcn(x,-2,3);
```

NKG.m

```
nkg  Nonlinear Klein-Gordon function

  u = nkg(P) returns the analytic solution of the nonlinear Klein-Gordon
  partial differential equation du/dy = s^2*(d^2u/dx^2) + a*u^n with
  initial condition u(0,0) = 1.

  The input structure P has the fields
  P.s = wave speed
  P.a = nonlinear dispersion coefficient
  P.n = degree of nonlinearity.
  P.lam = separation parameter

  Example:  P.s = sqrt(pi); P.a = 1/2; P.n = 3; P.lam = 1;
              u = nkg(P)
              ezsurfc(u,[-2,2,0,4],800)

See also ramanqi.m
```

RAMANCAUCHY.m

```
ramanCauchy  The Cauchy Data (initial conditions) for the Stimulated Raman
               Scattering Quasi-implicit method

  [f,g,x] = ramanCauchy(X,icFlag) returns the Cauchy data f(x) = u(x,0) and
  g(x) = v(x,0) for the steady-state stimulated Raman scattering model.
  These initial values should be functions with compact support over the
  input interval x = [-xo,xo].  There are three options as listed below.

  icFlag          f(x)            I(xo) = [-xo,xo]
  ------      --------------      ----------------
  'pulse'     1 for x in I(xo) and 0 otherwise
  'exp'       1-exp(- ((abs(x)-xo).^2 )/(s^2) ) for x in I(xo) and 0 otherwise
  'taper'     1 for x in I(xo/2), exp(- ((abs(x)-xo).^2 )/(s^2) ) for x in
              [-3xo/2,-xo] U [xo,3*xo/2], 0 otherwise

  The nominal value for the exponential beam has s = 25.  For the tapered
  beam, s = 5.

  g(x) = Co*f(x).  The nominal value of Co = 0.25 but it can be input
  as [f,g,x] = ramanCauchy(X,icFlag,Co).

  Note:  If the input domain is X = [-xo,xo], then output domain is
         x = [-2*xo,2*xo].
         The input domain should ALWAYS be symmetric about 0.

  Example:  xo = 55.36; X = linspace(-xo,xo,500);
               [f,g,x] = ramanCauchy(X,'exp');

See also ramanqi.m and eigenmatrix.m
```

RAMANQI.m

```
ramanqi  The Quasi-implicit finite difference method for Stimulated Raman
         Scattering

  [u,v] = ramanqi(IS) returns the full (complex valued) pump (u) and
  Stokes (v) electric fields as determined by the quasi-implicit finite
  difference scheme developed in Chapter 7.  The input structure contains
  the fields

  IS.time = T (the time integration endpoint from [0,T])
  IS.space = xo (the spatial variable extent over [-2xo,2xo])
  IS.dtime = dT (the time spacing) <= 2*exp(-T||f||^2)/||g||^2
  IS.dspace = dX (the spatial spacing)
  IS.Cp = = Kp/(2*Ks) = pump constant
  IS.Cs = 1/2 = Stoke's constant
  IS.CauchyData = Cflag (= 'pulse','exp','taper');
  IS.Co = the scale of the Stokes initial function g(x), nominally 0.25;

  [u,v,t,x] = ramanqi(IS) returns the Pump (u) and Stokes (v) fields as
  above along with the time (t) and space (x) domains associated with the
```

```
  computations.  In this setting, t = [0,T] and x = [-2*xo,2*xo].

  The electric fields are returned as u(t,x) and v(t,x).

  Example:  kp = (2*pi)/(3.5e-05); ks = (2*pi)/(4.1e-05);
            IS.time = 8; IS.space = 55.36;
            IS.dtime = 0.0225352; IS.dspace = 0.1845333;
            IS.Cp = kp/(2*ks); IS.Cs = 1/2;
            IS.CauchyData = 'pulse';

            [u,v,t,x] = ramanqi(IS)

  % returns the 365-by-1200 complex-valued Pump (u) and Stokes (v) fields
  % along with the 1-by-365 time vector (t) and 1-by-1200 space vector (x)

See also ramanCauchy.m and eigenmatrix.m
```

WAVEPLOT.m

```
waveplot  Time section plots of d'Alembert's traveling wave solution

  waveplot(U,X,T) returns the 2-by-2 plot of the solution U(X,T) at times
  T = 0, T = 1, T = 2, and T = 3.

See also dalembert.m
```

References

Note to the reader. If you access this text electronically, then the references have page links to their location in the book. These are set off to the extreme right and are bracketed. For those reference materials which have *digital object identifiers* (i.e., links to the publisher's website), these will be set off by the symbol DOI. For example, reference [15] has the internal page link [148] and DOI link 10.1017/CBO9781139167994. A mouse click on the page link [148] will take you to page 148 of this text. A mouse click on the DOI link will take you to the publisher's website for the specified reference.

[1] M. J. Ablowitz and H. Segur, ***Solitons and the Inverse Scattering Transform***, SIAM Books, Studies 4, Copyright © 1981, Philadelphia [137, 139]

[2] M. Abramowitz and I. Stegun, ***Handbook of Mathematical Functions***, Dover Publications, NY, Copyright © 1964. [72, 145]

[3] R. L. Anderson and N. H. Ibragimov, ***Lie–Bäcklund Transformations in Applications***, SIAM Books, Studies 1, Copyright © 1979, Philadelphia. [145]

[4] L. C. Andrews and B. N. Shivamoggi, ***Integral Transforms for Engineers and Applied Mathematicians***, Macmillan Publishing Company, Copyright © 1988, New York. [50, 85, 86]

[5] A. K. T. Assis and J. P. M. C Chaib, ***Ampere's Electrodynamics***, Copyright © 2015, Published by C. Roy–Keys, Incorporated, Montréal, Québec. [23] (https://www.ifi.unicamp.br/~assis/Amperes-Electrodynamics.pdf)

[6] P. Bamberg and S. Sternberg, ***A Course in Mathematics for Students of Physics***, Volumes 1 and 2, Cambridge University Press, Copyright © 1990, New York. [17]

[7] R. G. Bartle, ***The Elements of Integration***, J. Wiley & Sons, Copyright © 1966, New York. [49, 82]

[8] M. S. Berger, ***Nonlinearity and Functional Analysis***, Academic Press, Copyright © 1977, New York. [180, 183, 189, 192]

[9] M. S. Berger and M. S. Berger, ***Perspectives in Nonlinearity***, Benjamin Publishers, Copyright © 1968, New York . [130]

[10] M. S. Berger, P. T. Church, and J. G. Timourian, *Integrability of Nonlinear Differential Equations via Functional Analysis*, Proceedings of Symposia in Pure Mathematics, Volume 45, Part 1, 1986, pages 117–123. [173]

[11] J. Boussinesq, *Essai sur la theory des eaux cournates*, l'Académie des Science Institut National France, XXIII, pages 1–680, 1877. [190]

[12] W. E. Boyce and R. C. DiPrima, ***Elementary Differential Equations and Boundary Value Problems***, *Third Edition*, John Wiley & Sons Publishers, Copyright © 1977, New York. [41, 95]

[13] J. M. Burgers, *Further Statistical Problems Connected with the Solution of a Simple Nonlinear Partial Differential Equation*, Proceedings of the Academy of Science Amsterdam, Volume 57, Copyright © 1954, pages 159–169. [193]

[14] J. M. Burgers, *A Mathematical Model Illustrating the Theory of Turbulence*, Advances in Applied Mechanics, Volume 1, Copyright © 1948, pages 171–199. DOI: 10.1016/S0065-2156(08)70100-5 [132, 193]

[15] P. N. Butcher and D. Cotter, ***The Elements of Nonlinear Optics***, Cambridge Studies in Modern Optics 9, Cambridge University Press, Copyright © 1990, Cambridge, England. DOI: 10.1017/CBO9781139167994 [148]

[16] D. C. Champeney, ***Fourier Transforms and Their Physical Applications***, Academic Press, Copyright © 1973, New York. [86, 114]

[17] M. Cheney, *Inverse scattering in dimension two*, Journal of Mathematical Physics, Volume 25, Copyright © 1984, pp. 94–107. DOI: 10.1063/1.526003 [5]

[18] M. Cheney, *A rigorous derivation of the "miracle" identity of three–dimensional inverse scattering*, Journal of Mathematical Physics, Volume 25, Copyright © 1984, pp. 2988–2990. DOI: 10.1063/1.526050 [5]

[19] P. J. Costa, *The Linearization of Nonlinear Evolution Equations*, PhD Dissertation, University of Massachusetts, 1984. [173]

[20] P. J. Costa, *Raman Scattering, Lasers, and Convergent Finite–Difference Methods*, pages 185–236, from ***Bridging Mind and Model,*** *Papers in Applied Mathematics*, University of St. Thomas Press, Copyright © 1994, St. Paul, MN. [148]

[21] D'Addario String tension chart, [98]
https://www.daddario.com/globalassets/pdfs/accessories/tension_chart_13934.pdf

[22] J. W. Dettman, ***Mathematical Methods in Physics and Engineering***, *Second Edition*, McGraw–Hill Copyright © 1969, New York. [114]

[23] L. C. Evans, ***Partial Differential Equations***, Graduate Studies in Mathematics, Volume 19, American Mathematical Society, Copyright © 1991, Providence, RI. [32]

[24] L. D. Faddeev, *Three–dimensional inverse problem in the quantum theory of scattering*, Academy of Sciences of the Ukrainian SSR, Kiev, 1971, ITP–71–106E . [5]

[25] H. Flanders, ***Differential Forms***, Mathematics in Science and Engineering, Volume 11, Academic Press, Copyright © 1963, New York. [17]

[26] P. R. Garabedian, ***Partial Differential Equations***, Chelsea, Copyright © 1964, New York. [120, 123]

[27] C. S. Gardner, J. M. Greene, M. D. Kruskal, and R. M. Miura, *Method for Solving the Korteweg de Vries Equation*, Physical Review Letters, Volume 19, Copyright © 1967, pages 1095–1097.
DOI: 10.1103/PhysRevLett.19.1095 [4, 137, 139, 140, 173]

[28] C. S. Gardner, J. M. Greene, M. D. Kruskal, and R. M. Miura, *Korteweg de Vries Equation and Generalizations VI, Methods for Exact Solution*, Communications on Pure and Applied Mathematics, Volume 27, Copyright © 1974, pages 97–133.
DOI: 10.1002/cpa.3160270108 [4, 137, 139, 140, 173]

[29] I. S. Gradshteyn and I. M. Ryzhik, ***Table of Integrals, Series, and Products***, *Corrected and Enlarged Edition*, Academic Press, Inc., Copyright © 1965, 1980, London. [77, 86]

[30] D. M. Grobman, *Homeomorphisms of Systems o Differential Equations*, Doklady Akademia Nauk SSSR, Volume 128, Copyright (1959) pages 880–881. [173]

[31] K. E. Gustafson, ***Introduction to Partial Differential Equations and Hilbert Space Methods***, J. Wiley & Sons, Copyright © 1980, New York. [39]

[32] A. Haraux, ***Nonlinear Evolution Equations***, Lecture Notes in Mathematics, Number 841, Springer–Verlag, Copyright © 1981, New York. [157, 181, 195, 197]

[33] P. Hartman, *A Lemma in the Theory of Structural Stability of Differential Equations*, Proceedings of the American Mathematical Society, Volume 11, Copyright © 1960, pages 610–622. DOI: 10.1090/S0002-9939-1960-0121542-7 [5, 173]

[34] P. Hartman, *On the Local Linearization of Differential Equations*, Proceedings of the American Mathematical Society, Volume 14, Copyright © 1963, pages 568–573. DOI: 10.1090/S0002-9939-1963-0152718-3 [5, 173]

[35] P. Hartman, ***Ordinary Differential Equations***, *First Edition*, Wiley, Copyright © 1964, Hoboken, NJ. *Second Edition*, Birkhäuser, Copyright © 1982, Boston, MA. [5, 173, 177]

[36] O. S. Heavens, ***Optical Properties of Thin Solid Films***, Dover Publications, Incorporated, Copyright © 1965, 1991, New York

[37] D. Henry, ***Geometric Theory of Semilinear Parabolic Equations***, Lecture Notes in Mathematics, Number 840, Springer–Verlag, Copyright © 1981, New York. DOI: 10.1007/BFb0089647 [181, 184, 195]

[38] M. W. Hirsh, C. C. Pugh, and M. Shub, ***Invariant Manifolds,*** Lecture Notes in Mathematics, Number 583, Springer–Verlag, Copyright © 1981, New York. [177]

[39] D. Hoff, *A Finite Difference Scheme for a System of Two Conservation Laws with Artificial Viscosity*, Mathematics of Computation, Volume 33, pages 1171–1193, Copyright © 1979. DOI: 10.1090/S0025-5718-1979-0537964-9 [172]

[40] E. Hopf, *The Partial Differential Equation* $u_t + u\, u_x = \mu\, u_{xx}$, Communications on Pure and Applied Mathematics, Volume 3, 1950, pp. 201–230. DOI: 10.1002/cpa.3160030302 [4, 132, 173]

[41] E. Infeld and F. Rowlands, ***Nonlinear Waves, Solitons and Chaos***, Second Edition, Cambridge University Press, Second Edition Copyright © Infeld & Rowlands 2000, Cambridge, UK. [138, 139]

[42] M. C. Irwin, ***Smooth Dynamical Systems***, Academic Press, Copyright © 1980, New York. [177]

[43] F. John, ***Partial Differential Equations***, *Third Edition*, Springer–Verlag, Copyright © 1978, New York. [32, 88, 123]

[44] F. John, ***Plane Waves and Spherical Means***, *Applied to Partial Differential Equations*, Springer–Verlag, Copyright © 1955, New York. [119, 120]

[45] L. V. Kantorovich and G. P. Akilov, ***Functional Analysis***, *Second Edition*, Pergamon Press, Copyright © 1982, Elmsford, New York. [49, 191]

[46] T. W. Körner, ***Fourier Analysis***, Cambridge University Press, Copyright © 1988, Cambridge, UK. [86]

[47] D. Korteweg and G. de Vries, *On the Change of Form of Long Waves Advancing in a Rectangular Canal, and on a New Type of Long Stationary Waves*, Philosophical Magazine, Volume 39, Number 240, pages 422–443, 1895.
DOI: 10.1080/14786449508620739 [134, 190]

[48] E. Kreysig, ***Advanced Engineering Mathematics***, *Tenth Edition*, J. Wiley & Sons, Copyright © 1967, 2008, New York.

[49] M. D. Kruskal and N. J. Zabusky, *Princeton Plasma Physics Laboratory Annual Report*, MATT–Q–21, 1963, pp 301*ff.* [134]

[50] J. B. Marion, ***Classical Electromagnetic Radiation***, Academic Press, Copyright © 1965, New York. [23]

[51] R. E. Mickens, *Novel Explicit Finite–Difference Schemes for Time–Dependent Schrödinger Equations*, Computer Physics Communications, Volume 63, pages 203–208, Copyright © 1991. DOI: 10.1016/0010-4655(91)90249-K [172]

[52] T. Myint–U with L. Debnath, ***Partial Differential Equations for Scientists and Engineers***, *Third Edition*, North–Holland, Copyright © 1987, New York. [31, 52, 55, 110, 132]

[53] A. C. Newell and J. V. Moloney, ***Nonlinear Optics***, Addison–Wesley Publishers, Copyright © 1992, Redwood City, CA. [148]

[54] R. G. Newton, ***Inverse Schrödinger Scattering in Three Dimensions***, Springer–Verlag, Copyright © 1989, Berlin. [5]

[55] N. V. Nikolenko, *On the Reducibility of Nonlinear Evolution Equations to Linear Normal Form*, Soviet Math Doklady, Volume 27, Number 1, 1983, pages 119–122. [173]

[56] T. Nishida and J. Smoller, *A Class of Convergent Finite Difference Schemes for Certain Nonlinear Parabolic Systems*, Communications on Pure and Applied Mathematics, Volume 36, pages 785–808, Copyright © 1983. DOI: 10.1002/cpa.3160360605 [172]

[57] F. Olver, D. Lozier, R. Boisvert, C. Clark, ***NIST Handbook of Mathematical Functions***, Cambridge University Press, Copyright © 2010 by NIST, Gaithersburg, Maryland.
[72, 75, 80, 86, 103]

[58] J. M. Ortega and W. G. Poole, ***Numerical Methods for Differential Equations***, Copyright © 1981, Pitman Publishing Incorporated, Marshfield, MA. [156]

[59] A. Pazy, ***Semigroups of Linear Operators and Applications to Partial Differential Equations***, Springer–Verlag, Copyright © 1983, New York. [156, 157, 191, 192, 193, 195, 197]

[60] A. Penzkopfer, A. Laubereau, and W. Kaiser, *High Intensity Raman Interactions*, Progress in Quantum Electronics, Volume 6, pages 55–140, 1979. DOI: 10.1016/0079-6727(79)90011-9 [148]

[61] A. D. Polyanin and V. F. Zaitsev, ***Hand Book of Nonlinear Partial Differential Equations***, Chapman & Hall/CRC, 2004, Boca Raton, FL. [139]

[62] C. C. Pugh, *On a Theorem of P. Hartman*, American Journal of Mathematics, Volume 91, Copyright © 1969, pages 363–367. DOI: 10.2307/2373513 [177]

[63] S. Rabinowitz, *Private Communication*, 1993. [125]

[64] J. Rauch, ***Partial Differential Equations***, Springer–Verlag, Copyright © 1991, New York. [156, 181]

[65] A. C. Reynolds, Jr., *Convergent Finite Difference Schemes for Nonlinear Parabolic Equations*, SIAM Journal of Numerical Analysis, Volume 9, Number 4, pages 523–533, December 1979. DOI: 10.1137/0709048 [172]

[66] R. D. Richtmyer and K. W. Morton, ***Difference Methods for Initial Value Problems***, Second Edition, Interscience Tracts in Pure and Applied Mathematics, J. Wiley & Sons, Copyright © 1967, New York. [156]

[67] G. F. Roach, ***Green's Functions***, *Second Edition*, Cambridge University Press, Copyright © 1982, Cambridge, UK. [38]

[68] J. T. Schwartz, ***Nonlinear Functional Analysis***, NYU Lecture Notes, Copyright © 1965. [178]

[69] S. M. Selby, ***CRC Standard Mathematical Tables***, *22nd Edition*, CRC Press, Copyright © 1974, Cleveland, Ohio. [86]

[70] G. D. Smith, ***Numerical Solution of Partial Differential Equations***: *Finite Difference Methods*, Third Edition, Oxford University Press, Copyright © 1985, New York. [155, 161]

[71] S. L. Sobolev, ***Partial Differential Equations of Mathematical Physics***, Dover Publications, Copyright © 1989, New York (original copyright Pergamon Press © 1963). [123]

[72] M. Spivak, ***Differential Manifolds***, The Benjamin Cummings Publishing Company, Copyright © 1965, Menlo Park, CA. [17]

[73] M. Spivak, ***Differential Geometry***, Publish or Perish, Incorporated, Copyright © 1970, Berkeley, CA. [17]

[74] G. Strang, ***Introduction to Applied Mathematics***, Wellesley Cambridge Press, Copyright © 1986, Wellesley, MA. [72, 126]

[75] G. Strang, *Private Communication*, 1993. [126]

[76] J. A. Stratton, ***Electromagnetic Theory***, McGraw–Hill Book Company, Copyright © 1941, New York. [28]

[77] E. Tadmor, R. Sackler, and B. Sackler, *Semi–Discrete Approximations to Nonlinear Systems of Conversation Laws, Consistency, and* L_∞*–Stability Imply Convergence*, ICASE Report Number 88–41. [172]

[78] J. W. Thomas, ***Numerical Partial Differential Equations***, *Finite Difference Methods*, Springer–Verlag Texts in Applied Mathematics, Volume 22, Copyright © 1995, New York DOI: 10.1007/978-1-4899-7278-1 [151, 154, 156]

[79] A. N. Tikhonov and A. A. Samarskii, ***Equations of Mathematical Physics***, Dover Publications, Copyright © 1963, New York. [102, 103, 123]

[80] D. Vvedensky, ***Partial Differential Equations with Mathematica***®, Copyright © 1993, Addison–Wesley Publishing Company, Reading, MA. [139]

[81] G. N. Watson, ***A Treatise on the Theory of Bessel Functions***, Cambridge University Press, Copyright © 1966, Cambridge, UK. [72, 103]

[82] Wikipedia, *Airy Function*, https://en.wikipedia.org/wiki/Airy_function. [145]

[83] Wikipedia, *Peter Dirichlet*, https://en.wikipedia.org/wiki/Peter_Gustav_Lejeune_Dirichlet [17]

[84] Wikipedia, *Diederik Korteweg*, https://en.wikipedia.org/wiki/Diederik_Korteweg. [134]

[85] Wikipedia, *James Clerk Maxwell*, https://en.wikipedia.org/wiki/James_Clerk_Maxwell. [15]

[86] Wikipedia, *John Scott Russell*, https://en.wikipedia.org/wiki/John_Scott_Russell. [135]

[87] E. C. Zachmanoglou and D. W. Thoe, ***Introduction to Partial Differential Equations with Applications***, The Williams & Wilkins Company, Copyright © 1976, Baltimore, Maryland [27, 50, 98, 102, 117]

Index

Printed in the United States
by Baker & Taylor Publisher Services